THE POLITICS OF WATER
IN ARIZONA

THE

POLITICS

OF

WATER

IN

ARIZONA

Dean E. Mann

THE UNIVERSITY OF ARIZONA PRESS
TUCSON   1963

Preface

The history of the settlement of the United States is a history of individuals and small groups dissatisfied with their lot in their previous habitation moving into a wilderness in which they hoped for and generally found a more promising livelihood than the one they had left. This pattern was followed in Arizona as it was elsewhere. In small groups the Mormons moved into the river valleys of the Salt, the Gila and the San Pedro to establish an economy based on irrigation agriculture. With the discovery of gold and other precious metals, large numbers of fortune hunters flocked to Arizona to make their stake. When the grasslands of Texas no longer could support all of its cattle, the cattle were driven westward to the reportedly lush valleys of Arizona. The role that government played — national, territorial, or local — was relatively slight except for keeping the peace and, in view of Arizona's reputation derived from Tombstone gunmen, was less than adequate in that field.

But taking a longer view of Arizona history, it is clear that the role of government has been critical in its economic and social development. While individual impulse brought and continues to bring settlers to the state, government at all levels has played a critical role in making a basically inhospitable country suitable for human habitation in large numbers and at a relatively comfortable standard of living. In no realm of human activity is this more apparent than in the management of Arizona's natural resources. Arizona has been and continues to be dependent primarily on its resource wealth for the employment and well-being of its citizens. And even in the one major area of the economy which is not dependent upon Arizona's natural resources — manufacturing — governmental decisions and financial support have been critical in the location and maintenance of these manufacturing facilities which are dependent upon defense contracts.

In the arid Southwest, the location and development of water supplies has been the paramount necessity. Other resources — land, space, forests, and minerals—were found in abundance. But water, so necessary for other forms of economic activity and human and animal life, was in short supply, undependable and unevenly distributed in the few river valleys and their basins. Very early it was realized that to sustain even small communities it was necessary to undertake public works. By the turn of the twentieth century the federal government found itself a major participant in the development and management of water supplies for this arid country. Moreover, there was increasing recognition that for the protection of these water supplies public efforts were required to defend the land against misuse. These abuses took the form of wanton destruction of the timber resources and over-grazing which in turn had disastrous effects upon the sustenance and availability of water supplies.

There is a tendency to look upon the natural-resource problems as primarily problems of engineering and technology. Unquestion-ably, technological development has played an extremely important role in making the arid region habitable for large numbers of people. The possibilities inherent in desalinization of salt water and in con-trolling atmospheric vapor suggest the impact which technology may yet make in the field of water supplies. But modern technology with-out effective management, and *public* management in the provision of water supplies in the arid region, is useless in satisfying man's needs. For this reason, it is of equal importance to examine critically the ex-perience men have had in managing their water supplies through public institutions to determine the adequacy with which this techno-logy has been applied.

An examination of the public institutions involved in developing and managing water supplies requires an analysis of many facets of the political process. To understand the public policies that have been followed in the state of Arizona, it is necessary to look also at the political machinery through which the policy decisions have been rendered. Policy decisions are seldom made on the basis of uncontested objective evidence but are the result of contending interests which express conflicting values and which can look at the same set of ob-jective data and reach quite different conclusions. These interests of course contend in the marketplace through their economic means to buy and sell the resources that are necessary to sustain and improve their economic standing. But they must also operate through political parties, state legislatures, and state administrative machinery to realize their ends since these public mechanisms are able to make binding decisions which affect, if not determine, their well-being. Therefore it is necessary to examine such institutions as political parties and the methods of operation within the state legislature. Considerable attention must be devoted to the organization and operations of the various federal and state agencies which play such significant roles in this decision-making process.

The policy issues in water development and management and the public machinery through which decisions are made constitute the primary focus of this study. It is obvious that any discussion of policy leads inevitably to an analysis of the economic issues of resource policy. While not an economist, I have attempted to incorporate existing economic analysis into the study to indicate the alternatives that are posed by investment of one kind or another in water resource development. Similarly, a discussion of the water policy issues leads to the question of technological advances and these have been duly noted. In the final section of the book, some attention is devoted to the efforts and research being conducted in the state which are applicable to the water resource question.

A study as broadly conceived as this is necessarily limited in the depth to which it can go into each of the many facets of the resource question in the state of Arizona. Each of the chapters and some parts of chapters could easily constitute major studies in depth. For this reason, this study must be considered an initial effort to survey the general picture of water management in the state, which hopefully will lead to more detailed analyses of the various aspects of public water management and development.

In conducting this study, I received the assistance of innumerable individuals. I am particularly indebted to Professor Albert Lepawsky of the University of California for his encouragement in undertaking this study and for his helpful guidance in the process of its development and completion. Others who read the manuscript and provided useful counsel were Joseph P. Harris and Percy McGauhey, also of the University of California, and John Vieg of Pomona College. My colleagues at the University of Arizona — Paul Kelso, Rod Hastings, Bernard Hennessy, John Harshbarger and Russell Ewing — and Henry Caulfield, formerly of Resources for the Future and now Deputy Director of the Resources Program Staff of the Department of the Interior, read portions of the manuscript and gave me the benefit of their critical comment.

This study would not have been possible without the time and assistance given freely by the many public officials and private citizens in the state who provided much information and documentation.

The staff of the University of Arizona Library were particularly helpful in giving me access to the Arizona Collection and providing additional materials. The Department of Library and Archives in the State Capitol also provided useful information.

In editing the final manuscript I am greatly indebted to Miss Kit Scheifele for her long and patient efforts in putting the manuscript in presentable form. She was all an editor could be, making substantial improvements in every aspect of the manuscript. My gratitude is also expressed toward Jack Cross, the Director of the University of Arizona Press, for his encouragement and assistance at every turn. For the maps, I express my thanks to Mr. Don Bufkin.

Finally, and most importantly, I am grateful to my wife for her encouragement and support during the several years required in the completion of this study. Her patience, understanding, and assistance were of inestimable value and were a source of inspiration.

For all expressions of fact and opinion and for any errors, I am of course responsible.

Dean E. Mann

Washington, D.C.
February, 1963

Contents

Maps

East Phoenix-Tempe-
Scottsdale area.

Left: 1954

Below: 1962

Markow Photography
Courtesy of the
Valley National Bank

*Rapid urban growth of new communities and metropolitan areas
takes increasing amounts of the state's critically limited water supplies.*

Foreword

This book is a courageous account of the non-scientific factors relating to water management in Arizona. Dr. Mann's treatise provides an unbiased scrutiny of essential factors that must be considered in ultimate water-management plans. This succinct resume of past legal decisions and disjointed water planning reveals a record which should be of concern to every Arizona resident.

Because Arizona lies in the heart of the semiarid Southwest, it is being blessed with the arrival of many new citizens. Accompanying these recent residents is an increased demand for water for municipal, industrial, and recreational purposes. Agriculture and the mineral industry have long provided the fundamental economic base for development and growth in Arizona. Their acquisition of early water rights is only natural, and these industries remain essential for sustained economic stability. However, today there is keen competition for the available water to accommodate modern living demands and recreational enjoyment. While agriculture is the largest user of Arizona's water, it finds it increasingly difficult to pay the high cost for essential water supplies. In areas of short water supply, the dollar value per unit of water becomes the dominant factor in the control and ownership of water rights. Consequently, there is a new trend in water priority and demand in the Southwest — the command of water supplies is shifting to municipal, industrial, and recreational uses which can afford the higher cost.

Dr. Mann gives an excellent analysis of the many complex social, economic, and political factors as they have affected water policy and management for both the layman and water expert. He lucidly points out the inadvertent influence of the federal government on water planning in Arizona, and the reasons for the creation of diverse pressure groups. There is perhaps no other area in the world which has suffered such a myriad of interstate and international water problems compounded by the lack of scientific facts for understanding the physical environment controlling the water supply. Even more important, in the role of water management, is the effect of man's activities on the modification of the natural water system. Indeed, in places man has altered natural regimen of water supply to such an extent that it cannot be rectified by the application of concrete and labor. Leaders in water planning can no longer exercise imprudent measures that do not have the benefit of a realistic cause-and-effect analysis.

In the discussion of surface-water law in Arizona, the author shows how the common-law doctrine of riparian rights was found to be ill-suited for water development of the arid Southwest. This doctrine, although suitable for hydrologic conditions in a humid environment, was found by the early settlers to be impractical in an area where rainfall and streamflow is sparse and unevenly distributed. There could have been no other choice than the establishment of the doctrine of prior appropriation if the West were to be developed into the national asset it is today. These facts were clearly recognized in the early water history of Arizona and codes were established in keeping with the prior appropriation doctrine. Water rights under this doctrine are for beneficial use which permits transport of water to areas far beyond the original stream or watershed. Although there have been refinements, as documented by several cases, the state legislature has been careful to protect vested water rights. Today essentially all surface water is appropriated in Arizona and the present code appears to be adequate.

In comparison to surface-water law, the legal history of groundwater priorities is one of chaos and gross inconsistencies. As the author clearly points out, the primary reasons for this situation are: 1) a lack of utilization of scientific and technical information relating to cogent ground-water development, 2) the inability of the courts to foresee the large demand on subsurface water supply to accommodate the Arizona economy, 3) the rapid decline in water levels in certain areas, and 4) man's inability to distinguish between underground streams and percolating waters which allowed broad discretion in establishing rules of ownership. Dr. Mann provides an excellent documentation of many court cases which illustrate the inability of the courts to hand down consistent decisions.

After World War II the accelerated development of groundwater reserves began, and state leaders became concerned about safeguarding this resource. The U.S. Geological Survey was busy making intensive surveys on major ground-water basins to obtain factual data, and in 1948 the Arizona Legislature established a ground-water code. However, it did not attempt to solve the long-range problem and did little more than slow down development already underway. Several major lawsuits followed, and Governor Pyle appointed a commission to make intensive studies and recommendations to strengthen the 1948 code. In 1952 the commission report recommended the adoption of the correlative rights principle for regulation. This envisioned the closing of over-developed areas, the creation of districts to cut back pumping on a local basis, and the establishment of a commission to administer the law. This report was received with something less than enthusiasm by many residents. Further legal cases followed which debated the issues and police powers for enforcement. Since 1955 there has been little effort to establish a strong code, and today the economics of pumping costs are the prevailing influence on ground-water exploitation. Those enterprises which can afford the cost are withdrawing water, and priority rights are being purchased with land acquisition.

In the final chapter Dr. Mann has perceptively summarized water management in Arizona and factors which require the wisdom of Solomon for resolution. These factors certainly attest to the need for much public interest and talent. The Arizona State Land Department has received much criticism for its efforts to comply with its responsibilities in water management. Because Arizona includes so much federal land, state officials have had little choice but to comply with the diverse policies and pressure groups generated by the several federal agencies concerned with water management. On the other hand, more courageous leadership by state officials and politicians armed with scientific and technical information could have resulted in much progress in the resolution of long-range water problems.

The time is now for Arizona to establish basin-wide management-planning of water resources. The need is critical in several of the major valleys where intense development is depleting Arizona's most important resource. Planned management of water can not only alleviate a serious blow to local economy, but also provide the attraction for expansion. The time may be near when decisions to continue haphazard development or modify it with planned regulation will be made. Better understanding is needed of all the phases of Arizona's hydrologic system. All of the disciplines concerned with water need to be merged and harmonized if permanent occupation of the desert is to be achieved. It is indeed encouraging to note that there are trends in this direction. Such groups as the Arizona Water Resources Committee have recently undertaken objective analyses of the multitudinous factors relating to the optimum development of a limited water supply.

I heartily commend the author for his concise and cogent account of a controversial and much-debated subject. Dr. Mann's book is an outstanding contribution to water-management studies.

John W. Harshbarger
Department of Geology
University of Arizona

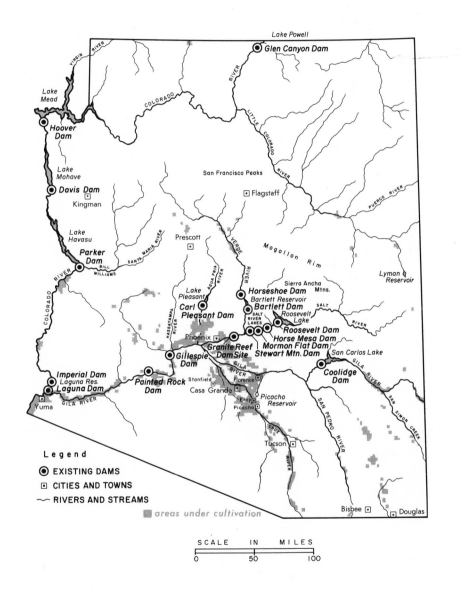

Lake Powell

◉ **Glen Canyon Dam**

Lake Mead

◉ **Hoover Dam**

Lake Mohave

◉ **Davis Dam**

⊡ Kingman

San Francisco Peaks

⊡ Flagstaff

Lake Havasu

Prescott ⊡

◉ **Parker Dam**

Mogollon Rim

Lyman Reservoir

Sierra Ancha Mtns.

Lake Pleasant

◉ **Horseshoe Dam**
Bartlett Reservoir
Bartlett Dam

◉ **Carl Pleasant Dam**

SALT RIVER LAKES
Roosevelt Lake

Phoenix ⊡

◉ **Roosevelt Dam**
Horse Mesa Dam
Mormon Flat Dam
Granite Reef Dam Site
Stewart Mtn. Dam

◉ **Gillespie Dam**

San Carlos Lake

Imperial Dam
Laguna Res.
◉ **Laguna Dam**

◉ **Painted Rock Dam**

Stantfield
Casa Grande
Florence ⊡

◉ **Coolidge Dam**

⊡ Yuma

GILA RIVER

Eloy ⊡
Picacho ⊡

Picacho Reservoir

Tucson ⊡

Bisbee ⊡   ⊡ Douglas

Legend

◉ EXISTING DAMS

⊡ CITIES AND TOWNS

⌇ RIVERS AND STREAMS

▨ areas under cultivation

SCALE   IN   MILES

0   50   100

River Systems and Irrigated Areas

—*Esther Henderson*

—*James R. Hastings*

—*Valdis Photos*

The three basic topographic regions in Arizona are the high plateau country to the north and northeast (top left), the desert basin and range area to the south and southwest (bottom left), and the belt of forested mountains and narrow valleys lying between them (above) which yields the major portion of the state's usable surface water.

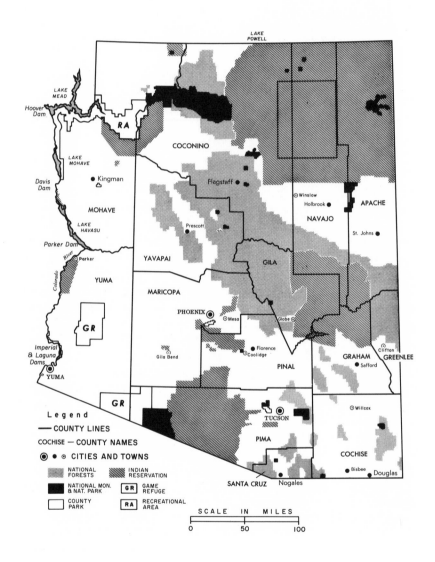

LAKE
POWELL

LAKE
MEAD

Hoover
Dam

RA

LAKE
MOHAVE

● Kingman

Davis
Dam

COCONINO

Flagstaff ●

⊙ Winslow
Holbrook ●

MOHAVE

LAKE
HAVASU

Prescott
●

NAVAJO

● Winslow
Holbrook ●

APACHE

St. Johns ●

Parker Dam

Colorado River

⊙ Parker

YUMA

YAVAPAI

GILA

MARICOPA

GR

PHOENIX ◉

⊙ Mesa

Globe ⊙

Imperial
& Laguna
Dams

◉ Florence
⊙ Coolidge

GRAHAM

Clifton ⊙
GREENLEE

YUMA

Gila Bend ⊙

PINAL

● Safford

GR

⊙ Willcox

TUCSON ◉

Legend
— COUNTY LINES

COCHISE — COUNTY NAMES

◉ ● ⊙ CITIES AND TOWNS

	NATIONAL FORESTS		INDIAN RESERVATION
	NATIONAL MON. & NAT. PARK	GR	GAME REFUGE
	COUNTY PARK	RA	RECREATIONAL AREA

PIMA

COCHISE

● Bisbee
Douglas ●

SANTA CRUZ   Nogales ●

SCALE   IN   MILES

0   50   100

Parks, Monuments, and Indian Reservations

CHAPTER 1

The Physical Setting

The aridity of Arizona is proverbial and legendary. One man said of this country, "Everything dries; wagons dry; men dry; chickens dry; there is no juice left in anything, living or dead, by the close of summer."[1] Some of the earliest prospectors were reported to have written home that in Arizona it was 80 feet to water and two feet to hell. Other stories, even more apocryphal, tell of those inveterate desert dwellers, the Gila Monsters, leaving Arizona in protest after one of its infrequent rainstorms asserting that they had to have a dry climate.[2] While the truthfulness of these stories may leave something to be desired, they attest to one situation in Arizona which conditions and circumscribes all human endeavor — the perennial shortage of water.

As the Hohokam Indian long before him, and the Spaniard in a later period, the Anglo-Saxon settler in Arizona has had to struggle to establish a permanent civilization in a country that seemed inhospitable to man. The Hohokam eventually was driven from his home by failure to solve, among others, his water problem.[3] He had developed a populous and a relatively prosperous society in central Arizona based on stored water and irrigation agriculture but found his desert home untenable, probably because of his inability to meet the water problem. The Spaniard had less difficulty, perhaps because he never colonized or settled the country as intensively nor engaged in agriculture to the extent that his predecessors or those who followed him did. But even the Spaniard was forced to recognize the limitations of the desert and to develop institutions designed to conserve the water supply. Today the Anglo-Saxon, his numbers growing rapidly, is trying to accommodate himself to the limitations which the shortage of water imposes, while at the same time attempting to improve Arizona's capacity to meet its expanding needs for water. He is facing the same problems that confronted his forerunners and "since water is the key

1

to the greater part of Arizona's modern development no aspect of its economic life merits greater attention."[4]

Although the state is not entirely within a typically arid zone, the major portion may be so considered, including the Painted Desert of the northeast, and the Sonoran Desert of the southwest. For example, the average annual rainfall at Phoenix, located in the south-central section of the state, is only 7.67 inches, while at Yuma in the southwestern corner the average is but 3.38 inches.[5] So arid is this latter region that it has been the source of many a joke, typical of which is the note sent to the central office in Phoenix in 1925 by a weather observer in Mohawk:

> Did not fill in [the rainfall column] as the only precipitation we had was a wild report that oil had been found in a dry hole at Stoval which report precipitated a large part of the population of Yuma with all the fervor of a Gold rush, and could not make up my mind, from information at hand as to proper classification, whether pure madness or plain damphoolishness.[6]

The major portion of the usable surface water in the state comes from precipitation which falls on the mountain ranges traversing the state from northwest to southeast. At Crown King in the Bradshaw Mountains, for example, the average annual precipitation is 27.73 inches, while at Workman Creek in the Sierra Ancha Mountains the estimated average is 31.44 inches, making it one of the wettest spots in the state.[7] The dryness of the state as a whole is indicated by the fact that, during an average year, approximately one-half the state receives less than 10 inches of precipitation.[8]

Coupled with this general scarcity of precipitation is the relatively wide temperature range in Arizona. Temperatures at Phoenix and Tucson, 125 miles south of Phoenix, can be expected to reach 110 degrees in the summer and seldom fall below 20 degrees in the winter. A much lower range of temperatures can of course be expected at higher elevations, such as at Flagstaff in the north where the summer high approximates the winter low at Yuma. The air is extremely dry, the annual average relative humidity ranging between 40 and 60 percent. The high temperatures, combined with the low relative humidity, produce a high rate of evaporation. At some stations the evaporation from free-water surfaces reaches 10 feet per year, although the state average is between 6 and 7 feet. Thus a considerable portion of the water supply is lost without ever having been put to use. The growing seasons are long, however, Phoenix having an average growing season of 304 days. Such a long growing season, of course, makes it possible to double crop, but this has the adverse effect of increasing annual water use.

Arizona rainfall is concentrated in two seasons, summer and winter. Summer storms contribute approximately 43 percent of the annual rainfall, the remainder coming in the winter. For long months during the spring and fall not a drop of rain will fall in many areas of the state. Today, nearly all of Arizona's streams are intermittent, because of the diversion of water for irrigation and the decline in runoff, although in the past the Salt and Gila rivers were perennial streams. The summer storms contribute little to the usable water supply because of the high proportion that is lost to evaporation and in restoring the moisture deficit in the soil. The winter storms, coming when temperatures are lower, and plants are not growing, are the most important sources of water for beneficial use.

Dendrochronologists, who utilize the analysis of tree-ring growth to estimate past rainfall, have concluded that Arizona has experienced during the past three decades a drought comparable to the worst droughts recorded by the tree rings. Stream-gauge records maintained by the U.S. Geological Survey demonstrate conclusively the downward trend in runoff in all of Arizona's major streams including the Colorado, Gila, Salt, and Verde rivers.[9] Surface runoff has been decreasing, therefore, during a period of greatly increased demand for water. The USGS provided the following estimates of increased use of water from the year 1950 to the year 1955. In million gallons of water per day, rural use declined from 36 to 14; public-supply use increased from 75 to 131; self-supplied industrial use increased from 45 to 157; and irrigation use increased from 4,628 to 6,910.[10] Total use increased from 5,784 million gallons per day to 7,212 million gallons per day.

The dependable supply of surface water has long since been appropriated. Nearly all of the increased water use has been dependent, therefore, on pumping from underground basins. In 1940 an estimated 1,500,000 acre-feet were pumped for irrigation; by 1953 pumping had increased to 4,800,000 acre-feet.[11] This figure declined slightly in 1954 and 1955 owing to decreased acreage in cultivation and above-average rainfall. The 1955 figure for pumping was 4,400,-000 acre-feet, about twice the amount of water diverted for irrigation from surface supplies, and pumping has remained over the four million acre-foot mark since.

The result of this greatly increased pressure on the underground water resources has been a steady decline in the water tables. In the most important agricultural areas of the state, those located in the broad central valley north and south of Phoenix, the decline in the water table has been precipitous. In the Maricopa-Stanfield area the table declined more than 75 feet between 1946 and 1955.[12] In the area between Picacho and the Casa Grande Mountains, pumping of underground water has been so great that a subsidence, or actual lowering of the land, has occurred.[13] The rapid decline of the water tables has affected the agricultural economy by increasing costs of pumping and also by decreasing yields of water in many instances. In some areas the costs have become prohibitive.

These water conditions require that the state of Arizona make a fundamental assessment of its economy. Barring some dramatic technological development in rainmaking or conversion of saline water, it appears that Arizona will have to rely only on its present surface-water supply — with all its variability — and the diminishing ground-water resources. It is clear that not all of the demands presently being made on the water supply can long be satisfied. Some water uses will have to decline to provide for others. It is incumbent on the state to make decisions in this matter in terms of its long-run interests.

This judgment rests on the assumption that the primary soil and water problems are at least in part man-made and therefore capable of solution by improved management and application of better techniques. There can be no argument with the fact that water use has increased tremendously and that this has been a causative factor in the state reaching its present limit on water supply. There is no doubt that the state can reverse the depletion of its water supply, if it so desires and if the long-run economic consequences are beneficial.

Some, however, assert that the water situation is not man-made but the result of a progressive desiccation of the area.[14] They attribute the decline in surface water and the erosion of the land and the depletion of the forage to the gradual changes in the natural processes rather than to misuse or over-use.

It appears certain, however, that man has at least hastened the erosive and depleting processes, whether or not these were dominant in the period prior to heavy immigration into the territory. The San Pedro River, for example, was described in 1846 as a river "an active man could jump across."[15] And near its confluence with the Gila River the grass and other verdure grew luxuriantly. Some early travelers found the rivers to be sparkling streams near which there were signs of beaver, quail, deer, turkeys, and javelina, and in whose waters fish played. One intrepid soldier even attempted to float provisions down the Gila River, a feat beyond the wildest imaginings today.[16] After examining the historical evidence, Thornwaite reports that in 1870 the valley of the San Pedro "had a shallow grassy bed and banks covered with luxuriant vegetation. Willow, cottonwood, sycamore, and mesquite timbers were abundant, and there were large beds of sacaton and grama grasses and sagebrush."[17]

Similar evidence was developed regarding other streams such as the Little Colorado where several explorers found a "narrow perennial stream lined with cottonwoods and willows" where "the surrounding hills once bore a good stand of grama grass."[18]

Early land promoters painted Arizona as a virtual paradise. Their testimony must be discounted to a considerable extent, but it may at least indicate that the territory was not always so poor and arid. In 1886 Patrick Hamilton, the Commissioner of Immigration, wrote: "Popular opinion has long considered Arizona a waterless region, but the truth is that few countries of the West are so abundantly supplied."[19] Both Hinton and Hodge felt that at least 2,800,000 acres could be irrigated while the former estimated that anywhere from 10 to 20 million acres could be irrigated from artesian wells.[20] Hamilton felt that all of the arable land in Arizona's borders could be watered with the existing water supplies.[21]

These same men had a similar optimistic view of the forage resources. According to Hamilton there was "a veritable paradise" of "millions of acres of fine grass lands . . . . "[22] There was no danger of "eating up the range" since, "after being grazed down to the roots, the sweet gramma grass shoots up next season with fresh vigor and luxuriance."[23] One traveler is quoted to the effect that "Arizona is decidedly the best grazing country on this continent, capable of subsisting millions of cattle without the aid of man."[24]

Since these travelers and settlers made their observations entire valleys have been virtually destroyed in their capacity to provide forage. Instead of small perennial streams, many streams have become rampaging torrents during brief periods of the year and dry washes during the remainder of the year. By a process of cutting and trenching, river channels widened and deepened, with resulting dissection of the bottom land. It has been estimated that at least 500,000 acres of the finest grazing land in the state were destroyed through this process and many other areas sadly depleted. In a careful study of the historical evidence James R. Hastings of the University of Arizona demonstrates conclusively the relationship between arrival of cattle in large numbers and accelerated erosion, while admitting that Arizona was something less than Paradise when the cattlemen arrived.[25]

Many farming lands have been badly damaged through improper methods of cultivation which allow wind action to remove soil from tilled fields. Water and more wind then cut deeply into the soil, resulting in land unsuitable even for grazing. The evidence appears incontrovertible that the acceleration of erosion is associated with increased use of land by man and the adoption of improper techniques in the management of resources.

The evidence would indicate, therefore, that remedial efforts could be efficacious in restoring the soil and water resources in order that they might better serve the needs of the people of Arizona. Undoubtedly it would be impossible to bring back the primitive conditions existing prior to intensive settlement nor would such a restoration be desirable. But it would appear to be in the long-range interest of the people of the state to maximize the production of usable water, reduce the losses attendant upon its use, protect the related resources — soil, forage — which contribute to a stabilization of the water supply, and make use of the best scientific information in extraction and appropriation of the water. It is to an analysis of the efforts made by state and federal agencies to plan for water needs of the people of Arizona that the following pages are devoted.

CHAPTER 2

Water Resources Planning

Perhaps the most popular concept in the literature of natural resource management is that of "planning." It is almost universally felt that orderly, wise, and maximum utilization of our resources is dependent on the creation of comprehensive and detailed programs, which are in turn based on the best available scientific data and agreement on the goals to be achieved. The lack of such planning, so absent in the past, will result, it is felt, in the United States becoming one of the "have-not" nations in many respects long before there is any necessity for that to happen. As Paul Sears stated:

... the allocation of resources to a population is a function of the culture of that particular group .... Fortunately, in the United States we are not under the pressures that exist in many parts of the world. We still have a great margin of safety. But this, rather than serving as an excuse for recklessness in the treatment of our resources, gives us the opportunity to think and look ahead toward the time when we shall have to come to some kind of terms with our own space and resources. These, however vast, still are certainly finite.[1]

Everyone engages in planning of one kind or another when he develops or utilizes our natural resources. Even the most rapacious destroyer of our resources engages in a kind of predetermined effort to maximize his benefit from the available resource. It is not, therefore, simply planning that is desired but planning that is based on the long-run needs of the society and in conformity with our best information regarding the inter-relationships of nature. We have evidence all around us of the limited planning that destroys the primary resources and often leads to the destruction of other resources in the delicate ecological balance. Too often, as Luna Leopold points out, those who express concern for conservation of our natural resources "concentrate on promoting development."[2] The real problem of conservation and the function of planning "are to identify those aspects of the resource which would be depleted or degraded if the economic forces were

7

allowed to operate unhampered."[3] The goal of the conservationist, therefore, is to protect nature's balance while at the same time providing for the satisfaction of man's multiple needs.

There is no dearth of evidence of planning in Arizona, dating all the way back to the time of the Hohokam Indians who populated central Arizona in the 1300's. The Indians developed an irrigation system that sustained a populous culture in the central valley of the state. Although the immediate cause of the decline of their culture is not certain, it is thought that it was directly related to their inability to control their water supply under adverse conditions. The Spanish period and the early American period both provide evidence of planning in the attempts on a communal basis to provide for an adequate water supply through *acequias* (irrigation ditches), the adoption of the prior appropriation system of water rights, and later, the construction of dams.

Increased pressure on our resource base has caused us to take another look at our resource planning because the limited and decentralized planning can no longer provide for the many varieties of usage and the amounts of usage demanded by our population. This is of course a nationwide problem resulting from a rapidly growing population, a higher standard of living, and the pressures of international politics. While it is undoubtedly true that technology will compensate for some of the predictable shortages, it is an act of sheer faith to rely entirely on such developments.

The need for more comprehensive and informed planning of our resources is nowhere more apparent than in respect to our water supplies. Although it is estimated that only one-fifth of the total water supply of the United States is being utilized at the present time, it is expected that the demands on the nation's water resources will double by 1980 and triple by the year 2000.[4] However, according to the President's Materials Policy Commission, "If a single word were used to sum up the water problems of the United States, that word would be maldistribution — maldistribution in regard to time and geographic areas."[5] The general aridity of the areas west of the 100th meridian is well known. Of these states in the arid region Arizona is among the most deficient in water supply. Nearly all of its area has a water yield of less than 12 percent of the national average. The average annual precipitation is 14 inches in Arizona, of which only 0.7 inch, or 5 percent, becomes stream flow or ground water.[6]

Coupled with the low precipitation is the extreme variability of rainfall in space and time. Storms are extremely localized, particularly those in the summer, and tend to be of short duration. Concentrated in a few summer and winter months, these storms always present the danger of flooding because of their high intensity.[7] Only with the construction of major works for the storage of water and protection against flooding has it been possible to provide a favorable habitat for a population now numbering over 1,300,000 people. These works, moreover, have been highly efficient in the sense that almost no water is allowed to waste from the major river system, the Gila, into the Colorado River. This does not mean, of course, that all the water is necessarily used efficiently. Seepage and poor water application, among other things, account for a good deal of waste.

Besides the uneven distribution of rainfall geographically and among periods of the year there are also wide variations over long periods of time, within recorded history. Total runoff in 1900 was only 290,000 acre-feet, while in 1905 it measured 5,200,000 acre-feet. In the past century there have been two serious droughts, from 1892 to 1904, and from 1942 up to the present time.

There are also problems with regard to water quality. The state is increasingly forced to use water of lower quality for irrigation. The most serious problem has occurred in the lower Gila Valley where the accumulation of salts has been severe. These salts, dissolved in the irrigation water, are deposited on the soil when the irrigation water is applied. Continued accumulation of these salts reduces the productivity of the soil and ultimately threatens to make it unsuitable for cultivation, unless the salts are flushed away. From 1951 to 1955 the streams entering the Gila Valley contributed more than three million tons of salts, for an average of 600,000 tons annually. The Safford Valley has also experienced similar conditions.

Agriculture has been the largest single user of water in Arizona, and in absolute terms, will undoubtedly continue to be for many years. In 1960, 95 percent of the water used was applied to agriculture.[8] Two conditions are forcing a change in the pattern of water use, however. As the water table has gone down, the expense of extracting the water has become almost prohibitive in some areas, causing acreage to go out of production. Reduced prices for agricultural products have naturally played an important role here. With the rapid growth of the state in population, and the gathering of these new settlers in the cities, there has been increased competition for the limited water supply. It is estimated that between 1949 and 1957 municipal use of water increased from 40,000 acre-feet to 120,000 acre-feet. Those using water for domestic or industrial purposes are able and generally willing to pay a great deal more for water than is the irrigationist. This competition for water has already affected the Salt River Project near Phoenix and it will undoubtedly affect the future of agriculture around Tucson. Of the 242,000 acres historically under cultivation in the Salt River Project only 190,000 acres are now devoted to agriculture, the remainder now being utilized for residential, commercial, and industrial purposes.[9]

A 1957 report on the water supply of the Tucson area, conducted by the Agricultural Experiment Station, indicated that some agricultural land would have to be taken out of production to provide water for the burgeoning population.[10]

The growth in population in Arizona has followed the trend found in most of the Western states. During the decade 1930-1940 the population increased only 14.6 percent. Between 1940 and 1950, however, the increase was a nation-leading 50.1 percent, with the population reaching 750,000.[11] By 1960 the population was more than 1,302,000, an increase of 73 percent over 1950.[12] Nor is the growth expected to diminish. In 1955 the Bureau of the Census projected the population of Arizona for succeeding five-year periods on the basis of varying assumptions concerning migration, fertility, and mortality, and estimated the following possible ranges of population: 1960 — 1,169,000 to 1,268,000; 1965 — 1,326,000 to 1,528,000; 1970 — 1,491,000 to 1,802,000.[13] On the basis of the 1960 census figures, these estimates for 1965 and 1970 are obviously low. However, at the very least the state can expect an increase in the population of 39 percent between 1960 and 1970. 2.5 million 1975

It can be expected, furthermore, that the greatest share of this increase will occur in the cities. During the 1940-1950 decade the populations of Phoenix and Tucson increased 78.2 and 93.9 percent respectively.[14] The state was only 35 percent urban during the period from 1920 to 1940, but by 1950 it had reached the 55 percent figure and the urban areas continued to grow rapidly during the 1950's. By 1960 the metropolitan areas around these two major cities comprised over 71 percent of the state's population.

The increased urbanization of the state is indicative of the trend toward economic diversification. New industry is coming into the state, attracted by the climate, relatively low costs of the factors of production, the existence of major government installations, and the individual company's desire for decentralization.[15] Most of this industry is classified as light industry, the best examples of which are the aircraft and electronics concerns. Between 1947 and 1958 over 400 new plants located in the Salt River Valley near Phoenix, providing over 20,000 new jobs or more than four times the number of factory jobs available there in 1947.[16] In 1960 manufacturing contributed $700

Rainfall in the desert is usually limited to intense but localized showers which contribute little to usable water supplies due to rapid evaporation and absorption by the arid soil (left).

—Esther Henderson

Most of the state's surface water originates in the high mountain belt, mainly from spring thaws of winter snow, and late summer rainfall. Small lakes, timberlands, and grazing for sheep, cattle, and wildlife characterize this important area (below).

million to the economy, ranking well ahead of mining and agriculture (including livestock), which contributed $415,776,000 and $435,554,000 respectively.[17]

Another rapidly developing area of the economy — tourism — must be considered in the context. Although this industry does not affect the water situation significantly in a direct way, it does affect the plans of those who continue to see agricultural pursuits as the predominant way of life in Arizona. The agriculturists look to the watersheds as a source of increased supplies of irrigation water through vegetation manipulation. The recreation industry has vigorously opposed such measures as destructive of the natural values of the state.

A survey by the National Park Service, published in 1950, asserted that "as the various sections of the [Colorado River] basin become better known and more accessible to the densely populated regions of the United States . . . catering to the recreationist should become a major industry."[18] The Park Service suggested that "to foster this industry it must be recognized that recreational use of land may, in certain places, be the highest or best use of the land . . . . The great stretches of open range, unobstructed by buildings, fences, transmission lines, and other signs of modern civilization, comprise one of the most important features of the basin. As other sections of the United States become more and more highly developed, this one feature of the Colorado River country if preserved, will have unusual appeal."

The tourist industry in Arizona has already established itself as a rival to other forms of enterprise. L.W. Casaday estimated in 1953 that the tourist industry provided a payroll exceeding those of trade, government, manufacturing, and services.[19] In some areas of the state, particularly in the north, primary dependence is on the tourist industry.[20] The Arizona Development Board and the Bureau of Business and Public Research at the University of Arizona are giving much attention in their publicity and research to tourism.

Increased recreational use of private and public lands may not seriously compete for water since human consumption is minor compared to irrigation use. However, under the Mission 66 program of the National Park Service, there is included in almost every expansion plan provision for the acquisition of water rights or the development of a water supply.[21] The most likely source of conflict lies in the management of public lands. The irrigationists have backed adoption of the practices proposed by the Arizona Watershed Program in the Barr report, *Recovering Rainfall,* which involve the cutting of timber and shrubs for the increased production of water for stream flow.[22] The sportsmen have resisted these proposals, arguing chiefly for the recreational values of what they claim is the pristine condition of the forest. There is also competition over the waters of the Colorado River. Both the Bridge Canyon Project and the Marble Canyon-Kanab Creek Project have come under National Park Service fire because they diminish the scenic beauty and geologic impressiveness of the Grand Canyon.[23]

The National Park Service sees many other areas as having considerable historical, recreational, or scenic value which are as yet unprotected or undeveloped. It strongly suggests the advisability of state action to ensure their beneficial use for the public.[24] The Park Service itself is in the process of planning the recreational programs for the Glen Canyon Dam area and the lower reaches of the Colorado River.[25]

In view of the extreme scarcity of water and the changing economic situation in the state one would expect state action to provide for orderly planning for management of the water supply. Certainly this would require the creation of some centralized machinery to pursue this task. Such machinery and planning have never developed, however, and it appears that there is a willingness to allow the economics of water use to dictate the purposes to which water will be put. The *Arizona Farmer-Ranchman* recently stated:

> The major road block in the way of statewide planning is that there is not enough water now to meet all the needs and demands for water. Under present conditions — until much more water is available — even statewide planning probably calls for sacrificing the full development of one area for that of another.[26]

But if it is difficult to achieve planning in times of shortage and maximum utilization it is even more difficult to conceive of it being achieved during periods of plenty.

Planning for the maximum development and utilization of water is replete with difficulty, rendering understandable the reluctance of the state even to attempt it. These difficulties are constitutional, legal, political, economic, and administrative. Each of these alone would be sufficient to hinder planning efforts, and the combination of the five makes such efforts virtually impossible.

The federal system, whatever its advantages in other fields, has complicated immeasurably the attempt to obtain basin-wide or even statewide planning.[27] Such a system depends either upon a clear-cut division of responsibility between the states and the federal government and among the states themselves, or on a willingness on the part of each to negotiate and reach a reasonable basis for cooperation between them. Frequently, none of these conditions is present and the management of water is controlled by the stronger party, either through acquiescence on the part of the weaker party, or through judicial decision. In the most significant instances in recent years, the stronger party has been the federal government.

The powers of the federal government in managing water are based on its constitutional powers to regulate interstate commerce and navigable streams and to manage its own property. The federal government, however, has separated its control over land and the supply of water thereon, allowing the states to determine the means by which and the purposes for which water will be used. In controlling navigable streams, however, it is not always possible to recognize asserted rights to water under state laws because of other demands made upon that water supply. "Rational" basin-wide planning and management consequently falter because of a failure on the part of the interests having a stake in the water supply to recognize a community of interest.

Illustrations of these basic conflicts are not hard to find in Arizona. Supposedly the states have the right to determine the basis for water use — usually either the riparian system or the appropriative system or some modification of the two. In the negotiations involving the support by the Bureau of Reclamation for the Central Arizona Project, the bureau virtually required that the state pass legislation seriously modifying the principles of ground-water use which had governed water use in the state since its earliest days. The demand was reasonable enough, but hardly an example of basin-wide planning in any real sense. In fact, it could be argued that it was planning in no sense, since the ground-water law passed in 1948 under bureau pressure was virtually ineffective. Such domination by the federal government causes one to question the extent of the "large field of water resources activities open to" the states, "including the 'right to choose their own form of water law.' "[28]

The states of the West, including Arizona, have become concerned over the asserted claims by the federal government that its agencies and its wards are not bound by state water law on lands withdrawn for their use.[29] During the protracted fight over the passage of ground-water legislation, frequent attention was drawn to the Indian reservations in the critical ground-water areas where no regulation was operative. The officials of the Bureau of Indian Affairs disclaimed any intention of abiding by state water law, asserting that reservation of the land resulted in reservation of the water supply sufficient to water the land. More recently this same problem has arisen in regard to surface waters. In a constitutional sense, the Indians and their federal attorneys may be quite correct, but such exemptions from state regulatory power make statewide planning and management farcical.

A similar problem has arisen on the Colorado River in the suit between California and Arizona. The federal attorneys have argued in behalf of their wards that the Indians have a first lien on Colorado River water, regardless of any appropriative rights already established under state law. Each state, it is asserted, must subtract from its share the amount of water claimed by the Indians.[30] In Arizona this could possibly mean the reduction in size of projects already in existence in order to satisfy the Indians' demands. The importance of these claims is indicated by a statement attributed to an attorney of the Salt River Valley Water Users' Association: "It might well be that Arizona and California will be required, to some extent, to forget their differences in order to jointly defend against Indian claims."[31] If these views that federal installations do not have to abide by state water laws are upheld, the consequences also could be serious because of the large number and the extent of the military installations in the state. There have been notorious examples of failure to abide by state game and fish laws on the part of the military on their extensive reservations. Military officials gave in only after protracted negotiations and argument.

The controversy which arose during the planning and construction of Glen Canyon Dam is another case in point. The Lower Basin states were concerned about the effect on water availability and power production. They demanded assurances from the Secretary of the Interior that in the construction and operation of the structure the United States "shall recognize the right of the Lower Basin to the consumptive use of water as superior to the Government's to accumulate or retain water at Glen Canyon," and that the United States would not "impair the performance of its Hoover Dam power contracts by the filling and operation of the Glen Canyon Reservoir."[32] For some time the Interior Department would give no guarantees to the Lower Basin states. As a result of the opposition to this doctrine of reserved rights in the federal government, the Western states caused to be introduced into Congress the Water Rights Settlement Act bill, which would provide that all persons and agencies, including the federal agencies, would have to acquire rights to the use of water in conformity with state law. It would further provide that no federal installation could interfere with any vested right to the use of water under state law unless authorized by law and upon payment of just compensation.[33] The National Reclamation Association, with Arizona members in agreement, supported the measure. The Senate held hearings, at which both the state attorney general and the state land commissioner testified in favor of the bill, but to date the bill has never been passed.[34] However, in April, 1962, the Interior Department announced the filling criteria for Lake Powell (the reservoir at Glen Canyon), which appear to satisfy the demands and calm the fears of the Lower Basin states. (See Chapter 7.)

The legal tangle of water law and water rights also presents a very real impediment to over-all planning. For planning requires and presupposes a favorable legal structure in the framework of which necessary adjustments can be made to meet the demands of the community. It is apparent that those who desire basin-wide development programs give too little recognition to the enormous problems of adjusting the already established rights in water under state and federal laws and constitutions. It may make economic sense to argue that "the West must soon decide whether its future must be sacrificed by its antiquated priorities system in water," but it is quite another matter to devise a legal system acceptable to all important interests in a basin.[35]

Striking illustrations of the legal problems involved in planning in Arizona are easily found. The U.S. Geological Survey suggested in 1951 that the state laws should "apply the same rule of law to surface water" that was applied to ground water, "recognizing the . . . interconnection between ground and surface water and the necessity of treating the common supply as a whole where such interconnection exists."[36] It is difficult to see how this end can be achieved in Arizona when the courts have asserted with apparent finality that contrary doctrines shall apply to surface and ground water. Barring an overturn of the rule of *Bristor* v. *Cheatham*,[37] which is not impossible but unlikely, the problem remains one of establishing a legal basis for regulation within the framework of the existing doctrines.

Others have suggested that there be a re-ordering of preferences for the use of water under state laws, recognizing new demands put on the water supply unforeseen by those who framed the appropriation statutes. While this suggestion may be useful in some states, its value in Arizona is extremely limited inasmuch as nearly all surface water has already been appropriated.[38] Ground water, of course, not being subject to the appropriation doctrine, is not governed by a system of preferences.[39]

It has been suggested that individual rights to water must be recognized as property, "but rights acquired decades ago should not serve as an estoppel to programs which provide a wiser use of water now from the standpoint of the overall public interest."[40] The end is undoubtedly desirable but the only solution advanced is condemnation proceedings which only the state or its agent may undertake. Uses of water other than those involving public agencies may have a high priority, yet no workable system has been devised to transfer rights from one owner to another except by willingness on the part of the present user to give up his right.[41]

Planning under Arizona water law has been made extremely difficult because of the rejection of the appropriation doctrine for ground water. It would have been possible to limit appropriations to annual recharge or to some reasonable amount proportional to annual recharge. In its *Bristor* v. *Cheatham* decision, the Arizona Supreme Court not only rejected this course but "overlooked its next best chance to halt the depletion of water," the correlative rights rule, which would have provided for equitable apportionment of available water among existing users in areas of limited supply.[42] Without such a rule, "it is doubtful whether the Arizona legislature can constitutionally define reasonable use strictly enough to stop the rapid drain of limited ground water supplies."[43] The Supreme Court of Arizona has indicated that the police power may be used to control ground-water use, and upheld application of the police power in the *Southwest Engineering Company* v. *Ernst* case. The extent of the police power is as yet undetermined, and it is not at all clear that it may be used to cut back the actual quantity of water pumped.

The economics of water conservation and use ranks high on the roster of problems facing the state. The question of the most economic use of water perhaps underlies other value questions inasmuch as it relates directly to the contest among agriculture, industry, municipalities, and recreation for their "fair" share of the water supply. It is, of course, recognized that other non-economic considerations do play important parts in the decisions ultimately made — the desirability of the family-size farm, the changes in demand for agricultural commodities, the demand that water be used at its place of origin — but it is nonetheless necessary to examine the economic problems without relation to other value questions.

In the past, economists have considered agriculture as the primary generator of economic activity. In a modern industrial economy, however, it is questionable whether this is the case, for allocation of water to industrial production may result in a greater increase in economic activity than would be possible under similar allocations to agriculture.[44] In most areas, where there is already a surplus of water, such allocation of water to industry would cause no deprivation to agriculture, but in the Southwest such an allocation would result in acres of land being taken out of production.[45] It has been suggested that the theory of marginal utility be used in judging the purposes to which water should be devoted. According to this theory the rule should be, "use water for the purpose that has the highest marginal value product," and a "maximum addition to the social product would be guaranteed."[46] There is no question that most reclamation projects in the past and many proposed for the future have financial feasability in that reimbursible costs will be returned to the government; a more important question is whether the water could not be better put to use in industry rather than in producing crops, thus contributing more to the national economy.[47] Congress has been noticeably more reluctant to authorize reclamation projects as the ratio of costs to benefits approaches 1:1. Traditionally, in reclamation projects it has been thought necessary and advisable to provide a subsidy to agriculture, but such assistance to the farmer is frequently based on social considerations rather than on strict economic arguments.

One of the serious roadblocks to reliable economic analysis arises from the difficulty of assigning economic values to all of the benefits of conservation practices. While the Bureau of Reclamation assigns dollar values to recreation and aesthetic enjoyment, others doubt that such assignation has significance. One observer suggests that aesthetic and ethical values should be "distinct and immiscible with economic ones," and that such values should not be forced to compete with each other.[48] He adds, "For resources which are principally noneconomic in value, let us decide whether we want them, but not by assigning a dollar sign to scenery and not by making the sale of hot dogs a measure of the worth of a park."[49]

Powerful attacks have been made in recent years on the more traditional economics of allocating water supplies for various uses. Some have argued, such as Hirshleifer, De Haven and Milliman, against sentimental attachments to particular interests in the economy with regard to allocation of water supplies.[50] They contend that the market pricing system is the best mechanism for allocating water supplies to their highest use, with a very minimum of public control. Others, such as Krutilla and Eckstein, using the concepts of welfare economics, have critically analyzed standards under circumstances in which the decisions must be made under budgetary restrictions.[51] These, among other studies, have emphasized the increasing attention being given to the justification of water-supply projects when there are only limited funds and conflicting claims on those funds.[52]

In the nationwide economy consideration must be given to alternative means of providing the agricultural products needed by the consumers. Serious objections have been raised to reclamation projects which result in the increased supplies of commodities already in surplus and supported by federal price supports. Obviously, there is no economic gain if the heavy investment by the federal government in a reclamation project is repaid in large part by governmental subsidies for the crops grown on the project. Congress has in recent years stipulated that water from reclamation projects could not be used on land planted to crops already in surplus.

The possibility of obtaining agricultural production from other lands rather than from proposed reclamation projects should be thoroughly examined. With the rapid growth of the population of the United States it is expected that large increases in agricultural acreage will be required to meet the demand. Department of Agriculture studies in 1953 indicated that the increases of population will require an addition of nearly 100 million acres of cropland to the nearly 400 million acres already in production, based on the 1945-1949 average level of consumption and average acre yield.[53] With increased yields per acre, this increase in cropland would be materially reduced although raised standards of consumption for the public would in part offset this reduction. This increased production must come either from improved agricultural lands already in production or through the development of lands requiring irrigation in the West. Citing Bureau of Reclamation figures, the Department of Agriculture calculates that 17 million acres of new land in the West with an adequate water supply, and 9 million acres with a supplemental water supply could be provided by present water supplies. The Senate Select Committee on National Water Resources saw a potential of 15,685,000 acres in the seventeen reclamation states, with 1,675,000 acres in the Colorado River Basin.[54] Although costs of development are high, productivity is estimated to be about 50 percent higher than on non-irrigated land.[55]

In the opinion of some, most of the increased production will come from expanded irrigation acreage in the West.[56] Others are of the opinion that the alternatives of restoring eroded land in other parts of the country (which might cost less than developing new lands in the West) or increasing the productivity of lands already in production offer more suitable lines of policy since water can be used more economically in the West for other purposes.[57] It is apparent that such decisions should be made before embarking on large-scale enterprises designed to bring lands into production or to "save" lands.

These economic considerations are applicable to Arizona. The contest over the waters of the Colorado River, while framed primarily in terms of legal rights, is also a dispute of great economic consequence. Presentations by Arizona and California, both before Congress and the Special Master for the Supreme Court, stressed the economic development that has come through efficient use of the water. Each side confidently predicted continued economic growth through the granting of the remainder of the stream's flow to its lands. The Special Master permitted the presentation of evidence on economic consequences and it is virtually impossible that this dispute will be settled without reference to the needs of the total region and, beyond that, of the entire nation.

Opinions differ concerning the responsibility of the federal government to "rescue" an area deficient in water supply. Michael Strauss, former chief of the Bureau of Reclamation, clearly favors federal assistance to water-short areas with the provision that protection be supplied against further agricultural expansion which would again deplete the water supply.[58] The alternative, he asserts, is to allow the deficient area to suffer from its own folly and reach an economic level commensurate with its limited resources. Arthur Carhart, on the other hand, considers such assistance to Arizona farmers not a justifiable claim on the federal government but a "big, fat handout to those boomers who gambled and mined water and when they had squandered their wealth, turned to Uncle Sugar to perpetuate them in their exploitation."[59]

There is no doubt whatever that Arizona farmers did expand their acreage beyond the long-term limits of the existing water supply under the influence of high prices. Considerable argument is engendered over the question of whether the expansion was the handiwork of land "boomers" or small individual farmers, but there is much evidence that the "suitcase" farmers played a large and significant role. Clearly some national policy is needed in regard to such expensive rescue operations lest the federal government be forced to bail out every irrigation project that overextends itself. It is very clear, in the case of Arizona, that too little has been done to protect against a repetition of the situation.

The nature of crop production in the state raises the aforesaid problem of deciding whether alternative means of production are not available. Although great emphasis is placed on the specialty crops such as citrus fruits and vegetables, the basic crops of the state are cotton, alfalfa, and grain sorghum.

These three crops utilize nearly two-thirds of the state's total cultivated acreage. A comparatively small proportion of the cotton has gone into loan under the Commodity Credit Corporation.[60] It would be an anomaly in the eyes of many if the federal government were to incur the expense involved in the Central Arizona Project if the additional production which resulted went into government loan, when the same or similar products are produced elsewhere without large-scale expenditures. It should be noted, however, that much of Arizona cotton is superior grade, commanding a high price in the market and not going into government loans.

A fundamental economic issue faces state and federal officials in regard to the proposals by the Arizona Water Resources Committee for watershed modification. The practices proposed for the watershed, if adopted, would occur on federal lands for the most part, and would entail large-scale expenditures. The purpose would be to remove uneconomic growth, such as chapparal and jack-pine, as well as considerable amounts of merchantable timber, in order to increase the water supply made available to agriculturists in the valleys. There is considerable uncertainty over the technical feasibility of such a program, those opposing it charging that the result would be accelerated erosion.

If the proposals were technically feasible, those supporting the vegetation modification proposals assert that the measures are economically sound since the return on an acre-foot of water at the headgate resulting from these management practices would be much higher than would be the return from management of the land for timber or forage production.[61]

There is little or no evidence concerning the costs of undertaking and maintaining the programs envisioned by the Arizona Water Resources Committee, although some work is apparently being done on this matter.[62] Furthermore, little has been said concerning who would bear the cost of such watershed treatment, although it is suspected that the federal government would carry a large proportion of the burden. It is extremely doubtful that the state would provide much money for such a program, although it did supply the money for the original study, *Recovering Rainfall*. Raymond Seltzer of the University of Arizona made a preliminary study of methods of allocating costs among those who benefit from watershed modification. He indicated that the money would probably have to come in the form of interest-free loans from the federal or state governments.[63]

The recreationists protest that the forest and chapparal regions should be protected in their virgin state because of the present and potential economic value of the watersheds for recreation. Furthermore, it is asserted, there will be a need for increased supplies of water for human consumption by those using the watersheds for recreation.

In regard to both the Central Arizona Project and the watershed modification proposals, it must be decided what share of the burden would be carried by the several interests involved. The state and local units of government have been singularly unwilling to assume financial responsibility for these projects. Not only would the state, the counties, the municipalities, or the irrigation districts be involved in the investment, but also the individual farmers who would be the direct beneficiaries of the management programs.

Our concern thus far has been entirely with the substantive problems of resource management and planning. It is necessary now to turn attention to the instruments devised from time to time to solve the substantive issues. It goes without saying that the planning of resource development requires an agency staffed with competent and trained personnel, supported by sufficient financial and moral backing, and organized in such a fashion that its will is carried into practice. The planning agency should be located so as to command the attention and respect of those whose responsibility it is to administer the natural resources of the state.

Long-range, integrated resource planning has been a hit-and-miss proposition in Arizona. The planning agencies that have existed have been short-lived, without adequate support, and have had relatively little impact on resource management. Also, there has been scant recognition that decisions made at one time might require additional study and research to determine suitability under changing conditions. It has been little realized that resource management demands coordinated and continuous planning of all of its aspects — land, water, plant, and animal. Harvey O. Banks, Director of the California State Department of Water Resources, recently listed the guiding principles for administration of water resources: (1) "each state should so organize its water problems on a comprehensive basis;" (2) water activities "must be tied into and must complement those with respect to other natural resources."[64] Agencies taking this broad and long-run point of view have never occupied a central place in the administrative machinery of Arizona nor have they survived long enough to have much impact on the thinking of public and private groups engaged in resource management. This condition bears out what one observer stated: "It is characteristically American to find the urge to build outstripping the urge to plan . . . ."[65]

Planning of a more involved and extensive nature was required as communities reached the limits of the dependable surface-water supply near the turn of the twentieth century. It was necessary to form cooperative associations for the purpose of financing and maintaining large-scale structures designed to conserve flood waters. These structures, some built by the federal government and others by private capital, greatly extended the agricultural possibilities of the arid lands in Arizona. State laws authorizing the formation of irrigation districts, soil conservation districts, water conservation districts, and power and electrical districts enabled local communities to take full advantage of the increased water supply.

The first endeavor to establish a resources planning agency in Arizona occurred with the passage of the Water Code in 1919. The Arizona Resources Board was created whose functions were to investigate and develop plans for the conservation and utilization of watersheds and water resources, to study plans for irrigation, drainage, flood control, water power, soil conservation, and storage and development of water. It was to recommend to the governor policies to protect public and private water rights.[66] This board was of such little consequence during its brief existence — it filed one report, in 1920 — that Griffenhagen and Associates reported, in studying Arizona administration, "Actually, the board has never been appointed and no funds have been made available for the work."[67] The board's report, while only a survey of possible areas of work and investigation, particularly in regard to the Colorado River, did suggest lines of approach of considerable potential value to the state if they had been followed up.[68]

It was not until the 1930's and under the stimulation and leadership of the National Resources Board that the state of Arizona again took interest in resource planning. Governor Mouer appointed the Arizona State Planning Board in 1934, consisting of the president of the University of Arizona, the heads of a number of state departments, some federal officers, and consulting engineers. This board, however, never had access to independently appropriated funds, and accomplished its work almost entirely through the voluntary cooperation of the officials involved.[69] When the legislature was given an opportunity "to promote more economical and orderly development of the State through the creation of an unpaid State planning commission," the House of Representatives turned it down after it had passed the Senate.[70] The most important accomplishments of the planning board were the compilation of a file of studies made on various resource topics,[71] and the publication, in cooperation with the Works Progress Administration, of a two-volume report on Arizona's physical plant and resource needs in 1936.[72]

In 1939 Governor Jones asked the legislature to create an agency to advertise Arizona, a theme later to be sounded by many other governors.[73] This the legislature turned down as it turned down Governor Osborne's request in 1941 for the establishment of a state resources board.[74]

In 1944, in response to the realization that some readjustment would be required in agriculture at the conclusion of World War II, a further attempt was made at statewide planning with the appointment of the Arizona State Resources and Planning Board. The Arizona Post-War Planning Committee, a division of the board, in 1944 reported its findings, which constitute a testimony to the inconsequential results of such discontinuous studies and reports.[75] This report noted the excessive withdrawals of ground water and recommended cutbacks in acreage. It recommended that serious attention be given the overstocked and depleted ranges. It suggested the need for attention to forest and recreational resources. Virtually none of these matters were given serious consideration at the state level. The overtaxing of ground waters continued at an even more rapid rate.

Like so many more states, Arizona responded to the work of the Hoover Commission at the federal level by the creation of its own "Little Hoover Commission." The state hired the management consultant firm of Griffenhagen and Associates to make a study of state administration. Its report, published in 1949, accepted the general tenets of administrative management popularized by the Hoover Commission, including executive leadership, responsibility, and control, the reduction of the number of independent agencies and departments, expansion of civil service, and the grouping of agencies by the tasks they perform.[76]

Of importance in regard to natural resources was the recommendation that there be established a department of natural resources. Included in this department would be 14 boards and commissions, offices, and departments which were operating independently. These would include the agencies responsible for land, water, power, mines, minerals, and game and fish. Only agriculture and the Arizona Interstate Stream Commission would be excluded, the latter resulting from its being "primarily a one-time activity of temporary nature. . . ."[77] The department would be headed by a commissioner appointed by the governor and confirmed by the Senate.[78] The numerous multi-headed agencies would be eliminated in such a reorganization, and they would become functional units in the department.

The consulting firm justified this reorganization plan on the grounds that "the broad administrative policies governing the conservation and development of the state's natural resources should be consistent and should be integrated for the various kinds of resources, and the general planning and programming of the work should be centralized for all activities that are closely related to the conservation and development of natural resources."[79] In its judgment the planning and programming were not consistent because of the large number of agencies and the improper organization for the flow of authority and responsibility.

It should be noted that the suggestions of the management consultants related primarily to administrative planning and program coordination rather than to broad, long-range planning involving the political, social, economic, and physical problems facing the state. There is little doubt that such integrated administrative planning would be beneficial in program execution, but it would not solve the problems owing to a lack of basic resource policy and an agency to devise that policy.

Even these modest suggestions were not acceptable to the legislature, however. The Special Legislative Committee on State Operations of the Senate approved in theory the report of the consultants regarding natural resources, but noted that some problems had to be worked out, notably "the very apparent and inherent conflicts of interest arising from the combination of power, water, land and the other functional authorities of the proposed board...."[80] Two problems noted were the long-term contracts negotiated by the Arizona Power Authority and the legal requirements regarding the use and disposition of land. As noted elsewhere, the important interest groups in Arizona have been reluctant to see these proposed reforms go into effect.[81]

Griffenhagen and Associates proposed more limited planning in respect to water with the creation of a division of water in the department of natural resources. This division

> would have general supervision and control of the waters of the state, of the appropriation of water, and of the transfer of water rights; control the construction of dams, reservoirs, structures; supervise underground waters; administer the soil conservation program of the state; coordinate flood control activities; conduct research in water development and control; develop comprehensive water programs; and cooperate with federal and state agencies in the conservation and development of water resources.[82]

Such a general program, except for construction, could be adopted under the present law, but there is little prospect of such happening in the State Land Department.

Other *ad hoc* studies have been made of the economy and resource base of Arizona. The House of Representatives' Committee on

Arizona Development published a report in 1950 in which it advocated much greater emphasis be put on the development of the tourist industry and suggested the creation of a public agency to advertise the virtues of Arizona as a recreational area.[83] In 1951, the U.S. Department of Agriculture, in cooperation with the University of Arizona, published a report which pointed out the limitations of agriculture owing to shortages of water and the precarious dependence on high prices for agricultural products.[84] The reports of the governor's Underground Water Committee in 1951 and the Underground Water Commission in 1953 are discussed more fully below.

Two other agencies having some interest and responsibility in planning are the Arizona Legislative Council and the Arizona Development Board. The council, created in 1953, has become increasingly important as a conduit through which proposed legislation is channeled before introduction into the legislature.[85] The Development Board was established in 1954 as the culmination of a drive to create an agency to publicize the attractiveness of Arizona for industry and tourism. The board directs its efforts primarily toward attracting new industry, but in so doing engages in resources studies. For instance, the board published in 1955 two studies describing the state's nonmetallic resources and the advisability of establishing a state parks and recreation board.[86] Although studies of this kind may be undertaken in the future, it is doubtful that the Development Board will ever occupy the position of a resource planning agency as such.

There have been several attempts to create private groups which would take leadership in the conservation of natural resources. These groups, however, have usually foundered soon after their inception for one reason or another. Their fate has usually been sealed by the inclusion within their membership of elements almost completely opposed to each other in their conception of what was necessary to protect the resource base.[87] In 1957 the Forest Advisory Research Council was formed, consisting of many of the leading figures in the research field in the state and leaders of several important interest groups. The council was established for the purpose of consultation and guidance in the experimental work in the forested regions which produce most of the state's water. Many of the members are also represented on the Arizona Water Resources Committee, which is dedicated to practices designed to produce more water from the forests for the agricultural users. The council meets periodically to assess the experimental work being done primarily by the Forest Service.[88] What impact such a group will have on resource planning is difficult to predict, although past experience would indicate that its influence will be slight.

E Pluribus Unem = "Look out for No. 1"

The Arizona Water Resources Committee has been extremely active in promoting watershed research. Working chiefly through Senator Carl Hayden, it takes credit for additional appropriations totalling around one million dollars which Congress has dedicated for watershed research in Arizona in recent years. Annually it sponsors a meeting of all federal, state, and private agencies contributing to Arizona watershed research and management, at which time papers are presented dealing with various aspects of watershed management problems and programs.

The governor of Arizona in 1956 recommended the statutory creation of a state planning board to give official status to a group he appointed unofficially. He stated in his address to the legislature, "I am convinced that the lack of long range planning in the past has cost the state untold sums and that this situation can be remedied only by sound study of the future, backed up by action."[89] The legislature showed little more interest in such a suggestion than it had in the past. It was willing in 1956, however, to authorize the Arizona Power Authority and the Interstate Stream Commission to cooperate through a coordinator in the joint prosecution of efforts to obtain water and power from the Colorado River. However, this arrangement is clearly deficient in terms of comprehensive planning, inasmuch as it is directed to a single facet of the water problem.

The same can be said of the Water Resources and the Ground-Water divisions of the State Land Department since they are concerned with narrow and limited aspects of the water problem. The State Land Department as a whole, in which the major part of the responsibilities for water management is located, indicates no desire to exercise leadership. It conceives its responsibilities to include administration of the statutes and not the promotion of policies. Policies, they assert, must come from the interest groups concerned with water use. The author was a consultant to an arid lands committee appointed by the governor in 1959 to study the needs of the state with regard to water planning and research. Its recommendation that there be established a permanent planning office in the office of the governor has not received much support.

The picture presented here, in summary, is evidence that the state of Arizona has been insufficiently interested in resource planning in the past, and demonstrates inadequate interest at the present time. Advances made in planning will have to overcome powerful obstacles that have heretofore prevented effective and consistent planning efforts.

Objective observers of the problem agree that efficient water management will require men of foresight, initiative, strong character, and technical ability, who are capable of comprehending the complexities of the situation and bringing order out of chaos. Moreover, such management will require closely coordinated support from state, federal, municipal, and private organizations. Without such a program it is inevitable that cultivated lands will revert to the desert, and critical water shortages occur.

Despite this, there are some who suggest that state efforts in research and planning might even be curtailed. In spite of the fact that the state is experiencing some reductions in acreage devoted to agriculture in certain areas, in spite of the continued downdraft on the underground water resources, and in spite of the problems relating to flooding and soil erosion, no state agency exists to assume the responsibility of relating the physical, economic, and political problems for the purpose of finding suitable solutions or adjustments. To be sure, efforts are being made along specific lines, particularly in soil conservation and forest research (with the major impetus coming from the federal government), but there is no agency to relate these varied fields in terms of the general economy of the state. It may well be, as one competent observer believes, that the operation of the economic system without planning and with all the attendant waste in water resources, will ultimately determine the nature of the productive economy of the state. As Luna Leopold stated in 1960, "Our problems arise principally because resources use involves people. Attempts to forecast 50 years in advance what is best for the people is even more difficult than talking logic to the burro."[90]

CHAPTER 3

Surface Water Law

The unique problems and the critical importance of water in the arid regions of the United States are nowhere more manifest than in the legal doctrines governing the appropriation and use of this vital element. In areas where there is an abundance of water the problems are mainly those involving floods, drainage, disposal of excess water, and stream pollution. In the arid regions the problems involve the conservation of all existing water and its maximum utilization.

When the earliest settlers of European origin came to Arizona, they adapted their laws and customs to meet the needs of the environment.[1] They found a territory with very inadequate rainfall, particularly at the lower elevations where the finest prospects for agriculture existed. Precipitation alone could not possibly provide sufficient moisture to sustain an agricultural economy. (Dry-farming has been tried extensively in Arizona but has remained marginal.) The settlers were also faced with the problems of periods of drought alternating with wet periods — both of which might be of many years duration. Further, they found that precipitation was not evenly distributed, but varied widely within a single basin or even within a few miles. From the earliest times, reaching far back into the Spanish period, the agriculturists found it necessary to manage the water supply communally, primarily by means of *acequias,* or irrigation canals, which provided water both for riparian lands and those far-removed from the banks of streams. Later it became necessary to create means of impounding runoff to provide water for those periods of the year when streamflow was inadequate or nonexistent.

The Anglo-Saxon settlers who came to Arizona in the middle of the nineteenth century quickly adapted themselves to the conditions of the arid West and were among the first of their culture to adopt irrigation practices in the United States. In so doing, it was necessary to modify and in some cases reject entirely the legal theories that had

29

obtained in the humid area of the United States — the common-law doctrine of riparian rights.[2]

The doctrine of riparian rights established public ownership of stream water. It provided that private persons could lay claim to rights of use of the water but not to ownership of the corpus of water itself. Rights to use stream water depended upon location, the primary requirement being that the user own land abutting on the stream. The riparian landowner had a right to use water for domestic and stock-watering purposes and in some Western states even for irrigation, but he was enjoined from diminishing either in quantity or quality the flow of the stream. There was no requirement that the water be put to beneficial use and there was no lapsing of the right to use the water because of failure to use it over a period of time.

The law of prior appropriation, while maintaining the public character of all water, bases the right to the use of water upon the application of that water to some beneficial purpose. The right is dependent on a demonstration that the water is to be applied to some beneficial use, not on the location of the land whereon the water is to be applied. Rights in stream water do not in any way depend upon ownership of land bordering on a stream or land through which a stream runs. Since the right depends upon use, the water may be transported to land far removed from a stream, even as far as an entirely different watershed.

The right to water also depends on the time at which the water was diverted for beneficial use. A water right is valid only when it does not interfere with or damage the right of another who has a prior claim to the water. He who diverts water for beneficial use earliest has the better right and other rights of subsequent appropriators are subject to his. In case of shortage of water the latest appropriators are required to relinquish their water to the prior appropriators in order that the latters' claims may be fully satisfied. The aphorism "first in time, first in right," is an accurate transcription of this doctrine.

There is a clear assumption under this doctrine that all water is to be used and none is to be wasted. There is no intention to maintain the flow of water in the stream. Every available drop is to be utilized, even to the extent of drying up the stream itself. One has a right only to that amount of water that can be used beneficially. Surplus waters are subject to appropriation. If one fails over a period of time to use water to which he has laid claim he loses his right and another may appropriate that water.

It is readily apparent that the riparian rights doctrine ill-suited arid-land agriculture. If water rights depended on riparian land ownership, then only those who owned land bordering on the stream could engage in agriculture. Precipitation was too sparse and undependable to support agriculture on the desert. The riparian rights doctrine would have led to monopoly of agriculture by a few strategically located landowners. These might use the water beneficially or allow it to waste at their discretion. The prior appropriation doctrine allowed the water to be put to use beneficially regardless of location and did not permit waste or disuse. As Kinney stated it, "Nature clearly designed, in spite of . . . the inequality of precipitation, that the rain should still be permitted to shed its blessings on all. . . ."[3]

In Arizona the common-law doctrine of riparian rights was never in force and has been uniformly rejected in constitution, statute, and court decision since the creation of the Territory in 1863. The Spanish recognized Roman civil law, and consequently the common law was not accepted in any of its forms. Spanish customs clearly postulated the need for irrigation for agriculture. As stated by the Supreme Court of the Territory of Arizona in 1888: "Up to about a third of a century ago . . . the territory of Arizona had been subject to the laws and customs of Mexico, and the common law had been unknown; and that law has never been, and is not now, suited to the conditions that exist here, so far as the same applies to the uses of water."[4]

The modification of riparian rights appeared in the Kearny Code, the organic law for the new Territory of New Mexico which was established in 1846 and included Arizona. The code provided for the construction of public or private *acequias* to supply water for non-adjacent lands for consumptive use.[5]

The first Territorial legislature of Arizona, in passing the *Howell Code* which served as the Territorial constitution, declared:

> All streams, lakes, and ponds of water capable of being used for the purposes of navigation or irrigation, are hereby declared to be public property; and no individual or corporation shall have the right to appropriate them exclusively to their own private use, except under such equitable regulations and restrictions as the legislature shall provide for that purpose.[6]

Article LV of the *Code* further provided that "all rivers, creeks, and streams of running water in the Territory of Arizona are hereby declared public. . . ."

The legislative determination of the public character of stream water has gone unchallenged. The most significant development has related to the question of what constituted water that was subject to the law of prior appropriation. Additions to the surface water which could be appropriated were made in the state Water Code of 1919 and in amendments passed in 1921. The 1919 code declared public "the water of all natural streams, or flowing in any canyon, ravine or other natural channel, or in definite underground channels, and of springs, and lakes. . . ."[7] The desire to have all waters utilized for a beneficial purpose prompted the legislature to amend the code so that "the waters of all sources . . . whether perennial or intermittent, flood, waste, or surplus water" in addition to the other categories of water in the code were public.[8]

The courts have declared that drainage water does not come under the code and is not subject to appropriation.[9] Under the Drainage Act of Arizona drainage districts could be organized with power to construct works and obtain legal title to property "including all waters collected in, controlled, or handled by means" of such works.[10]

While the public character of surface water has gone unchallenged, the basis for use of water has led to a considerable amount of litigation. The *Howell Code* did not specifically abrogate the riparian rights system but gave a number of indications that such was its intention. The *Code* guaranteed all rights in *acequias* and the right of Arizona settlers to construct *acequias* and "obtain the necessary water for the same from any convenient river, creek, or stream of running water."[11] The *Code* also provided that in times of water scarcity those having the oldest title to land would have precedence for the water for irrigation.[12] These provisions clearly assume the prior appropriation doctrine since only under these doctrines could water be transported to non-riparian lands and preference be given to earlier users. The *Code* also provided, however, that the "common law of England, so far as it is not repugnant to or inconsistent with the constitution and laws of the United States, or the bill of rights or laws of this Territory, is hereby adopted. . . ."[13] Some argued as a result that the riparian rights system was not repugnant to the laws and constitution of Arizona or the United States.

In *Clough* v. *Wing* the Supreme Court of the Territory dismissed the riparian system in sweeping language:

> . . . the right to appropriate and use water for irrigation has been recognized longer than history, and since earlier than tradition . . . . The riparian rights of the common law could not exist under such systems; and a higher antiquity, a better reason, and more beneficent results have flowed from the doctrine that all right in water in non-navigable streams must be subservient to its use in tilling the soil.[14]

The Territorial legislature further gave evidence of the rejection of riparian rights when in 1885 it amended the *Howell Code* by stating that the common law was adopted "so far only as it is consistent with and adapted to the natural and physical condition of this territory, and the necessities of the people thereof. . . ."[15] In 1887 the legislature specifically rejected the riparian rights doctrine in language that later became part of the Arizona Constitution. "The common law doctrine of riparian rights shall not obtain or be of any force or effect in this territory."[16]

In the case of *Boquillas Land and Cattle Company* v. *Curtis et al.*, final confirmation was given the prior appropriation doctrine.[17] In this case, involving the right of a non-riparian landowner to construct a ditch across the land of a riparian landowner to bring water to his land, the Territorial Supreme Court reiterated the primacy of the appropriation rule and pointed out that even if the doctrine of riparian rights had at any time applied, this doctrine was abrogated by legislative rule. The Supreme Court of the United States sustained the arid region doctrine for Arizona, contending that the provision for adoption of the common law in Arizona "construed with the rest of the Code" is "far from meaning that patentees of a ranch on the San Pedro are to have the same rights as owners of an estate on the Thames."[18]

The federal government early extricated itself from the problem of deciding the basis for use of water, leaving the field to the states. By an act of July 26, 1866, the Congress recognized and acknowledged local customs, laws, and court decisions governing rights to the use of water.[19] The Desert Land Act of 1877 authorized the adoption of the prior appropriation doctrine by providing that "the right to the use of water by the person so conducting the same, on or to any tract of desert land . . . shall depend upon bona fide prior appropriation; and such right shall not exceed the amount of water actually appropriated, and necessarily used for the purpose of irrigation and reclamation."[20] Congress reserved all non-navigable water for the use of the public under the laws of the states and territories named in the act, one of which was Arizona. The Reclamation Act of 1902 specifically provided that that act was in no way intended to interfere with any state or Territorial laws relating to the use of water for irrigation.[21]

For nearly three decades the Territorial government made no attempt to regulate the manner in which a valid appropriation should be made. The *Howell Code* did nothing more than establish the fact that those rights already established in irrigation canals were confirmed and that inhabitants had the right to construct such works in the future. The determination of the methods for a valid appropriation was left entirely to the courts. The courts were also required to adjudicate cases involving a conflict of rights in their first instance.

The courts of Arizona, without legislative guidance, have been directed consistently by the following principles, as stated in the case of *Stewart* v. *Verde River Irirgation and Power District*:

> ... we have consistently held that appropriations of the use of the public waters of the state and the rights acquired thereby were regulated by three principles, (a) appropriations were made by actual application of water to a beneficial use, (b) the prior appropriator in point of time was prior in point of rights, and (c) the time the right of appropriation accrued was the day of the actual beneficial application of the waters to the purpose for which they had been appropriated.[22]

In the early case of *Clough* v. *Wing* the court, citing decisions by the courts of the states of California, Colorado, and Utah, asserted that a valid appropriation involved an intention to take water, accompanied by an open, physical demonstration that the purpose was being pursued. It further required that the water be put to some beneficial purpose and that the actual appropriation be pursued with diligence.[23]

In 1883 the legislature passed the first statute governing the appropriation of water by granting the right to all persons, companies, and corporations to appropriate for beneficial purposes. Such persons or corporations were granted the right to construct dams, reservoirs, ditches, and other waterways to meet their needs. Those first appropriating the water were declared always to "have the better right."[24] In 1901 the Supreme Court of Arizona refined the definition by stating that an appropriation involved the use or application of water to a beneficial purpose either under conditions prescribed by statute or custom "so as to vest a permanent right to such use and application." In order to constitute a permanent right there was also required permanent ownership or control of the means by which the use of the water might be enjoyed.[25] Control of the means of diversion did not mean that an appropriator had to have ownership of the means of diversion for such works could be owned in common by a number of appropriators or by another person who had contracted to serve those with rights of appropriation. This decision, confirming the roles of the irrigation companies and cooperative associations such as the Salt River Valley Water Users' Association in supplying water to farmers, was reiterated in the 1904 decision of the Supreme Court in *Gould* v. *Maricopa Canal Co.*[26]

Ownership of the means of diversion by persons or by a corporation other than those actually applying the water to the land raised a number of problems. Some of the corporations distributing water maintained that they had rights in the water itself, that they were bona fide appropriators of the water, and thus could determine the distribution of the water to shareholders as they saw fit. The Supreme Court of Arizona disagreed on the grounds that the irrigation companies did not possess irrigable lands on which to apply the water. The irrigation companies were mere agents of the landowners; if they obtained ownership of land they could acquire a right to appropriate water in the same fashion as any other landowner.[27]

There were many abuses associated with the delivery of water by irrigation companies. The companies often contended they were private associations obligated only to provide water for their own members. The irrigation companies tended to consider water rights as divorced from ownership of land and which thus could be sold separately from the land. In some instances water rights "floated" from place to place, causing insecurity in land values and manipulation of the water supply during periods of drought.[28] The companies were accused of depriving older canals of water "rightfully theirs by priority" and treating public water supply as corporate property. Water "instead of being strictly appurtenant to land, was sold to users as a separable commodity."[29]

In 1892 there began a series of court decisions designed to settle water rights and determine the prerogatives of irrigation companies. The so-called Kibbey Decision arose out of a dispute over the proper allocation of waters to the various canals operating in the Salt River Valley during the periods of shortages of water.[30] The surface water was nearly all appropriated and many early appropriators were feeling a serious pinch. The court rigorously applied the doctrine of prior appropriation in attempting to determine the legal claim upon water by all sections of land with the Salt River Valley. It attempted to determine when each parcel of land was brought into cultivation under each canal and then allocated the water for each canal on the basis of the amounts necessary to cultivate such land under proper methods of cultivation. The court affirmed the doctrine that water rights were permanently appurtenant to land and that there were no such things as "floating" water rights which could be bought and sold except in regard to a specific plot of ground.

The Kibbey Decision was not effective. Even while the decision was pending, the canal companies made agreements to circumvent the decree. The companies divided the water supply among themselves without regard to the Kibbey Decision and as a result "a majority of the Salt River Valley farming population was unlawfully dominated by the water companies which controlled the water supply as though it were their own instead of acting as distributing agents only."[31]

With the *Slosser* and *Gould* decisions in 1901 and 1904 respectively the Supreme Court of Arizona upheld the Kibbey Decision. In the Slosser case it ruled that "a water right, to be effective, must be attached to and pertain to a particular tract of land, and is in no sense a 'floating' right."[32] The companies could not discriminate in favor of their shareholders and thus deny water to persons with prior rights. A water right could be sold, however, as long as the water was used only on the land of the person to whom it was sold. In the Gould decision the Supreme Court invested the canal companies with a "quasi-public" character by requiring that "to the extent . . . that such a canal company has diverted and carried water from a public stream, and to the extent to which this water has been applied by appropriators for the necessary irrigation of their lands, the canal company must continue this service so long as such service is required by said appropriators and the water is available from the common source."[33] In contrast with the Kibbey Decision, these decisions were enforced and helped to break the arbitrary control exercised by the canal companies over the farming population.

In the considered judgment of many observers, the state had virtually reached the limit of its agricultural development under the conditions existing at the beginning of the twentieth century. The dependable flow of the major streams had long been appropriated.[34] The inefficiency of pumps made the use of ground water uneconomic. The only possibility for increasing the water supply was to improve the storage facilities along the rivers in order to provide for those long periods when stream flow was at a minimum. Although some small projects had been successful in capturing flood waters, major structures required more capital than the local water-users could amass. The answer to their needs was supplied by the National Reclamation Act of 1902.[35]

The passage of this act was an unmitigated boon to the West, but it also provided its share of legal problems in regard to the distribution of the stored water. It was necessary to establish a basis for water rights in the flood waters impounded behind the dams, and to quiet title to these waters. This need led to the case of *Hurley* v. *Abbott et al*,[36] in which Hurley, an early appropriator, brought suit against a large number of landowners in the Salt River Valley in order that he might quiet title to the water which was his owing to his early appropriation. The result of this action was the Kent Decree which determined the water rights for every parcel of land in the Salt River Valley.

The job of the court was to determine the rights of each individual defendant and each parcel of land to water in the river and to establish the various dates of appropriation of water by each landowner. This the court did by taking exhaustive testimony regarding the historical development of the valley and by establishing the date at which each section of land was brought into cultivation and the amounts of water used thereon. The court followed the doctrine of prior appropriation in giving priority of normal stream flow to those who had early rights in the stream. It determined the standard amount that should be allotted to each plot of land of equal size.

In regard to the stored waters, it was impossible to establish prior rights since the water would be available only if stored. In allocating stored water, the court divided the lands receiving the water into three classifications. Class A land, having preferential rights, was that land which had been farmed and irrigated continuously or nearly so. Class B land, with rights second only to Class A land, was that land which had been brought into cultivation during periods of above-normal stream flow and discontinued for lack of water. Class C, with no established rights, was that land which was in irrigation districts but not cultivated or irrigated. The court decreed that, according to an agreement entered into between the United States and the Salt River Valley Water Users' Association, the members of the association, regardless of the class of their land, were to receive the stored water "proportionately according to the acreage of the land, and irrespective of any priority of irrigation or cultivation of such land."[37] The rights to normal streamflow were of course left undisturbed.

This consent decree, from which no appeal was taken, settled firmly the rights to water in the Salt River Valley and made possible the construction of Roosevelt Dam. The same pattern was followed in the 1920's with the construction of Coolidge Dam on the Gila River. In each case a water commissioner was appointed to carry out the terms of the decree; these water commissioners continue their duties in spite of other procedures for determining water rights under the State Water Code.

Before 1893 the government of Arizona did little in the way of regulating the manner in which surface water was appropriated or of prescribing rules for settlement of conflicts. The *Howell Code* established the public nature of water and established rules regarding the management of irrigation canals but was silent on the methods or standards to be used in making an appropriation. In 1893 the Territorial legislature required the publication and notice of an intent to divert water. Such notice posted at the place of intended diversion was to state the amount to be appropriated and the location of the dam for diversion. Such appropriators were then to use all diligence in constructing the necessary works for failure to do so would cause a forfeiture of the right.[38]

From 1893 to 1919 nothing was done to improve the method

of establishing rights in surface water. Many appropriators ignored the legal requirements concerning filing notices of intention to appropriate. The courts concurred in bypassing these requirements by deeming the application of water to beneficial use sufficient to establish a water right. The failure by the courts to require adherence even to these minimum requirements led to considerable litigation over rights. The courts were often not competent to make judgments since the decisions depended on factual information which the litigants were frequently unwilling or unable to provide. There was often confusion concerning the land to which the rights applied, and seldom was there any hydrographic data which was necessary for a proper judgment.[39] It was this outmoded and inadequate system that gave rise to the passage of the State Water Code in 1919.[40]

The State Water Code wrote into the legal structure of the state the principles which had governed the decisions of courts for many decades. It declared that all surface water and water flowing in underground streams belonged to the public and was subject to appropriation and beneficial use. Beneficial use was to be "the basis, measure and limit to the use of water."[41] When the owner of a right failed to use the water for five successive years he lost his right and the water reverted to the public and was subject to appropriation.[42]

Water could be diverted for several purposes which were deemed beneficial: domestic, municipal, irrigation, stock watering, water power, mining, for personal use, or for delivery to consumers.[43] The person first appropriating was to have the better right. Authority to appropriate water for fish and wildlife was granted in 1941 but was relegated to an inferior position when and if the State Land Department was required to determine the relative values to the public of proposed uses of water.[44] Authority was given to construct and maintain works necessary to effect the beneficial use of the water for these purposes. The code centralized the administration of the water laws in the state water commisioner who at the present time is also the state land commissioner. Persons desiring to make an appropriation of water must make application to the water commissioner, giving pertinent information regarding the water supply from which the water is to be taken, the nature and amount of the proposed use, the location, point of diversion, necessary works, and the time needed to complete the works.[45]

The commissioner is given broad discretionary powers in granting or rejecting applications for permits to appropriate water, although review by the courts is guaranteed. He shall approve applications except when such applications conflict with vested rights, menace the public safety, or are contrary to the interests and welfare of the public. He may approve an application for less water than applied for but not for more. He may approve applications for municipal use to the exclusion of all subsequent applications if the need exists.[46]

The granting of a permit to appropriate water does not guarantee the appropriator that he has obtained an absolute right. "A junior appropriator with a certificate of water right may be restrained from using water if it can be established that the prior appropriator's rights are being impaired."[47] Moreover, the commissioner has no power to transfer a water right from one and give it to another.[48] The commissioner is guided in his actions, however. He may not grant an application for the use of water to generate hydroelectric power over 25,000 horsepower without specific legislative authority.[49] When there are pending conflicting applications for the use of water from a given source, the commissioner must give preference to certain uses. According to the Arizona Revised Statutes, "The relative values to the public for the purposes of this section shall be: 1. Domestic and municipal uses; 2. Irrigation and stock watering; 3. Power and mining uses."[50] Actual construction of the necessary works must begin within one year and the works must be completed within a reasonable period determined by the commissioner, usually not to exceed five years.[51]

Any applicant whose rights are affected by a decision of the commissioner may appeal to the Superior Court of the county where the proposed appropriation is located and the court may modify the commissioner's decision if he "has abused his authority."[52]

The commissioner is also made responsible for the determination of conflicting rights to water. He may make such determination only on petition signed by one or more water users on a stream. Water users may bring suit in Superior Court and thus avoid any administrative decisions; a court may transfer a suit to the State Land Department for its initial determination.[53] The department must publish notice of the investigation, examine the stream or supply of water and the lands upon which the water is used, and take testimony. The findings of the department are filed in the Superior Court, the court affirming the decision if no exception is taken within a time set by the court. If exception is taken, the court hears the parties, including the state, on the exceptions. The court may affirm or modify the order of the department or remand for further action.[54]

The legislature was careful to protect vested rights. The state water commissioner was to have "general control and supervision of the waters of the state . . . except distribution of water reserved to

special officers appointed by courts under existing judgments or decrees."[55] The code did not affect relative priorities determined by the
judgment of a court, impair the right of eminent domain when conferred by law, or in any way affect appropriations initiated under and
in compliance with prior existing statutes except when it was necessary
to adjudicate rights.[56]

In its general control and supervision of the surface waters of
the state, the State Land Department is empowered to "make surveys,
investigations and compilations of water resources in the state and
their potential development, and may cooperate for such purposes
with the United States."[57] The department has contracted with the
U.S. Geological Survey to obtain information on both surface and
ground water.

When necessary, the department may divide the state into water
districts with reference to watersheds both for the protection of claim-
ants to water and for the general supervision by the state. The department may appoint water superintendents in these districts whose
responsibilities involve the division of water according to established
water rights and the conservation of water by economical and careful
distribution of the water.[58] No action has been taken under this provision.

The Water Code has not gone unchallenged from water users
who felt that their interests were adversely affected by either the code
or its administration by the water commissioner. In 1924 the Supreme
Court was called upon to determine the constitutionality of the provisions which allowed the state water commissioner to make decisions
in regard to relative rights of water users in a stream.[59] It was argued
that the power of the water commissioner to make these decisions
vested in him judicial power contrary to the state Constitution which
vests the judicial power in the courts. The court rejected this contention, arguing that in undertaking his administrative duties "the commissioner merely paves the way for an adjudication by the court . . . .
The duties of the commissioner are not unlike those of a referee. . . ."[60]
The water commissioner brought a recommendation to the court for
its determination, and the court could inquire into the reasonableness
of the evidence supporting the commissioner's decision.

Objections were also made to the charging of fees proportional
to the number of acres for which an appropriation of water was sought.
In a case decided by the Supreme Court in 1937 the court upheld the
fee schedule as not unreasonable in view of the services rendered to
the claimant to the water, and stated that the money exacted did not
constitute taxation.[61]

It is clear that the State Water Code made no radical departure from the practices that had obtained since the earliest days of Arizona history. Its primary advantages lay in the codification of existing law and precedent and in the establishment of centralized machinery for the administration of water law. It provided at least the form for administrative determination of water rights, although in fact the water commissioner has been severely hobbled in his work as the result of the niggardliness of the legislature in providing funds.

Although the authority of the state water commissioner appears clear in the code, regarding the determination of water rights and the granting of a permit to appropriate, in recent years the administration in the Land Department has appeared to doubt its authority. It has been disclaiming any discretion in granting or denying a permit to appropriate water. Such an interpretation appears to be in conformity with a general policy of retrenchment occurring in the State Land Department in relation to its water responsibilities.

The laws governing the appropriation of surface water in Arizona are settled and are little disputed either as to their constitutionality or their wisdom. Nearly all surface water has been appropriated and what little remains is not suitable for irrigation. Relying on the fact that 1,200 permits to appropriate water were granted from 1955 through 1959, Struckmeyer and Butler contend that "it is not now correct to assume that there are no appropriable waters left in Arizona . . ."[62] They admit that new appropriators face difficulty because they will frequently be in conflict with vested rights. It is probable, however, that the state land commissioner grants many more permits than are justified by the available supply.

While the appearance of new legal doctrines concerning the appropriation of water may be doubtful, there will continue to be litigation over water rights. Recent years have witnessed a scramble to appropriate every drop of water available in the state. The state land commissioner related that a leak occurred at one of the new steam power plants in central Arizona, and a small trickle of water flowed out of the power company's property for a brief distance. Before the leak could be repaired, three persons had filed for appropriation with the State Land Department. The construction of a dam for flood protection, erosion control, or fishing always brings the possibility of suits by those who claim prior rights along a watercourse.

—Chuck Abbott

More than 90 percent of all the water used in Arizona is applied to agriculture. Canals and ditches transport the water to the fields from surface-water reservoirs and pumps. Important crops include cotton (below), safflower, feed grains, alfalfa, and produce (left) such as lettuce and cauliflower.

—Esther Henderson

CHAPTER 4

Ground Water Law

In contrast to the settled nature of surface-water law, ground-water law has been confused and inconsistent. Use of ground water in significant amounts did not occur until the 1920's and since that time the state has attempted to establish a legal system to govern its use while protecting the supply. The period from 1948 through 1955 in particular witnessed a considerable amount of legislative and judicial activity in this field without solving the basic issues that confront the state.

One of the serious difficulties in dealing with the legal aspects of ground water has been the lack of scientific and technical information regarding the water beneath the surface of the earth. Although the geologic sciences have made notable advances in the past fifty years, the legal doctrines upon which the courts as well as the public have come to rely were made at a time when the knowledge of ground water was relatively slight.

Furthermore, the early decisions in regard to ground water were made without considering the heavy reliance which would be put upon this resource in later years. The courts could not foresee that in the middle of the twentieth century well over 50 percent of the water used for agricultural purposes in Arizona would come from this source. In the early cases the courts were dealing with water which apparently had little potentiality for uses other than domestic or stock watering. There appeared to be little necessity of vesting it with a public character similar to surface water. It was generally agreed, moreover, that there was no relationship between the surface and underground water flow, and water that percolated through the soil.

Owing to full utilization of the available surface water supply and the increasing need for water in an expanding economy, subsurface water has come to supply the preponderance of the water presently used in Arizona. Despite the rapid growth of the urban areas, acreage under cultivation has also increased enormously, with a resulting increase in demand for water for irrigation. Approximately 95 percent of all water — surface and subsurface — used in the state is used for agriculture.[1]

In 1889 only 65,821 acres were irrigated. By 1909 irrigated land reached 320,051 acres — the largest acreage ever cultivated in the state's history to that date.[2] Demand for water rose accordingly. In 1940 an estimated 1,500,000 acre-feet of water was pumped for irrigation; by 1950 this had risen to 3,410,000 and by 1953 to 4,800,-000 acre-feet.[3] Of the 4,450,000 acre-feet of ground water used consumptively in 1953, 4,225,000 acre-feet were used for irrigation.[4] These figures for consumptive use differ from pumping totals because a certain portion of pumped water eventually is returned to the underground table.

The rapid extraction of water from underground reservoirs has had a profound effect on the underground water tables. From 1946 through 1955 the water table declined in some areas between 55 and 70 feet. The average lift for pumped water for the entire state in 1958 was 280 feet, with the pumping lift varying from 50 feet in some areas to over 450 feet in the Maricopa-Stanfield area of Pinal County.[5] In the Queen Creek-Higley-Gilbert area the water level declined more than 60 feet in the period between 1946 and 1955.[6] In the same period the decline in the Tempe-Mesa-Chandler area was around 55 feet, while in the Eloy area it was nearly 60 feet.[7] The increasing depths to water in some areas have made it economically infeasible for some operators to continue in agricultural production because of the excessive costs of bringing water to the surface.[8] The problem is serious and a solution is not readily obtainable. Constitutional, legal, political, and economic limitations of rather staggering proportions make any attempted solution difficult to achieve.

For the reasons stated above the Territorial government did not vest "percolating" water with a public character as it did with surface water. The *Howell Code* of 1864, which was Arizona's Territorial Constitution, ignored entirely the subsurface water in declaring that "all streams, lakes, and ponds of water capable of being used for the purpose of navigation or irrigation, are hereby declared to be public property."[9] In 1919 the legislature for the first time took action involving subsurface water when it included within the State Water Code a provision stating that water flowing "in definite underground channels . . . belongs to the public, and is subject to appropriation."[10] The legislature did not define "definite underground channels" but it was assumed that there was a distinction between water flowing in such channels and water that percolated through the soil into relatively unmoving "reservoirs."

This distinction between water in underground streams and water that percolated through the soil was supported both by the common law and the decisions of courts throughout the West. Water in underground channels was generally governed either by the riparian rights doctrine or the arid region doctrine of prior appropriation, while percolating water was governed by the English common law rule, which provided that "all percolating waters are considered a part of the soil where found, and, therefore belong absolutely to the owners thereof."[11]

The legal doctrines on which ownership and use in Arizona were based did not receive court sanction until 1904 in the case of *Howard v. Perrin*.[12] In this case, a squatter claimed the right to appropriate water from what he claimed was an underground stream. The landowner claimed title to the water on the grounds that it was percolating water and part of the realty. The two litigants agreed on the legal distinction between water flowing in subsurface channels and percolating water. The court had to decide the factual question of whether the water was from an underground stream or was percolating water, and not whether the legal distinction was valid.

The decision of the Supreme Court of Arizona, later affirmed by the United States Supreme Court, declared the water percolating and not subject to appropriation.[13] It stated that surface and underground stream water was treated the same under the doctrine of prior appropriation. The distinction lay "between all waters running in distinct channels, whether upon the surface or subterranean, and those oozing or percolating through the soil in varying quantities and uncertain directions."

Arizona courts thus formally accepted the distinction between percolating and flowing water without regard to its ultimate impact on the economy of the state. One of Arizona's leading hydrologists, G. E. P. Smith of the University of Arizona, stated of this decision that "it was not the studied conclusion of the court that such division was in the interest of justice or was best suited to the physical and economic conditions in the Territory," since the case did not involve extensive property or the welfare of the entire community.[14] Smith contended that the court, in its responsibility to rule on the suitability of the common-law doctrine governing percolating waters for Arizona, merely accepted what it found in other Western states and territories. In his estimation, since the Territorial legislature had laid down no rules, the court could have used the broadest discretion in establishing rules for the ownership and use of this valuable commodity.

The court continued to apply the rule of *Howard* v. *Perrin* in subsequent cases. In 1918 in the case of *McKenzie* v. *Moore* the Supreme Court of Arizona decided that a subsurface spring of water was not one of the sources from which water might be appropriated.[15] The court found such water percolating in nature and therefore part of the realty. Since there had been no mention of springs in the statutes the doctrine of prior appropriation did not hold. Without proof that the water flowed in a definite channel, it was assumed to be percolating. Two months later the legislature, in passing the State Water Code included springs in the category subject to appropriation and in 1921 changed it to "springs on the surface."

In 1926 the Supreme Court decided the case of *Proctor* v. *Pima Farms Company*, a case involving an entire river basin and agricultural lands totalling 14,500 acres.[16] The Pima Farms Company had purchased considerable land along the Santa Cruz River and had developed a water supply for the land from pumps of large diameter and capacity. The company sold the land to farmers and then supplied them with water from its wells. The pumping by Pima Farms imposed a heavy draft on the ground-water supply, resulting in a significant dropping of the water table and in the need to deepen wells. Proctor, an early farmer in the area, sued to prevent the company from lowering the water table, claiming the rights of a prior appropriator. He asserted the right to have such stream's level remain so that his "means of capture and diversion as originally installed will not be impaired . . . or to have the later appropriators deliver to him his water in such manner as to make it available for his uses."[17] Both Proctor and Pima Farms agreed that the source of water was an independent underground stream despite the fact that the water supply from which they pumped was over a mile in width and ran in no discernible channel or within any defined banks.

The court accepted the assumption of the parties that the body of water was an underground stream and therefore applied the doctrine of prior appropriation. The court stated, "If the first appropriator's rights are superior under the law, they should be made so in fact."[18] Pima Farms was required, therefore, to make available to Proctor water undiminished in quantity and quality and in a place equally accessible to him for diversion.

This decision reinforced the distinction between flowing and percolating subsurface water but appeared to extend the concept of an underground stream to water which was clearly not within banks and channels. Smith observed: "The description would fit most of the rivers of the state and would include most of the ground water supplies."[19] The decision further demonstrated the inability of the courts to deal with a subject concerning which they had little technical information or training.

In 1931 another opportunity to review ground-water law arose in the case of *Maricopa County Municipal Water Conservation District* v. *Southwest Cotton Company et al.* which involved millions of dollars in property values.[20] The court recognized the importance of its decision for the future of Arizona and therefore treated the matter "as though it were of first impression in all aspects."[21] After rehearsing the water-law doctrines from the early Spanish period and comparing them with the common law, the court found neither system of law determinative and considered the government of Arizona able to make whatever rules it so desired regarding ground water.

The court first found that the *Howell Code* and subsequent legislation and constitutional provisions had not applied to percolating waters. Only surface waters were included in statutory enactments. On the other hand, since the *Howell Code* specifically provided that the common law was adopted when not inconsistent with the Constitution of the United States and the Bill of Rights of the Territory, the common-law doctrines governing percolating water did apply and therefore made such water "the property of the owner of the land, subject to the rules of the common law."[22] The court held that the legislature never "specifically made percolating waters subject to appropriation, and, if we apply the usual rule of 'expressio unius,' has very carefully excluded them therefrom."[23]

In thus reasserting the doctrine of *Howard* v. *Perrin* the court contended that whether the doctrine of that case was *dicta* or not, "It has been accepted as the law of this jurisdiction for so long, and so many rights have been based on it, that only the clearest showing that the rule declared was error would justify us in departing from it."[24] Since the legislature had not upset the judicial finding, it was to be upheld.

The case involved the damming of the Agua Fria River by the Maricopa County Water Conservation District. Southwest Cotton Company argued that such damming caused a decline in the water available from its well drilled on land bordering the stream below the dam. The company asserted that the water in question was all part of a large subterranean stream and therefore subject to appropriation. The court, however, rejected the view of the company, agreeing with the conservation district that the water underlying the cotton company's land was percolating water and thus not available for appropriation like surface water. The water district could dam the river and divert its flow regardless of the effects of such diversion on agricultural developments, unless the company could prove that it had prior rights to appropriate water.

In contrast with the decision in the Pima Farms case, the court took a very restrictive view concerning what was necessary to prove the existence of an underground stream. It required clear evidence of a channel, with well-defined bed and banks, and current, and a certainty of location. The presumption would be, without evidence to the contrary, that underground water was percolating and therefore subject to ownership by the overlying land owners. The court refused to consider the question of the relative rights of users of percolating water coming from the same basin of underground water, but suggested two alternatives: the strict rule of ownership by the overlying owners of the land, or the doctrine of correlative rights.

Subsequent cases involving ground water did little to alter the legal situation. In *Fourzan* v. *Curtis* the Supreme Court decided that a spring which did not naturally provide water to the surface of the earth was not subject to appropriation since the Water Code made only springs on the surface subject to that law.[25] In *Campbell* v. *Willard* artesian water was considered as percolating water unless there was definite proof that the water had its origin in an underground stream.[26] In the case of *Parker et al.* v. *McIntyre et al.*, the court declared that springs on the surface providing sufficient water to be put to beneficial use (in this case for stockwatering) were subject to appropriation.[27]

It was during the 1930's that concern for the conservation of underground water supplies first developed. Owing to the increased efficiency of pumps, higher prices for cotton, and lower costs for power, ground water began to play an important part in Arizona agriculture.[28] Some saw the early need for settlement of the legal and policy questions surrounding ground water but their hopes for settlement were in vain.[29]

In 1939 the water commissioner was able to win from the legislature an appropriation for $10,000 for ground-water investigations, to be undertaken in cooperation with the United States Geological Survey. The commissioner recommended no substantive action on a code until a survey was completed, although he indicated that "quite a little interest" was shown in a state code.[30] During the early 1940's proposals were introduced into the legislature calling for the establishment of study committees for the writing of a code, and even for the passage of a ground-water code, but each of these measures died along the thorny legislative path.[31] As early as 1942 the Arizona Farm Bureau Federation called for a code.[32]

In 1944, after five years of investigation — extremely limited owing to inadequate financial support — by the U.S. Geological Survey, the state land commissioner (who had assumed the duties of the state water commissioner in 1943), informed the governor that without a ground-water code "the agricultural development of the state can never be safeguarded against overdevelopment that will always threaten the return of certain areas to the desert."[33] He pointed out that speculators would open up marginal lands and threaten the existence of established irrigation districts unless a restrictive code was imposed. In 1944 the Arizona Agricultural Post-War Planning Committee of the Arizona State Resources and Planning Board, after considering the sizable growth of agriculture during World War II, confirmed this view, saying that "abandonment of developed acreage appears inevitable."[34] The committee recommended the adoption of a ground-water code, and the adjudication of all existing rights to water.

Governor Sidney Osborn time and again asked the legislature for funds to finance an extensive underground water survey beyond what the Geological Survey was then engaged in, but he was consistently turned down until 1945. It was the Central Arizona Project that finally tilted the scales in favor of legislative action. By 1945 the pressure for enactment of some kind of code increased with the first expression of views on the Central Arizona Project by the Bureau of Reclamation. The bureau found the project economically feasible, and sound from an engineering viewpoint, but was reluctant to approve such a project unless the state took action to control its own ground-water depletion problem. The governor warned that the project would not receive Bureau of Reclamation support without a state ground-water law.[35] This argument, in addition to the previous arguments, convinced all but those who maintained that the state could obtain water from the Colorado River without having to contract with the federal government that a law would be passed. (There were some who had maintained from the earliest attempts to utilize Colorado River water that early filings on the river were sufficient to give the state the right to divert water, without the approval of the federal government.) The result was the passage of the Ground Water Act of 1945.[36]

The primary purpose of this act was to provide information on wells and to require data concerning the nature and extent of ground water in the state.[37] The act required that all persons owning and operating wells report such wells to the state land commissioner with certain information regarding their operation — depth, whether cased or not, the land on which the water was used, amount of water produced, etc.[38] No wells were to be drilled thereafter without a notice of intention to drill having first been filed with the commissioner.

While the act did make the first faltering steps in providing information about the rate of depletion, it did nothing to lessen the speed with which the water supply was being exploited. During the post-war period agriculture continued to expand under the stimulus of high prices; new lands were brought into production causing ever-increasing demands on the water supply.

Between 1945 and 1948 several attempts were made to write and pass a restrictive ground-water code. The proposed codes took various forms. One would have declared all ground water public property and would have given the state land commissioner the power to regulate the appropriation and use of water except in established irrigation districts.[39] Another would have adopted the correlative rights principle and would have restricted the pumpage of water on that basis.[40] Under this doctrine, "owners of overlying lands have equal rights to the ground-water supply for use on such lands and each is entitled to an equitable apportionment if the supply is not enough for all."[41] A third bill would have declared as public all ground water not already subjected to use and would have restricted the utilization of these waters. All vested rights would have been protected, with the public waters subjected to management by the state land commissioner on a sustained-yield pumpage basis.[42] A fourth measure would have restricted public control to areas below 2,500 feet in elevation.[43]

There were several bases for opposition to these proposed codes. Perhaps the most important opposition came from those who opposed any code at all. In spite of the declaration of the state land commissioner that the irrigation farmers were "predonderantly in favor of legislation which would lead to the most beneficial utilization and protection of ground water resources," a large segment of these farmers opposed a code in any form.[44] Indicative of this attitude was the exclamation by one farmer in a legislative committee hearing in 1947. He said, "Who is going to tell me what to do and how to do it? If my land is destroyed through lack of water I want to destroy it myself; I don't want you [presumably the state legislators] to do it."[45]

Those who opposed any and all codes were generally the more recent developers of land through use of ground water. They would have had inferior rights under any scheme based on the doctrine of prior appropriation. While undoubtedly some were interested in a permanent livelihood on the desert, many were of the "suitcase" variety, willing to make an investment for short-term profits with full knowledge that the resource eventually would play out.

Some opposition developed to any arrangement in which farmers were treated in a discriminatory fashion with respect to administration of a code. In one form this meant that irrigation districts were in effect exempted from the code and in another form this meant that farming areas above a certain elevation would be exempted. It was felt that all should be regulated or all should be left alone. There were those who objected to what they called "dictation" by the Bureau of Reclamation in urging passage of a ground-water code in return for bureau support of the Central Arizona Project.

There were serious legal and constitutional problems involved in the proposed codes. Opinions ranged from those who thought the legislature could declare all ground water in public ownership, and therefore subject to public control, to those who believed the legislature was powerless to act to regulate what the courts had declared to be private property.

The primary support for the code came from the older irrigation farmers who saw their investments jeopardized by those whom they viewed as land speculators.[46] The *Arizona Farmer*, chief spokesman for the irrigation farmer, accused the opposition of being "ignorant and selfish," and called for early passage of a code.[47]

Meanwhile the Geological Survey continued to compile its survey data, which demonstrated the serious depletion of the underground water supplies in central Arizona.[48] The USGS found that under the conditions then existing, in most of the heavily developed areas the annual safe yield was being exceeded — and in at least one locality by at least 18 times.

Governor Osborn was an ardent supporter of a ground-water code and was determined to see one passed. He assailed what he called the "forces of greed and destruction" who resisted passage of a code.[49] Failing to obtain action from the regular session of the legislature in 1947, he called the legislature back into session three times for the specific purpose of writing a ground-water code; finally he was successful.[50]

The result of the legislature's action, taken more in desperation than from conviction, was the Ground Water Act of 1948 which everyone admitted was a stop-gap measure designed to slow down the rapid depletion of ground water but certainly not designed to solve the long-range problem of balancing agricultural development with the available water supply.[51] Representative Murphy of Maricopa County, in a beautifully mixed metaphor, asserted the code was "as weak as restaurant soup and should have been sent from the Senate with crutches."[52]

The act declared it a matter of public policy "in the interest of agricultural stability, general economy and welfare of the state and its citizens to conserve and protect the water resources of the state from destruction, and for that purpose to provide reasonable regulations for the designation and establishment of such [critical ground-water areas] within the state."[53] The act defined critical ground-water areas as "any ground water basin . . . not having sufficient ground water to provide a reasonably safe supply for irrigation of the cultivated lands in the basin at the then current rates of withdrawal."[54] The state land commissioner, after obtaining information about the safe annual yield and the use of water in each basin, was authorized to designate critical ground-water areas. He could do so on his own initiative or under petition by users of ground water.[55] He was required to hold hearings and then make a decision which was conclusive as to facts unless appealed within a stated time.[56]

The importance of the designation of an area as critical lay in the power of the commissioner to refuse to permit the construction of irrigation wells in a critical area except under certain conditions. He was required to grant the permit "except that no permit shall be issued for the construction of irrigation wells within any critical groundwater area for the irrigation of lands which shall not at the effective date of this Act be irrigated, or shall not have been cultivated within five years prior thereto."[57] Clearly the intent of the act was to give the state land commissioner administrative control over the development of ground water in critical areas and to restrict its use to that land which was cultivated within a previous five-year period, and to control the drilling of wells. Only the replacement and deepening of wells which had been in operation during the previous five years were to be permitted in critical areas.[58] Wells used for the purpose of domestic supply, stock watering, domestic water supply, industry, or transportation were exempted.[59]

The code was designed to limit the acreage irrigated by means of ground water and to restrict the number of wells in these critical areas. Powerful opposition in the legislature prevented the imposition of any restrictions on the quantity of water pumped from wells already in operation. The act specifically stated that "nothing in this Act shall be construed . . . to affect the right of any person to continue the use of water from existing irrigation wells or any replacements of such wells."[60] The act attempted to prevent expansion of agriculture by use of ground water and thus prevent more serious overdrafts, but did nothing to reduce existing overdrafts or to forestall "critical" water situations in other locations. Although some weakening provisions had been removed, such as local option, the obvious weakness of the law lay in the fact that the bill preserved the concept of private property for underground water and imposed no restrictions whatever on the pumpage of water on land having a five-year-old history of cultivation prior to the effective date of the act. The doubts concerning the constitutionality of the act led to elimination of any provisions which would impose limits on pumping.

The objections on constitutional grounds were serious. Some contended that the act violated due process of law in that it was arbitrary use of the police power to interfere with private property. If the water was denied landowners, their land was worthless also, so that the act in fact diminished the value of their property.[61] The act was also charged with being in violation of the equal protection of the laws clause of the U.S. Constitution and a comparable provision of the Arizona Constitution in that it failed "to regulate the use of underground water by present irrigation pumpers, while preventing land owners who do not have wells or pumps from drilling new wells."[62] It was further argued that the act conferred legislative authority on the state land commissioner because of its indefiniteness. The legal profession apparently felt that the law would be declared unconstitutional on any one or all of the above grounds.

The passage of the Ground Water Code occasioned a great deal of activity on the part of those who wanted to bring into cultivation new land prior to the operation of the act. Restrictions on drilling were not to apply until the state land commissioner had declared areas critical after public hearing. The commissioner reported that the notices of intention to drill new wells "showed a material increase" much of which occurred "during the week preceding the effective date of the Ground Water Code of 1948."[63] The first critical area, near Eloy, was not so designated until April 4, 1949, and other areas in the state were not designated critical until 1951 and later.[64] Meanwhile, landowners could continue to construct wells in water-short areas.

The ineffectiveness of the code is amply demonstrated by the statistics on pumped water and land under cultivation during the years immediately following passage of the code. With the continuation of the drought, ground water became continually more important to Arizona each year. The state land commissioner reported the following amounts pumped for the years 1949-1954: 1949 — 3,250,000 acre-feet; 1950 — 3,500,000 a/f; 1951 — 3,680,000 a/f; 1952 — 3,750,000 a/f; 1953 — 4,800,000 a/f.[65] The decline in surface water and the practice of double-cropping accounted for some of the increase. But the effect of new lands on pumpage can be measured by the increase in irrigated acreage from just under 1,000,000 in 1949 to nearly 1,300,000 acres in 1953.[66]

The land commissioner was given insufficient money to provide adequate supervision of critical areas. Violations were reported but apparently went unpunished. Ben Avery of the *Arizona Republic* asserted the only attempts at enforcement involved the writing of letters.[67] It was quickly realized that there was little hope for Governor Osborn's expressed belief that it would be necessary only to spread the ground-water supply over a ten- or twelve-year period until supplementary water was obtained from the Colorado River. The Central Arizona Project had twice failed in Congress and litigation appeared to be the only means of securing water from the Colorado. If the ground-water base were to be protected there would have to be more stringent restrictions imposed on use of ground water. The U.S. Geological Survey continued to warn of the rapid decline in the water tables in central Arizona and the higher costs, lower yields, and poorer quality of water resulting from these declines.[68]

Soon after the passage of the 1948 code another attack was made on the theory of private ownership of percolating water in the case of *Bristor* v. *Cheatham,* filed in the Superior Court of Maricopa County. This case involved two landowners who pumped from a common underground water supply. Bristor, the earlier user, had used his water for domestic purposes. Cheatham and others later installed powerful pumps to extract water for agricultural purposes, causing a decline in the water table; in fact, Bristor's well went dry. Bristor sued Cheatham on the grounds that Cheatham was pumping from a common supply and transporting the water from the lands where it was pumped to other lands for agricultural purposes, in violation of what Bristor argued was the rule regarding ground waters. It was contended that the water could be used only on the land from which it was pumped and that one user could not do injury to another who shared the same supply. Bristor, in effect, argued for a form of the correlative rights theory as it had been adopted in the neighboring state of California. Cheatham contended that the absolute English rule applied — that the owner of the overlying land could do with ground water as he pleased.

The case excited a great deal of attention throughout the state and occasioned the entrance of several important legal firms who represented some of the most important interests in the state, such as the Salt River Valley Water Users' Association, Goodyear Farms, and the Central Arizona Project Association. These firms filed briefs as *amici-curiae* on both sides of the case. Significantly, a number of these friends of the court rejected both legal doctrines presented by the litigants and argued in lengthy briefs in favor of the doctrine of public ownership. They took the position that the ground-water situation could be reversed only if ground water was declared public property and if the legislature were provided a firm legal base to apportion the remaining water supply in some equitable way.

On November 14, 1949, the Superior Court of Maricopa County granted Cheatham's motion to dismiss Bristor's causes of action. On December 29, 1949, Bristor filed a motion for appeal; the case went to the Arizona Supreme Court on Feb. 10, 1950. The difficulty the court had in deciding this case is indicated by the fact that the court did not reveal its decision until January, 1952, nearly two years after receiving the case.

In the fall of 1951 Governor Pyle appointed a committee to study the ground-water situation and to recommend methods of strengthening the code, which, he said, "everyone admits is shot full of loopholes and virtually ineffective."[69] This committee consisted of some of the leading figures in the agricultural industry in the state. The group met several times with technicians in the field of hydrology, studied available information, and held several open hearings during the fall months. After much deliberation, the committee recommended to the governor a program which called for a reversal of previous policy on ground water. Their report advocated declaring ground water in public ownership and adopting the rule of prior appropriation with some modifications toward the principle of correlative rights.[70] All ground water basins would be classified on the basis of the relationship of total pumpage to total supply of the basin. Some would be closed to further drilling of wells when the overdraft was excessive; some would be restricted; and others would be unregulated when there was no existing or potential imbalance in pumpage and supply. A complicated schedule operating on the basis of a combination of rights based on the prior appropriation principle and the correlative rights principle would determine the amounts of water to be withdrawn, depending on the condition of the ground-water basin.

The recommendations were transmitted to the 1952 session of the legislature by Governor Pyle, but without his approval and with his suggestion that the legislature was free to make its own decision.[71] It was in this situation that the Supreme Court delivered its opinion in *Bristor* v. *Cheatham*.[72] The court, by a 3-2 margin, declared all ground water public property and subject to the rule of prior appropriation.

The court majority took the remarkable position that previous decisions of the court in distinguishing between percolating and underground stream water were erroneous. It dismissed the doctrine of *Howard* v. *Perrin* as *dicta* and the doctrine of the Southwest Cotton Company case (including the rule of *stare decisis*) by stating that it would be "more harmful to the public at large" to follow the rule, "than to overrule precedent and establish a sound principle."[73] The court held that the nature of percolating water had been decreed by the Desert Land Act, which declared the "right to the use of water by the person so conducting the same, on or to any tract of desert land . . . shall depend upon bona fide prior appropriation. . ."[74] The Supreme Court maintained that by this act Congress dedicated "to the public all interest, riparian or otherwise, in the waters of the public domain," and abrogated the common-law rule in respect of riparian rights as to all lands settled upon or entered after March 3, 1877. It cited the decision of the Supreme Court of the United States in upholding the constitutionality of the Desert Land Act in which that court stated that the act held that all water not navigable was severed from the land "free for the appropriation and use of the public. . ."[75]

As long as the public character of the water supply was maintained, stated the court, the states and territories were free to make their own rules for using the water supply. No legislative enactment was necessary to invest ground water with a public character. Inaction by the legislature in failing to specify the steps necessary for the appropriation of ground water did not divest the water of public character, even though the legislature had prescribed steps for other waters.

The majority was concerned with the impact of its decision on the economy of the state and rights of individuals. Noting that counsel for Cheatham had argued that "citizens throughout the state relying upon the decision in that [*Howard* v. *Perrin*] and subsequent cases, have spent large sums of money and that they therefore have vested rights which must be recognized," the court replied that "the vested rights of the users of percolating waters since the decision in the *Howard* v. *Perrin* case are more fully protected under the law of prior appropriation than under the so-called common law rule."[76] The court saw "inevitable exhaustion of all underground water in the

State of Arizona if the rule of private ownership . . . is still held to be law," since such a rule would put pumping beyond the power of the legislature to regulate.[77] The court believed the only feasible way to control the excessive pumpage was to give preference to those with priority of use, and thus make all later appropriators subject to the availability of water to prior users. Public ownership would permit the writing of reasonable rules by the legislature.

The dissenting opinions disagreed on the law and on the predicted effects of the decision. One dissenter compared it to the "dropping of a gigantic atomic bomb in our midst" which would "destroy and wipe out all rights and investments that have been acquired by the expenditure of millions of dollars and industry of thousands of citizens."[78] The later appropriators would be put at the mercy of the prior appropriators who would be fully protected by the decision. Persons who had drilled wells within the last 5, 10, or perhaps 15 years would have to close down operations whenever a prior appropriator demonstrated that the ground-water supply had diminished to his detriment. The dissenters held out the prospect of endless litigation and literal destruction of a good share of Arizona agriculture since almost every well had some effect on the ground-water level in an area.

The dissenters believed the court had no right to upset what had been settled law since the Southwest Cotton Company case.[79] They argued that the court had misconstrued the Desert Land Act because that act, they claimed, did not dictate the method of appropriation for the Western states, but gave each state the option of adopting the prior appropriation doctrine for all its water, or the common-law doctrine, or any combination of the two it might choose. Arizona, by omission on the part of the legislature, and by positive declaration by the courts, had adopted the common-law rule for percolating waters. The dissenters asserted there was power under which the legislature could enact regulatory legislation. They suggested various lines of approach, including the doctrine of reasonable use, and the exercise of the police power. One justice suggested the correlative rights doctrine as a method of sharing ground water among those in the same basin.

The decision of the Supreme Court met with a violent reaction among those who felt they faced ruin from the decision, even including threats of personal harm to the justices of the Supreme Court. The court was subjected to attacks by various spokesmen for the opponents of public ownership, particularly in Pinal County where the water situation was most serious. The farmers there contended that effects of the decision were immediate in that cotton-ginning companies, banks, and other institutions were refusing financing.[80] Luncheon clubs and Chambers of Commerce meetings heard predictions of impending doom for Arizona's agricultural economy.[81] Meanwhile, the attorneys for Cheatham prepared briefs in hope of obtaining a rehearing of the case.

The advocates of a strong ground-water code took heart from the decision and felt the way cleared for positive action during the 1952 legislative session.[82] Heads of prominent banks cautioned against unjustified "scare" talk, asserting that their lending policies remained the same. They feared that the "campaign of fear and hysteria" would frighten away outside investment capital on which Arizona banks depended.[83] The officials of the Salt River Valley Water Users' Association called for support of the decision and ordered its attorneys to prepare for a continuation of the fight for public ownership of underground water.[84] It also pressed for immediate passage of a ground-water code. Many of the leading irrigation districts held meetings and instructed their attorneys to prepare new briefs upholding the court's decision.[85]

With the appeal for a rehearing before the court and rumors rife that one justice had had a change of heart, the legislature was reluctant to take up a ground-water code in the 1952 session.[86] Governor Pyle's committee on ground water, finding support for its recommendations in the Supreme Court decision, modified its original recommendations to fit the decision.[87] Governor Pyle sent the revised recommendations to the legislature with a notable lack of enthusiasm.[88] Pyle previously had championed a strong ground-water code, and one observer saw in Pyle's attitude an attempt to maintain good relations between himself (a Republican) and the Democratic leadership of the legislature which was known to oppose a strong code.[89]

The leadership of the legislature apparently had decided already against passage of a new code, with or without the support of the governor. The Speaker of the House was reported to have said as early as January 26, 1952, there would be no new code.[90] The opponents of a new code introduced a bill which authorized the appointment of a commission by the governor for the purpose of making a study of the underground water situation.[91] The supporters of this bill argued that the previous committee had done its work too hastily and that it based its conclusions on inadequate and inaccurate information.

It was at this point that the Supreme Court granted a rehearing in *Bristor* v. *Cheatham* and stated that a new opinion was being written. The granting of the rehearing in effect nullified the original decision and indicated that the court was going to reverse itself. It was assumed that the court was going to declare percolating water in private ownership. This change of heart made impossible passage of a code in 1952.

Hearings (described by Pickrell of the Salt River Valley Water Users' Association as "more of a farce than anything else") were held by the legislature on a new code, and on the study commission bill.[92] Hydrologists, geologists, and representatives of farm organizations testified. The opponents of a new code and of the Supreme Court's original decision vigorously supported the study commission bill. As a result, the code bill died in committees but the study commission bill was passed with heavy majorities in each house. One amendment was added, which bowed toward the ground-water problem in that it prohibited pumping from wells in critical areas which were not completed prior to the effective date of the ground-water act.

It was in this confused situation that the Underground Water Commission took up its duties. No one was quite sure of the legal nature of ground water. Restrictions were in existence in regard to the development of new land by ground water and the pumping from illegal wells in critical areas, but there was little real possibility of enforcement. Officials of the State Land Department were dubious about its enforcement. One said the code would require a policeman by every well. Another said it could be enforced but that no one really wanted it enforced. There were serious doubts in many circles, particularly in the legal profession, that any code would be constitutional in view of the Supreme Court's expected reversal.[93] The *Arizona Farmer* lamented:

> But precious little sense has ever been displayed in dealing with ground water. Every move toward conservation, toward recognizing and confirming the rights of the first ground-water users, has been blocked by the pump-and-run boys. They are able to make more noise and throw up a bigger stink than anybody else. Apparently they are to have their own sweet way until all the water is gone. Then what?[94]

The act creating the Underground Water Commission stipulated that all members of the commission were to be farmers on the presumption that the farmers then would not have any excuse for failing to pass some kind of code the next year. The Speaker of the House stated, "I don't think they are going to want a code any more next Jan. 1 than they do now, so let's not leave them any loopholes to complain."[95] In the appointment of members the governor was apparently favorably disposed toward those who opposed a stringent code. The Farmers' Protective Association, representing opponents of previously proposed codes, made available to the governor a list of names from which he might choose members of the commission.[96]

The commission held numerous hearings throughout the state during 1952 and gathered voluminous testimony from farmers on their water problems. The testimony indicated a wide disagreement on the nature of regulation desired, and even on the question of whether there should be any regulation at all. Many expressed dissatisfaction with the 1948 Ground Water Act. A large number of farmers wanted the economics of agriculture and pumping to determine the manner in which the underground water resources were exploited, untrammeled by "bureaucrats" or "politicians" appointed by the governor. If some regulation was desirable, a large number wanted it by local option and under local administration. A few called for regulation on a priority basis with strong enforcement powers, but these were clearly a small minority.[97]

The Underground Water Commission made its report on January 1, 1953, recommending the adoption of the correlative rights principle as the basis for regulation, in order to provide "an equitable apportionment of water among all present legal users in over-developed areas."[98] It recommended the closing of over-developed areas to further pump irrigation, the creation of districts for local determination of necessary cut-backs in pumping, provision for industrial or municipal acquisition of water rights by means of purchase, and the establishment of a commission to administer the law.

The reaction to the proposals of the commission varied. The *Arizona Daily Star* said the report indicated that the commission had bowed to the heavy pressure of the agricultural spokesmen. It said the local option arrangement "would be like the patient deciding to what extent the surgeon should operate. There would be the very human desire to avoid surgery entirely . . . ."[99] The *Arizona Republic* called the report "realistic" and felt that it should "convince the legislature that no stone has been left unturned to find out what situation the state faces or what should be done about it."[100] Everyone agreed with Governor Pyle that nothing concrete could be done until the Supreme Court gave its new decision on ground water.[101]

The Supreme Court finally made public its reversal of opinion in March 1953.[102] The reconstituted majority asserted that the common law regarding percolating water had prevailed in Arizona and had not been contravened by federal law. It again pointed out the police power as "possibly the only source of power the legislature possesses. . . ."[103] It did not specify the manner and extent to which the legislature might exercise that power since that question was not before the court.

In dealing with the common-law rule, the court adopted the doctrine of reasonable use and specifically rejected the doctrine of correlative rights. The reasonable use rule required that all water pumped from underground be used on owners' lands to the extent necessary to improve those lands; the water must be used in a manner reasonable to the needs and requirements of the land.[104] The reasonable use doctrine required no apportionment, but only reasonable use, even though pumping might damage another water user. The court withheld judgment regarding what constituted reasonable use in order that decisions might be made in individual cases in terms of particular circumstances.

The new dissenters derided the confidence of the majority in the use of the police power, contending there was no legal authority anywhere in the country for using this power in statewide regulation of the water supply. They predicted "that the mad race to 'mine' percolating waters which are our greatest natural resource will continue unabated until such times as these waters are declared to be public in character and suitable regulatory measures are adopted."[105]

Newspaper opinion was sharply divided on the reversal. Many felt that the new decision virtually precluded regulation of ground water. The *Arizona Republic* opined that economics would be the means of control thereafter.[106] Some felt it a blow to the Central Arizona Project since regulation was required in order to get supplementary water from the Colorado River. Others felt the decision made all existing laws on ground water unconstitutional.

With the end of the 1953 legislative session near, there was little hope for passage of a new code. Therefore, the legislature passed Senate Bill 107. This established restricted areas which were closed to agricultural development by means of ground water, and prohibited the drilling of new wells in these areas until March 31, 1954. These restricted areas were virtually the same as those that had already been declared "critical" under the 1948 code by the state land commissioner. The life of the Underground Water Commission was extended to the above date for the purpose of recommending definite measures to solve the water problem in light of the court's new opinion.[107]

The commission deliberated during the next months in preparation for the 1954 legislative session. During 1953 there was a tremendous increase in pumping, amounting to over one million acrefeet more than in 1952. Opinion continued divided on the question of means of controlling the exploitation of the ground-water basins. This division crystallized further with a Superior Court decision late in 1953 which ruled that the Ground Water Act of 1948 was unconstitutional.[108] Further doubt was thus thrown on legislative efforts to curb pumping.

With the opening of the legislature in 1954 the Underground Water Commission recommended a code to the governor which followed closely its recommendations of the previous year. Water conservation districts would be established to determine the necessity of, and the extent of, reductions in pumping. Vested rights would be guaranteed. All critical areas then existing would be closed for drilling wells. New critical areas could be designated by a newly established commission or on petition by the landowners in an area.[109] The governor gave only tentative support to these recommendations in view of the unsettled legal situation.[110] He urged the continued support of the Underground Water Commission so that it might make whatever alterations were necessary in light of subsequent court decisions.

Like its predecessors, this version of a code received indelicate treatment from the legislature. Each interest sought exemption from any code that might be adopted or demanded the right to construct supplementary wells when others played out.[111] Strategically located legislators prevented action in legislative committees.[112] Efforts by the governor, Lewis Douglas (former ambassador to Great Britain and one of the state's leading citizens), and some legislators availed nothing. A complete impasse was reached.

The only hope of preventing a further deterioration in the situation lay in extending the prohibitions already in existence. House Bill 367 was the answer. This provided for the continued life of the Underground Water Commission for one additional month, and an extension of the prohibition on drilling wells in critical areas for one year more. This passed both houses by heavy majorities. The governor signed this bill, but severely rebuked the legislature for its unwillingness to correct the ground-water law. He accused the mining companies, the farmers, and the municipalities of each contributing to the defeat of the code by their demands for exceptional status. He alleged that a "few rail-perched lobbyists . . . are more effective in wrecking useful legislation than the 99 members of the 21st legislature are in passing it."[113] With this message ringing in their ears and with threats of a special session being given widespread currency, the legislature went back to work to appease the governor.

The renewed effort again proved in vain. Only to keep the governor from calling a special session, Senate Bill 135 was passed by large majorities in both houses.[114] This bill transferred the duties of the Underground Water Commission to the state land commissioner, confirmed existing critical areas, provided for the establishment of enlarged critical areas after hearings by the state land commissioner, prohibited the issuance of permits for new wells for the purpose of irrigating lands not already under cultivation in critical areas, and prohibited the use of ground water for irrigation in violation of either the 1948 code or the 1953 amendment. The governor allowed this bill to become law without his signature, declaring to the legislature:

> I could never sign such legislation as this, representing as it does, a sorry, weak, and confused ending to a two-year struggle for an adequate underground water code to protect our entire economy against the dangers of dwindling water supplies.[115]

Ironically, that same month of April, 1954, when the legislature completed its rout of ground-water legislation, was declared "Conservation Month" by Governor Pyle.[116]

Legislative action on ground water virtually ceased in 1954. Governor McFarland, coming to office in 1955, barely mentioned water problems in his address to the legislature and no bills of any general significance were introduced.[117] The state land commissioner continued to hold hearings and designate new and enlarged critical areas during 1954 and 1955,[118] but none has been created since that time. Hearings were held regarding the creation of a critical area in the McMullen Valley, where considerable agricultural development has occurred, but no action was taken.[119]

It was in 1955 that the case of *Southwestern Engineering Company* v. *Ernst* reached the state Supreme Court.[120] Southwestern Engineering Company had applied to the state land commissioner for a permit to construct a well on land within a critical area which did not have a history of cultivation prior to the passage of the ground-water code. The commissioner denied the application and the company then sued to enjoin the commissioner from preventing the construction of the wells.

The primary argument of the company was that the waters under their property were percolating and therefore part of the soil and subject to whatever use might be made of them. It argued that the Ground Water Code amounted to a deprivation of property without due process of law and without just compensation. It further argued that the classification involved in the code was arbitrary and unreasonable and therefore violated the equal protection clause of the 14th Amendment to the federal Constitution since the classification was not reasonably related to the purpose for which the code was passed. The code allegedly discriminated among persons within a single class, the distinction between present and potential users being unwarranted.

Finally, the company contended the act was unconstitutional for want of definiteness and because it gave the land commissioner law-making power.

In considering these arguments, the court found it necessary to review the meaning of the police powers of the state in regard to the rights of property. After reviewing numerous cases, the court said:

> We are of the opinion that there is a preponderant public concern in the preservation of the lands presently in cultivation as against lands potentially reclaimable, and that whereas here the choice is unavoidable because a supply of water is not available for both, we cannot say that the exercise of such choice, controlled by considerations of social policy which are not unreasonable, involves a denial of due process.[121]

The police power, then, could be used to restrict property rights in water because of the needs of "social policy."

Although admitting the distinction between present and potential users was an unusual classification, the court said that this was not sufficient to invalidate the code since the classification had a rational basis. All occupations called by the same name did not have to receive the same treatment under this constitutional standard since there were sound reasons for the distinction, reasons involving protection of the community against economic loss. The court further asserted that the application of the restrictions of the code to people within critical areas and not to those outside these areas was constitutional and a reasonable classification related to the conditions existing within the state.[122]

The court accepted the argument that a law must not be so "vague, uncertain and incomplete" that reasonable men could not agree on the law's meaning and application, but found that there was sufficient certainty concerning the principles to be used in the determination of critical areas and the procedures to be followed. The court stated that it was a well-settled principle that the legal consequences expressed in the law could take effect upon the determination of a fact or condition by an administrative agency. The complexity of the ground-water problem required that the responsibility for the determination of the facts be delegated to an administrator.

The lone dissenter in this case was the writer of the original decision in *Bristor* v. *Cheatham*. He contended that the law violated the due process clause in denying the exercise of rights of property. He maintained that the classification was unreasonable since it did not in fact promote the stated purpose of conserving ground water. He argued that the law in no way restricted pumping so that the alleged purpose of the law was not served by its classifications.

"Social Policy" = extra rights for Sodomites

More recent developments, while not challenging the constitutionality of the code, have further raised questions concerning its usefulness. The prohibitions against the drilling of new wells in critical areas, except for replacement and deepening, expired in 1955 when the legislature failed to extend them in revising the Arizona State Code. In November, 1956, the Supreme Court, in the case of *Ernst* v. *Collins*, decided that the state land commissioner had the authority to issue licenses to construct new wells only in the event that it was necessary to replace a failing well in a critical area.[123] In May, 1957, however, the court permitted the construction of a new well in a critical area even though it was clearly not a replacement of an existing well. In the case of *Vance* v. *Lassen* the court said that since the legislature failed to re-enact the prohibition on well drilling, a well could be drilled to provide water for lands in cultivation during the qualifying period.[124] The land commissioner was required to issue the license. There was no limit, therefore, on the number of wells which could be dug in the critical areas which had a history of cultivation five years prior to 1948.

In 1960 the Supreme Court further weakened the code in the case of *Arizona* v. *Anway* in which the court stated that a landowner could transfer the application of ground water from a parcel of land having a history of cultivation prior to the effective date of the 1948 act to a parcel of land not having such a history.[125] While it appeared to the State Land Department that the code forbade expansion of the acreage developed by ground water, the court said this was only an implication drawn from the code and the court could not expand the statute.

On the legislative side, the legislature, with the support of the State Land Department, passed an amendment to the code which permitted land to be irrigated by ground water in critical areas when it had had a history of cultivation five years prior to the creation of the critical areas.[126] Thus, land brought into cultivation as late as 1955 in some instances can be irrigated with ground water, in spite of the original purpose of limiting agricultural development in critical areas to that land in cultivation prior to 1948.

The irrigated acreage in the state declined between 1955 and 1957 — in part because of the increased costs of pumping water from greater depths. In 1958, however, agricultural acreage began to expand again, partly as the result of new areas brought into cultivation such as Harquahala Plains, Salome, Wenden, and others. These areas are almost entirely dependent on ground water and have not yet been declared critical since they are only recently developed. The 1948

Ground Water Act does not afford these areas protection against overdraft.

Perhaps the most important change in the picture has been the rapid urbanization of Arizona in some of the most important agricultural areas. Phoenix, together with the surrounding complex of cities, has grown rapidly, causing reduction in agricultural lands through subdivision for residences. Substitution of residential development for agriculture may relieve temporarily the pressure on ground water and increase the supportable population since residential water use is markedly less than agricultural use. In connection with the Tucson area, a 1957 study concluded that "if the city continues to grow at its present rate, some water will have to be diverted from agricultural use to meet future municipal needs, or new sources must be found."[127]

Some geologists have suggested the use of recharge wells to capture much of the flood water now lost through evapotranspiration. The feasibility of this technique, either technically or economically, has not been proven, and experimentation with recharge wells has not received adequate support by the state or its cities. The U.S. Geological Survey makes an annual report on ground-water levels in Arizona, but owing to financial limitations imposed by its matching arrangements with the state can do little basic geologic investigation of water-resource structures.

With the legal issues apparently settled there is little interest in altering basically the existing legal and administrative arrangements involving ground water. Farmers will continue to pump until it is economically no longer feasible to do so, or until they receive offers sufficiently attractive to induce them to sell their water rights. Meanwhile, new lands are opened up without restriction and with the eventual danger of over-development. Ground-water laws have perhaps prevented the expansion of agriculture and further overdevelopment of land dependent on ground water, but they have not redressed the serious imbalance of withdrawal and supply that existed before the laws were put on the statute books.

The relentless increase in population in Arizona cities apparently will continue, and with such increase, the competition between water for agriculture and domestic use will intensify. The U.S. Geological Survey has warned both the state and its municipalities that they are drawing on a water bank account of uncertain size; yet they are making few attempts to ascertain what the size of the account is.

CHAPTER 5

The Politics of Water

The physical, economic, and social limitations which the shortage of water imposes upon the people of Arizona causes the management of water to be a matter of primary interest to the general public and the various interest groups composing it. Public interest in the management of the water supply was displayed immediately upon the creation of Arizona as a Territory when all surface waters were declared public in the *Howell Code* in 1864. Public funds have been and continue to be vital in providing water at the proper place and time. Public agencies regulating land and water usage maintain, with varying degrees of success, their historic responsibility of ensuring maximum utilization of water supplies for the public interest.

The very concept of the public interest, however, tends to gloss over the underlying and deep-seated conflicts that exist over the management of water. While administrators concerned with over-all planning may pretend to see the proper methods and techniques for managing water and proper purposes for which it should be managed, there is no such agreement among the major interests whose very existence depends upon obtaining sufficient quantities of this "white gold." In times past this lack of agreement has caused many an Arizonan to stand by his headgate with revolver in hand to prevent another from taking one drop of that water he considered due him. In 1935 it caused the governor of Arizona to send national guardsmen to the Colorado River to prevent the federal government from proceeding with its developmental work there. Mining corporations, with little benefit of law, diverted water from watershed to another in order to ensure an adequate supply. The statement that "water is more precious than gold and more explosive than dynamite" certainly applies to Arizona.[1]

Basic economic issues have a very real relationship to the issues concerning the management and utilization of the water supply.

Whatever increases of water use occur in one segment of the economy will ultimately result in a deprivation of that water for use in another, barring some technological breakthrough in such fields as rain-making or desaltation of sea water. The expansion of the cities, for example, has resulted in a reduction of cultivated acreage in the areas around the cities. Although this has temporarily alleviated the water problem in some areas, since domestic uses demand far less water than agriculture, the problem will occur again as the cities continue to grow. The increased importance of tourism and recreation may prevent all that the farmers want done in the way of managing the forested and chaparral areas for increased water production. Industry, which has been entering the state in increasing magnitude in recent years, also draws on the limited water resources.

Perhaps the most important question affecting the basic resource policy in Arizona is that concerning the long-term direction of the state's economy. In this regard, water is a most important limiting factor. Some provision must be made for the sustenance of the rapidly growing population. In the past, agriculture and mining have been the mainstays for the economy, constituting 20 percent of the total income of the state in 1953.[2] But manufacturing has recently overtaken both of these, and agriculture will probably continue to decline in relative importance. In the long run, mining will probably also diminish in importance as the vital ore resources are depleted.[3] Furthermore, both these industries are greatly dependent on the world market which has a history of marked instability and fluctuation.

Nearly 20 percent of the state's income is derived from government payments. Such income is highly dependent on the international political situation and the willingness of the national Congress to support reclamation projects and the like. Some economists foresee the major opportunities for expansion in the fields of manufacturing and tourism, both of which have shown decided increases in the last decade.[4] The advantage seen in diversification based on tourism and light industry is an economy more self-sufficient and less dependent upon outside economic forces.

The dependence of the state on water, as well as on other primary resources — metals, timber, crops — has created a political situation having its own peculiar characteristics, although shared by other states having similar backgrounds. A sense of individualism permeates the thinking of the members of many of the most important interest groups. On the whole, this attitude promotes an almost instinctive negative response to governmental activity, in spite of the remarkable degree to which these very same interest groups have been and are dependent on governmental policy for their well-being.[5] It is this which produces the anomaly of Arizonans asking the federal government to finance such things as the Central Arizona Project, while at the same time electing and re-electing people who express continued fears about "the federal government taking over."

While integrated effort and governmental support have been a necessity, the almost mythical notion of self-sufficiency has a decided staying power, often with dire results as in the case of water utilization. The influx of new population from other areas, and the introduction of new industry may alter this picture in the long run, but it appears certain that the now-entrenched economic groups will continue to dominate the political scene in the foreseeable future.

In a sense, the present power of well-organized economic interests is a testimony of the failure of liberalism which contributed to the thinking of those men who were instrumental in the writing of the Constitution of Arizona. That document was the product of the two dominant political forces operating in Arizona in the early twentieth century — progressivism in political organization and management, and the labor movement. The stamp of Progressivism is clearly in evidence in the emphasis on popular government through such devices as primary elections, the initiative, referendum, and recall, and the various restrictions imposed on elective and appointive officials. The influence of the labor movement is found in provisions allowing the state to enter directly into business, directing the legislature to provide for workmen's compensation, and establishing a corporation commission with wide powers of regulation over industry and utilities in the state.[6]

This so-called "liberal" constitution was written in reaction to the domination of the Territorial government by corporate and frequently nonresident interests. It was generally understood that the Territorial legislatures had been virtually the property of the large corporate interests — particularly the mines and the railroads. One student observed:

The late former Governor George W.P. Hunt has declared that every territorial legislature was under the control of the corporations . . . . It was charged that if an anti-corporation bill managed to be pushed through both the House and the Council, $2,000 would secure the governor's veto. Both of the major parties were under the influence of the corporations.[7]

It was common knowledge that these interests could and did exercise their influence through the locally elected representatives as well as through the patronage appointees from Washington.[8] The antagonism of the farmers and workingmen of the Territory was directed particularly toward the prevailing methods of taxation, methods which permitted the valuation of mining lands on the same basis per acre as the farming and grazing land in the same area.[9] It was felt that only by throwing off the limitations imposed by "carpet-bag" rule and permitting the "public" to gain control of the legislative process would government serve the ends of others than the powerfully organized corporate interests.

The legislative history of the state since 1912 indicates the inadequacy of the devices incorporated into the constitution to break the power of these economic groups. Only occasionally have the referendum and initiative proven useful in overcoming the irresponsibility or intransigeance of the legislature, and never in a matter relating to water or natural resources policy. On the other hand, these special-interest groups have used these devices, or the threat of their use, to promote or defend policies as the occasion demanded.[10] The legislature has been described by one observer as "a reactionary, hand-picked legislature," dominated by a "cabal" consisting of the cattlemen, copper companies, utilities, commercial farmers, and railroads.[11] The liberalism with which Arizona began its history may not have vanished entirely, but its luster has dimmed considerably with the reassertion of power by these well-organized groups. In the early 1950's Stocker professed to see a renascence of the liberal movement in collaborative efforts by labor and education groups, but as yet this union has not been very effective.[12]

The inverse relationship between the direct influence of pressure groups in the legislature and the strength of political parties is borne out in Arizona. There has been little in the way of genuine competition between the two major parties for most political offices or for control of the legislature. The Democratic Party is the primary vehicle for election at nearly every level of government, although in state-wide elections Republicans have achieved much greater success. There has been a Republican Senator since 1952, and from 1950-54 and 1958 to the present time the governor has also been a Republican. But only during brief periods has the Republican Party gained influence in the legislature through force of numbers, and only four Republican governors have been elected over a 48-year span from 1912 to 1960. At times there have been no Republicans in the state Senate. This condition promotes divisions within the Democratic Party based primarily upon "the influence of varied economic organizations of the state."[13] The absence of genuine competition has led to a situation in which "cliques" (as Waltz calls them) can control the legislative proceedings without fear of political reprisal. Some observers see a changing situation here also, due to the large numbers of Republicans coming into the state who are settling in the two major metropolitan areas around Phoenix and Tucson.[14] The Phoenix area, for example, has had a Republican Congressman since 1952. But there are few real signs of a Republican trend at the county level save in the two metropolitan areas.

The factionalism inherent in the modified one-party system has provided the means by which a conservative majority has been able to control the legislative process during most of the recent sessions. Only occasionally does the liberal bloc in the Democratic Party actually gain control of the House and almost never in the Senate. For this reason the so-called "Pinto" Democrats, the conservatives in the party, generally are able to ally themselves with the Republicans to form a majority. The conservative majority, composed mostly of Democrats, has effectively ruled the legislature. Compounding this internal party division with the bicameral system which provides its own devices for inhibiting majority rule, one has a picture of a legislature characterized by deadlock, misunderstanding, or spite, all of which seem to prevail during a normal session. The distrust and confusion resulting from two houses, one of which is dominated by the "cow" counties and the other by the two metropolitan counties, is amply demonstrated by nearly every legislative session.

The ends sought by private interest groups are furthered by various other means, some embedded in the constitution and others the result of policy. The state constitution requires an absolute majority vote in each house of the legislature for passage of all legislation; occasionally this provision has played into the hands of minority groups which can delay or prevent action by refusing to participate or vote on an issue. The threat of the referendum has often forced the proponents of legislation to seek the two-thirds majority necessary to prevent the use of the referendum device. This threat has been a potent one considering the differences in economic resources among the interest groups in conflict, since the referendum requires large expenditures of money and can be used only by those willing and able to bear the cost.

The majorities in each house of the legislature, which are rigidly controlled by a caucus system, are normally conservative. The conservative majorities usually can control all of the internal procedures of the legislature, including the committees, thus reducing the power of legislative minorities to obstruct legislative action. The committee system, however, occasionally does play into the hands of veto groups that can prevent action which would detrimentally affect the interest they represent. A bill may be assigned to as many as four committees, each one of which has the power to prevent the bill from being considered on the floor. With each committee composed of representatives having particular concern for the subject matter of that committee, powerful interest groups are in excellent positions to protect their own well-being.[15] Representative Abel of Maricopa County, a member of the minority Democratic bloc in 1955, reportedly stated in exasperation:

> We have this great number of committees in the House so you can trade chairmanships for votes for speaker. The judiciary committee is for the sole purpose of finding any bill unconstitutional that is not for the benefit of the mining companies and cattlemen. The efficient government committee is a typographical error; it should have been inefficient government. Ways and means was created to seek ways and means to kill bills in committee.[16]

It is not only in the legislature that the important interest groups have found the means of protecting themselves. The system of popular election for executive officials and the limited powers of the executive over his administration have produced a similar situation. The public agencies, in many instances, have become little more than spokesmen for the private interest groups with whom they must deal. The close union between the Arizona Game Protective Association and the Arizona Game and Fish Department has long been recognized. The cattle growers for a long period considered the State Land Department a protector of their interests and even protested when the state land commissioner, who appeared to be especially sympathetic with their goals, was removed for engaging in shady land operations.[17] The alliance between the Arizona Interstate Stream Commission and the Central Arizona Project Association is not only accepted but widely praised. The Arizona Corporation Commission has been looked upon as a protector of the private utilities rather than a regulator of them.

This fragmentation of executive power, while perhaps not a primary causative factor in the influence of private groups, has operated to the benefit of these groups nominally subject to control by these executive agencies. The unwillingness of the legislature to accept administrative reform, so much talked of in Arizona, cannot be understood without recognition of this fact.[18] The failure to accept resource planning at the administrative level may also be considered in this context.

While fragmentation of political power at the formal levels of government is the pronounced characteristic of Arizona government, this centrifugal force has been counterbalanced in part by a system of private alliances which have welded the major entrenched units into a formidable machine. "These forces are headed up by the three 'C's

— copper, cattle, and cotton — but they include also a 'U' for utilities and a double 'R' for railroads."[19] The first three of these groups have a vital interest in water management and utilization.

They work together in a tightly-knit, efficient and generously financed combine, often behind fronts with such guileless-seeming names as the Arizona Tax Research Association. With one or two exceptions they have the support of the state's press. Their principal instrument is the Arizona legislature, which consistently plays McCarthy to their Bergen.[20]

Undoubtedly, the Tax Research Association is the most potent lobby in the state legislature, at times even having had a formal role in the proceedings of the legislative committees, particularly in the Ways and Means committees. In 1947, after Governor Osborn characterized the Tax Research Association as the "fourth department of state government," the legislature barred the association from sitting in on committee deliberations.[21] In what was either bravado or disgust, the Speaker of the House, who was also vice-president of the association, reportedly said, "Anyone who says the association did not run that session doesn't know what he is talking about."[22]

The cattlemen, it is asserted, operate as the front men for the other interests, notably in the Senate, because of the extreme favoritism given the rural areas in that chamber, composed of two senators from each county.[23] The Tax Research Association, concerned above all else with preventing increase in the tax rates affecting their members, resists most welfare legislation, and has in recent years taken on the teachers' association in its fight for higher salaries,[24] the labor groups in their quest for the establishment of a labor department, and the cities in their desire to remove budget limitations.[25] The association also works with executive agencies toward keeping a minimum level of taxation — with apparent success. In a remarkable bit of self-revelation about its operations, the association expressed its satisfaction with the way things were running in the following words:

We, of the staff of the Arizona Tax Research Association, desire to take this opportunity to express our appreciation to the many public officials who have been so cooperative and courteous to us during the budgetmaking period just ended . . . . This year marked the seventeenth year we have all worked together in the process of budget making, and we can say . . . that our association with those officials charged with the duty of making budgets and setting our tax rates, has been most pleasant at all times.[26]

The history of resource management in Arizona must be considered in the light of this atomistic political system. Each of the major interest groups has a direct interest in resource policy; some, such as the cotton, copper, and cattle groups because of their utilization of a natural resource; others such as the railroads, because of a concern for taxation. These groups, singly and in cooperation, use their power in the legislature and the administration to prevent adjustments in resource policy which would be damaging to their positions. These groups are occasionally at odds, but generally they work together to promote maximum usage of the resources with a minimum control by the public, or public control firmly in their own grasp. Those who seek adjustments in resource use favorable to other interest groups have to contend with these political facts of life.

The conservation movement harbors many different and often mutually antagonistic elements within it, giving rise to the question whether there is such a thing as "conservation" at all. These extreme elements range all the way from those who desire to preserve an area as nearly like its primitive conditions as possible to those who want "conservation" but only in terms of the single resource in which they have a vital stake. Both have their spokesmen in Arizona. While this diversiveness within the ranks of the conservationists is not peculiar to Arizona, it of course reduces their political effectiveness.

The former groups take a dim view of the so-called developmental work in Arizona, contending that the "developers" have lost sight of the true place of Arizona in American culture. They contend that the "developers" are ripping up the soil, denuding the forests, and generally disturbing the natural conditions which flourished in the past, and thus depriving man of his community with the earth. Perhaps the most vocal members of this group are found in and around Tucson, with leadership supplied by Arthur Pack of the Pack Foundation and national publicity by Joseph Wood Krutch. The latter has written: *How about a RUBBER CRUTCH?*

> To live healthily and successfully on the land, we must also live with it. We must be part not only of the human community, but of the whole community; we must acknowledge some sort of oneness not only with our neighbors, our countrymen and our civilization, but also with the natural as well as the man-made community.[27]

Man, Krutch contends, is too egocentric to see this natural community and conceives of the world only as a place to serve his own interests.

Krutch, and others with this point of view, assert that the dreams of the Chambers of Commerce in the larger cities for more people, more irrigation, and more industry will be denied because of the lack of water and that "the demand for living space will have to be frustrated in other, more easily over-exploitable regions before the dreams of the boosters are realized as the nightmares such dreams have a way of turning into."[28] What is the desert good for? For space, because "room is becoming one of the scarcest things on earth for most people because it is one of the things which no economy of abundance seems to plan to supply abundantly."[29] Space is necessary because not to have known the mountains or the desert is not to have known one's self.[30]

This almost reverential attitude toward nature and particularly toward the desert is shared by many Arizonans. Said one conservationist recently, "To the true lover of the desert . . . the relinquishment of any wilderness area is a tragic thing that amounts to personal defeat."[31] Many who have come to Arizona for its natural beauty support these views. Said Frank Lloyd Wright, in justifying his proposal for a new state capitol building (which was rejected by the men of practical affairs),

> Arizona, youngest of the United States, is also youngest in geological time. Therefore outlines are sharpest and colorful; contours are most picturesque. Her terrain is unique in the world; destined, in spite of the obtuse insistence upon industry and agriculture, to become the playground of these United States of America.[32]

These conservationists have not been very influential in curbing the growing tide toward maximum development of the state's resources for human use. Realization of this comparative lack of influence on public thinking may be the reason for the establishment of a program for the study of conservation education at the University of Arizona. There is considerable evidence that these groups would like to have a larger share in the decision-making on resource policy.

When the naturalists join with their almost natural allies, the recreationists and the sportsmen, they form a more imposing bulwark against the onrushing movement toward increased industrial, commercial, and agricultural development. The recreationists protest that they do not demand non-use of the desert and forested areas but only that these should be managed for maximum value for all interests, and that such management should be well planned. They are particularly concerned about protecting these areas from commercial exploitation which will eliminate recreational values. "Arid lands are not expendable lands. They can be valuable lands from a recreational standpoint. Our growing populations and ever-increasing margin of leisure time demand more such areas."[33]

The recreationists are unwilling to accept a secondary position in terms of economic importance to the state. An *ad hoc* group of state and federal officials in 1941 stated the following concerning the importance of the recreation industry to the economy:

> . . . if we add to the most conservative estimates of *the amount* spent by tourists, the total amount spent by Arizona residents, and the various private and governmental agencies and organizations for outdoor recreation, the sum will reach an amount which will unquestionably place the recreation industry in first place.[34]

In their interpretation of figures developed by the National Wildlife Federation concerning the economic importance of recreational activities, the Arizona Game Protective Association again claimed preeminence for recreation in 1956.[35]

The issue is clearly joined between the naturalists and the developers, with the water-giving watersheds of the state in the center of the battle at the present time. The two sides have tended to obfuscate the issues and many have taken such extreme positions that it is difficult to find reasonable discussion, let alone a basis for resolution of the conflict. For example, Rich Johnson, former editor of the *Arizona Farmer-Ranchman*, and now president of the Central Arizona Project Association, and vice-chairman of the Arizona Water Resources Committee, in speaking to a meeting of the Arizona Section of the Soil Conservation Society is reported to have said the following:

> Conservation is an economic problem. If we could amputate sentimentalism, romanticism and hobbyism from the body of conservation, progress would be faster and more certain than it is.[36]

Calling the recreationists "useless appendages," he argued that forest lands must be managed for water and forage; he asserted that "worshippers of every leaf and twig that grows" must be convinced that "woody plants — even trees — may be worthless weeds."[37] He did not consider the erosion problem critical, although he admitted that there was insufficient information on the subject. However, he believed that liquid soil — silt — could be managed to create new farm lands "more or less where we want them."[38]

This view, expressed by an important official of the groups supporting watershed modification programs, could be expected to bring a violent retort from wildlife groups. The editor of the *Arizona Wildlife-Sportsman* responded that the above views were that of the lunatic fringe and would be laughable except for the fact that the people supporting them were in deadly earnest.[39] He argued that it might make just as much sense to eliminate cotton and cows as trees since they also were declining in relative importance to the state economy. In his opinion, however, none of these alternatives made any sense. The management of the forested areas demanded by the sportsmen, he said, was management for multiple use rather than for special interests which Johnson represented.

Most of this is sound and fury since the resolution of the conflict over the management of the state's desert and forested lands can come only through the federal government. The state could, of course, embark on a modification program on its own lands, but these generally are not high water-producing lands. Also, it is doubtful if the state will undertake *any* management, let alone the expensive procedures involved in watershed modification. The federal agencies managing these lands cannot ignore the pressure put on by organized interest groups that use these lands. There is much evidence that they are paying careful attention to all that is being said and done without committing themselves to any policy except the present one of management for multiple use. These federal agencies are reluctant to embark on any watershed modification programs without greater research evidence. They too are under attack by the cattlemen, farmers, and some municipalities for their unwillingness to take up the cause of vegetation modification. These groups, led by the Arizona Water Resources Committee and the Water Resources Division of the State Land Department, are keeping up a steady promotional effort designed to convince the land-management agencies to accept their position. In this effort, the Land Department has enlisted the support of Senator Carl Hayden who has been able to obtain increased appropriations for the establishment of experimental projects and for speeded-up silvicultural practices on the national forests where there is the greatest expectation of water-yield increases. The Arizona Farm Bureau Federation and the Arizona Cattle Growers' Association have both indicated approval of watershed modification and the Arizona Section of the National Reclamation Association pressed for favorable resolutions at the national meetings. Meanwhile the sportsmen have been firmly opposed to all schemes for vegetation modification.

The politics of a water-short state are a mixture of science, economics, esthetics, and emotion. The amount of each element going into this political potion varies from time to time but seldom is one of these elements entirely absent. The verdict of science is called upon to support each side on water problems. The economic value to the state of each of the contending interests is calculated in such fashion as to make each the leading industry of the state. The need for less tangible things such as space or scenery is rallied to the cause against such mundane things as crops or industry. The decisions that are made regarding water, soil, and land management, or the failure to make decisions, will decree the nature of society in Arizona for the decades to come.

—*Charles W. Herbert, Western Ways Photo*

Competition for the use of available water is keen among the widely varied interests. Claims of each group have merit, and finding a mutually acceptable apportionment is extremely difficult. Artificial lakes for recreation like the Apache's Hawley Lake (above), and the stock-watering catch basins built by cattlemen (below) conflict...

—*Naurice Koonce, Ray Manley Photo*

—*Ray Manley Photo by Carroll*

. . . with other requirements such as agricultural irrigation (above) and the reservoirs and power plants (below) essential to the growth and development of more populous areas. Continually increasing. . .

—*Phoenix Chamber of Commerce*

—*Phoenix Chamber of Commerce*

... amounts of water are being required for municipal and domestic use in growing urban areas (above) and for the development and growth of the many new industries which are springing up around the state.

—*Pete Balestrero, Western Ways Photo*

CHAPTER 6

The Colorado River

It is impossible to discuss the politics of water in Arizona without giving attention to the Colorado River controversy — especially to the political and legal issues involved. The solution to the problem of how to obtain and utilize Colorado River water in the state has been one of the dominant political issues, making or breaking politicians from county sheriffs to governors. It has been said that during the 1920's and 1930's one could not be in favor of the Colorado River Compact and hope to win an election in Arizona. In the same way, it is virtually impossible for a candidate to win an election now unless he favors the Central Arizona Project.

The people of Arizona have come to look upon the officials of California, and particularly those of the Imperial Irrigation District and the Metropolitan Water District of Los Angeles as diabolical schemers who are dedicated to the robbery of Arizona's birthright. Others, particularly from the Upper Basin states, have come to a similar position owing to the uncompromising opposition of Southern California to any and all proposals for Upper Basin development.[1] With the Colorado River running through Arizona or along its border for 688 miles, the longest stretch in any single state, it has become a source of much frustration that little of the water has been put to use within Arizona's borders. The psychological impact of finally having to accept defeat on the Colorado River Compact after twenty-two years of fruitless resistance has further contributed to the intensity of these feelings.

THE COLORADO RIVER COMPACT

The seven states that are within the drainage basin of the Colorado River are divided into two groups: the Upper Basin states — Colorado, New Mexico, Utah, and Wyoming; and the Lower Basin

81

states — Arizona, California, and Nevada. The division point between the two is Lee's Ferry. The Colorado River Compact, concluded in 1922, divided the waters between the Lower and Upper Basin states, but it did not touch the equally vital question of the apportionment of waters to the states within each basin. The compact envisaged apportionment of waters to individual states by the terms of agreements subsequently to be entered into by the states in each basin. It was written at a time when the available records showed much larger quantities of water in the Colorado than have been shown by records of longer duration compiled since that time. Variations in streamflow range from 10 million acre-feet to over 26 million acre-feet measured at Yuma.[2] The prolonged drouth of the 1930's and 1940's forced a downward revision of the figures from those used in determining the Colorado River Compact. The 48-year streamflow record from 1897 to 1943 measured at Lee's Ferry was an average of 16,270,000 acre-feet per year, while the 10-year record from 1931 through 1940 indicated an average of only 11,800,000 acre-feet per year.[3] Harold Schwalen of the Department of Agricultural Engineering at the University of Arizona has contended that the 16,270,000 figure should be scaled down to 14,900,000 acre-feet per year.[4]

The impetus for an agreement between the two basins came from the Upper Basin states which looked with some uneasiness at the voracious water and power developments in Southern California. They feared — and they feel that their fears have had considerable justification in light of events in recent years — that the early development of the river for the benefit of Southern California, without some legal protection for their water rights, would threaten irrigation and power projects in their areas at a later time.

Although the Colorado River Compact was signed by the representative from Arizona, the legislature of the state refused to ratify it. Arizona's commissioner signed the compact only after he obtained agreement — at least he thought he had — that the beneficial consumptive use by the Lower Basin could be increased by one million acre-feet in order to compensate Arizona for the inclusion of the Gila River under the terms of the compact.[5] He also sought for a definite division among the states of each basin as well as between the basins but was overruled.[6]

THE COMPACT AND STATE POLITICS

During the time that the compact was being negotiated, Arizona was in the throes of an election campaign for governor and other state officials, a campaign in which the approach to a solution of Colorado River problems became involved. The Republicans, led by their incumbent Governor Campbell, in general supported the reclamation policies of the federal government, while the Democrats, led by the redoubtable George W. P. Hunt who had formerly been governor for three terms, expressed a suspicion of developmental projects except on such terms as Arizona laid down.[7] Hunt won a victory of landslide proportions and immediately indicated that he would give his consent to no part of the agreement then being completed at Santa Fe which in any way compromised the rights of Arizona.[8] He questioned the adequacy of the engineering data on which a compact could be based, expressed fear of allowing any water to cross the border into Mexico for irrigating land for "asiatic colonies," and suspected the demands of California for power developments. He asserted,

> We have at least two million acres in this state that shall be irrigated with the water of the Colorado River. Arizona must have every bit of the power out of that river that she can use in any of her industries.[9]

Governor Campbell, now a lame duck, backed the compact concluded at Santa Fe and led a campaign to get it ratified at the January 1923 session of the legislature. The dispute took on decided party colorations inasmuch as the Republican administration in Washington, and particularly Herbert Hoover, supported the compact as did Republicans generally in Arizona.[10] With Governor Hunt opposing the compact, and calling for its defeat in his message to the legislature in 1923, the legislature followed his lead and refused ratification of a compact even with reservations.

The reasons for the failure of the compact in the 1923 legislature and thereafter have been variously assigned. Some attributed the opposition of the mining, agricultural, and power interests as the most important factor.[11] Undoubtedly these groups were by and large opposed along with a large proportion of the general public. The farmers feared they would not receive a fair share of the water supply. The power interests opposed public power development of the Colorado. The mining companies objected to federal development since the installations constructed by the federal government would contribute nothing to a relief of the tax burden they were carrying.[12] Uncertainty, however, played a most important role. The failure, according to Parsons, represented a "suspicion of any ill-considered move which might jeopardize the interests of important groups within the state as such . . . ."[13] For, "the full implications of any policy on Colorado River development for any political or economic group within the state were not yet wholly clear."[14] Viewing the Colorado River as the state's most important resource, the spokesmen for Arizona did not wish to take precipitate action they might later regret. They did not believe the river could be developed without the consent of Arizona and that Arizona's only bargaining power lay in refusing to ratify the compact.[15]

This fateful decision taken in 1923 dictated the form of politics on the question for the next two decades. Although there was considerable sentiment in favor of the compact in 1923, this sentiment was virtually eliminated by the opposition. Rejection of the compact and opposition to the Boulder Canyon Project became the only fashionable position to take for Democrats and Republicans alike. Even the Republican governor elected in 1928 opposed ratification of the compact. It could be said, with little qualification necessary, that the "state" position was that of opposition to the compact and any development made under its terms. According to one writer, politicians "were using the issue for political purposes," posing as the defenders of Arizona's birthright against the "octopus" — California.[16]

Arizona continued its opposition through the period when the Swing-Johnson bill was being considered by Congress, by the terms of which the Colorado River Compact became effective without the consent of Arizona. The various interest groups solidified their opposition, particularly the power interests and the farmers. The farmers were even more adamant since they recognized that no water could be made available from Hoover Dam for Arizona lands and that such a structure could provide water for lands in Mexico growing crops using cheap labor and competing with Arizona crops. The worsening farm depression only increased their opposition since any water on new land would increase the competition of farmers already in desperate circumstances.[17] The sole support for the ratification of the compact and the acceptance of the Swing-Johnson bill came from Yuma and Mohave counties. These counties had a natural interest in these matters inasmuch as the construction of the dam would regulate the river and protect the lands under cultivation near Yuma and Parker, and also provide a steady water supply for irrigation. The support from Mohave County was based primarily on the economic advantage accruing from having the structure located within the county.[18]

ARIZONA vs. CALIFORNIA — EARLY LITIGATION

Arizona lost its battle in the halls of Congress with the passage of the Boulder Canyon Project Act in 1928. The state then shifted its attack to the courts. In each of three attempts, it was unsuccessful. The first case involved an attempt to obtain an injunction to prevent the construction of Hoover Dam on the grounds that the Boulder Canyon Project Act constituted an unconstitutional invasion of the *quasi-sovereign* rights of Arizona, that Congress could not regulate the Colorado River since it was a non-navigable stream, and that construction of the dam would cause grave injury to the state. It was argued that construction of such a dam, partially within the borders of Arizona, depended on the consent of the state engineer of Arizona under state law and such consent had not been obtained.[19]

The Supreme Court rejected these contentions, asserting that the court could only accept the judgment of Congress as to navigability, and that developments envisioned by the Boulder Canyon Project Act related to navigation. The federal government did not have to rely on the consent of a state official in carrying out its constitutional powers since to do so would negate the delegated power of the federal government. Arizona could appropriate water from the Colorado River; until its right to water was actually interfered with, there was no cause for complaint.

In a second case, decided in 1934, Arizona sought to perpetuate testimony before the Supreme Court of the United States so that this testimony relative to Arizona's rights to Colorado River water might be available at a later time when Arizona's rights might be interfered with by California. In this action Arizona claimed rights under both the Boulder Canyon Project Act and the Colorado River Compact. It sought particularly to submit testimony concerning the negotiations involved in the Colorado River Compact in order to show that the one million acre-feet of water added to the Lower Basin's portion was intended for Arizona to compensate for inclusion of the Gila River in the compact.[20]

The court once again denied Arizona's claim, stating that Arizona could not assert a claim based on the Colorado River Compact when it had not ratified the compact. The court further argued that Arizona had failed to show the relevance of Article III (b) dealing with the one million acre-feet to the Boulder Canyon Project Act under which it claimed rights. The court said the evidence which Arizona sought to perpetuate was not of the character that would be admissible in any proceeding related to Article III (b) and therefore denied Arizona's suit, stating, "If Arizona's rights are in doubt it is, in large part, because she has not entered into the Colorado River Compact, or into the suggested sub-compact."

In 1934, with the beginning of construction on Parker Dam,

the structure to impound water for diversion to California, desperate measures were taken. Governor Mouer sent the state militia to the dam site to prevent any work on the dam, asserting that there was no statutory authorization for the dam. The federal government sought an injunction against the use of troops but the Supreme Court of the United States upheld the validity of Arizona's contention.[21] With the passage of legislation specifically authorizing the construction of Parker Dam, Governor Mouer was forced to withdraw the troops.

The third and final attempt by Arizona involved a suit calling for the Supreme Court to apportion the unappropriated waters of the Colorado River equitably between Arizona and the other basin states in order that the title to its share be quieted. It also called for judicial limitation of California's share to that stipulated in the Boulder Canyon Project Act and California's Self-Limitation Act. The defendants claimed that the court had no jurisdiction because, under the law of appropriation which governed in all the basin states, no rights could be claimed until an appropriation was made. Since Arizona had made no appropriation of the water for which she sought judicial determination, she had no rights to defend.[22]

The court refused to rule on the question of whether it could decree an equitable division of unappropriated waters. It contended that the rights of the United States in such a proceeding were so superior that no determination of the rights of the states could be made without also determining the rights of the federal government. Since the United States had not agreed to be sued and had not been made a party to the suit, Arizona's petition was denied.

During the period of litigation there were numerous attempts to obtain a settlement of the disagreements between Arizona and California by negotiation but each attempt ended in failure owing to the uncompromising positions taken by each side. As a result, the Colorado River Commission of Arizona stated as early as 1935, "Your Commission is convinced that it will be impossible to arrive at an agreement respecting the division of waters of the river between the States of Arizona, California, and Nevada."[23]

RATIFICATION OF THE COMPACT

The apparent exhaustion of legal remedies and the seeming impossibility of coming to agreement with California over the apportionment of water seemed to leave no way out for Arizona except to ratify the compact and obtain whatever benefits were possible under its terms. Opposition to such a course remained strong, however, and it was not until circumstances became desperate that legislation to that end was enacted.

During the late 1930's Arizona found itself facing an extreme drought situation. The years 1938 through 1940 were among the driest recorded in modern times; the dams were virtually empty and the diversion for agriculture was much diminished.[24] Increased pumping compensated for some of the decline but it was recognized that underground resources were not unlimited. With impending disaster to their agricultural investments, the farmers in the central valley in particular began to reconsider the merits of the Colorado River Compact and the Boulder Canyon Project Act.[25] With the decline in streamflow there was also a decline in power output. During the year 1940 there was a critical power shortage, relieved only by the rapid completion of transmission facilities from Parker Dam to the Phoenix area which transmitted power from Hoover Dam. This shortage and the impending demands resulting from the war effort led to the Arizona power survey discussed elsewhere, in which the Federal Power Commission recommended that the state use to the utmost the power developed along the mainstem of the Colorado.

These conditions altered the attitude of some of the major groups in the state, particularly the farmers who were facing grave water shortages. The opposition of the power utilities remained since water and power matters were inextricably interrelated in Colorado River questions. Some argued that the only way to get Colorado River water was to negotiate with the Upper Basin states for an increased amount of water from their future allotments under the Colorado River Compact.[26] A small group known as the Colter Highline Canal Association vigorously opposed a change in Arizona's position, arguing as it had for 20 years that Arizona had valid claims on Colorado River water not subject to the terms of the compact or federal laws. By the early 1940's, however, its ranks were greatly diminished. Even in 1956, however, there remained in the Arizona legislature one or two members dedicated to this position.[27]

At the height of the water shortage in 1940 the Department of the Interior even began surveys of the possibility of bringing Colorado River water from Parker Dam, directly to the west of Phoenix, to central Arizona. This investigation was discontinued when 1941 proved to be a wet year.[28] The next year found the drought resumed and new demands being made for use of Colorado River water. In 1943 the state legislature appropriated $200,000 to be made available to the Bureau of Reclamation for conducting surveys of various routes for bringing Colorado River water to the central valley.

Governor Osborn, who had become convinced that ratification of the Colorado River Compact and negotiation of contracts with the Department of the Interior for delivery of water and power were the only hope for Arizona industry and agriculture, became the leader in the fight for this reversal of position. In 1943 he offered his water and power authority bill to the legislature, stating,

> With the passage by Congress of the Boulder Canyon Project Act in 1928, the era of theorizing about the Colorado's riches ended. Whatever our previous opinions about the best place or the best plan for utilizing its water or the fairest basis for dividing its power, we now can recognize that the decisions have been made . . .[29]

However, he was unable to martial sufficient strength in the legislature, his bill failing the House after passing the Senate.

The year 1944 brought to fruition the governor's efforts. After authorization was given by the legislature, a contract was negotiated with the Secretary of the Interior for the delivery of 2,800,000 acre-feet from the Colorado River. This contract was ratified by the legislature early in 1944 after which the Colorado River Compact was ratified also.[30] Virtually the only opposition remaining was the Colter Highline Canal group. Many, including the governor, looked forward to a tremendous quantity of water coming into the state as a result of their action and through the Central Arizona Project then under study by the Bureau of Reclamation. Governor Osborn estimated in 1944 that it might be possible to obtain as much as four to five million acre-feet per year if the compact were ratified, depending on the amount of water in the available surplus to be apportioned in 1963.[31]

CENTRAL ARIZONA PROJECT

As noted elsewhere, the Bureau of Reclamation recommended the construction of an aqueduct from Lake Havasu to the central part of the state through which water could be pumped for agricultural purposes primarily. There would also be constructed a dam at Bridge Canyon for the production of power, a large proportion of which would be used for the pumping of the water into Arizona.[32] This Parker pump-plan has been adopted officially by the state of Arizona and supported through two attempts to get the project authorized by Congress. Nearly every important organization in the central valley has been mobilized in support of the Central Arizona Project, led by the Central Arizona Project Association, and the Arizona Interstate Stream Commission. All of the major newspapers in the area have given the project full support also. There is virtually no dissent, except for some misgivings on the part of the people around Parker and Yuma where there is fear that the Central Arizona Project would deprive them of water already being used on their irrigation projects.

In spite of the almost complete unity in the state on the question of bringing Colorado River water into the central areas of the state, the project has foundered on the unwillingness of Congress to give its consent. Serious objections were raised over the economic justification of the project. One commentator asserted that the primary reason for Congressional reluctance to authorize the project was economic: excessive subsidization to be paid for by the taxpayers and the power users.[33] The hearings over and over again raised the question whether Arizona had title to the water necessary for the project and whether adjudication of the dispute between California and Arizona over the Colorado could be made a justiciable issue without authorization and appropriation of money by Congress.[34] Congress at length decided that no authorization would be made until the question of title to the water had been settled.

ARIZONA vs. CALIFORNIA — THE FINAL LITIGATION

In 1953, after California had made diversions from the Colorado in excess of the 4,400,000 acre-feet limitation imposed by California on itself as a condition prior to the passage of the Boulder Canyon Project Act, Arizona sued in the Supreme Court to restrict California to that amount. It is this suit which currently holds the fate of the waters of the Colorado River for Arizona.

Originally there were basically three issues involved in this litigation. Two of them involved the place of the Gila River under the Colorado River Compact and the other the charges for evaporation losses on the Colorado itself. The compact provided that the Lower Basin states could increase their beneficial consumptive use by one million acre-feet above the average 7,500,000 acre-feet which was to be allowed to flow past Lee's Ferry annually.[35] Arizona had maintained from the beginning that this one million acre-feet was intended for Arizona to compensate for the inclusion of the Gila River system in the terms of the compact.[36] Arizona had objected to its inclusion on the grounds that the Gila entered the Colorado at a point where it could not contribute to usage elsewhere in the United States and that its waters had already been appropriated in their entirety. Unfortunately for Arizona, however, such an understanding was not made explicit in the compact and California now contends that this water is surplus water unapportioned and subject to appropriation.[37] Authorities range on both sides of the question.[38]

A second important source of conflict arose out of the meaning of the term "beneficial consumptive use," a term used in the compact but nowhere defined therein. It is Arizona's contention that this term refers only to man-made depletions of the river and not to all water used consumptively. California, on the other hand, asserts that beneficial consumptive use refers to diversions minus returns to the river. The Gila River system is a wasting system inasmuch as a great amount of the water in the system is lost through seepage, evaporation, and other natural causes. Only a part of its virgin flow ever reached the main stream under natural conditions. Arizona therefore maintains that it should not be charged for the consumptive use of water which never would reach the main stream anyway; its charges should consist only of that flow which would reach the river if it were not diverted. In other words, Arizona claims it should not be charged for salvaged water. California contends that the use of all water consumptively from any stream in the Lower Basin must be charged against that state wherein it is used. Again, the quantity involved in this issue is more than one million acre-feet.

The third central issue concerns the charges for evaporation losses in the reservoirs along the main stem of the Colorado River. It is Arizona's contention that these losses should be ratably charged to Arizona and California in proportion to the consumptive use of water for which each state is charged. California asserts that it is bound by the California Self-Limitation Act already and that its contracts with the Secretary of the Interior do not provide for reductions for evaporation losses while Arizona's contracts do. If California were charged with evaporation losses the losses might run as high as 600,000 acre-feet annually.

There are other issues of no little importance. It is Arizona's contention that the surplus waters referred to in the compact are not subject to apportionment until 1963. California, on the other hand, asserts the right to appropriate such waters before that date, and has done so by means of contracts with the Secretary of the Interior, which provide ultimately for the delivery of 5,362,000 acre-feet of water annually. California contends that the water described in Article III(b) was surplus water and therefore to be divided between the states of California and Arizona.

Arizona charges that the contracts entered into between the various California districts and the Secretary of the Interior in excess of the 4,400,000 acre-foot limitation are invalid. Arizona further charges that the contracts entered into by Arizona provide for 2,800,-000 acre-feet to be diverted from the main stream, subject to the availability of such water under the Colorado Compact and the Boulder Canyon Project Act. With California exceeding its limit, a loss of water available to Arizona under its contract has resulted.

California claims that its appropriations above the 4,400,000 acre-feet figure were prior to the Boulder Canyon Project Act and were guaranteed by the compact and the act, and that the above figure applies only to water apportioned by Article III (a) of the compact, thus permitting California to exceed that figure from the III(b) and III(c) water. California further charges that Arizona's contract with the Secretary of the Interior is invalid on the grounds that such a contract was dependent upon ratification of the Colorado River Compact. Arizona's ratification, it is alleged, was invalid since it came too late and involved interpretations of the compact which were unacceptable to California.

For several years Arizona placed great stress on the equity of its claims, asserting that a denial of access to the Colorado River water would result in great harm to her economy, "necessitating a substantial reduction in planted acreage with all of its attendant evils."[39] California contended that Arizona's arguments on the basis of equity were without merit.

> The primary factor in an equitable apportionment case is priority of appropriations. If relative needs and equities are to be weighed, California's evidence will show that for any water taken from California's existing projects for use in Arizona, the possible gain to Arizona will be more than offset by the certain detriment to California . . . . The crisis in Central Arizona is largely self-created, the result of uncontrolled expansion of acreage dependent on pumping from the underground.[40]

California claimed that its appropriations from the river have priority and that its uses of the water are as economic and productive as any found in Arizona.

The course of litigation resulted in some victories for both sides during the first five years. California strove to involve the Upper Basin states in the litigation — a delaying tactic, said Arizona. Both Arizona and the Upper Basin states opposed this move vigorously and it was denied. On the other hand, Arizona sought to prevent the admission of evidence regarding the claimed priority of appropriations on the Colorado River by California, contending that all rights to Colorado River water in Arizona and California were dependent on contracts made under the compact and the Boulder Canyon Project Act. In this view Arizona was supported by the Justice Department of the federal government. In this instance, the Special Master ruled in favor of California in admitting evidence of prior appropriations.

In August, 1957 Arizona, with a change of counsel, struck out on a new tack, giving less emphasis to its claims to the water on the basis of equity and more to its rights on the basis of contract under the Colorado River Compact.[41] In particular, this has meant a reinterpretation of the compact clause allowing the Lower Basin to increase its allotment of water by one million acre-feet. Instead of contending that this referred to water in the Gila River system, Arizona now argues that Arizona's rights to the Gila were already perfected and protected by the compact. The III(b) clause is now conceived of as referring to the waters of the various tributary streams entering the Colorado River below Lee's Ferry. Rights had been established, it is asserted, to the waters of these streams long previous to the compact and provided a basis for Arizona's claims to that water.

Arizona's final position is that it has title to 2,800,000 acre-feet of mainstream water under Article III(a) of the compact. It also has title to an additional one million acre-feet under Article III(b) of the compact, subject to any claims by Nevada to this water. It has right to use one half of the surplus in excess of that amount of water apportioned by Article III. It asserts that all water entering the Colorado River below Lee's Ferry was excluded from the terms of the contract. Under these terms, Arizona would be entitled to as much as 3,800,000 acre-feet of water from the mainstream of the Colorado, plus one half of whatever surplus might be available. Under this position the water Arizona had already appropriated from the rivers within Arizona, exclusive of the Colorado, would not be charged against Arizona under the compact.

California saw the issues quite differently at the conclusion of the trial. It argued that the Colorado River Compact did apply to the Lower Basin tributaries. It further urged that no decision could be made without an interpretation of the Colorado River Compact and without the Upper Basin states being made parties to the dispute. Both of these contentions hinged particularly on the question of where the water was coming from to satisfy the guarantee of 1,500,000 acre-feet to Mexico. Since it appeared that Arizona and Nevada both expected the Upper Basin states to deliver more than the amount required by the compact, California argued that the Upper Basin states had to be parties and the compact would have to be construed. Arizona contended that the Mexican water would come from return flows to the Colorado at a point so far downstream that they could not be used in the United States.

California argued that no decision could be made without a

determination of the water supply available for beneficial consumptive use in the Lower Basin. Arizona claimed that it was impossible to determine the future flow of the Colorado River and was unnecessary to do so in any event. California had maintained from the beginning that there was not adequate water to satisfy all of the claims to water, since the Colorado River Compact had assumed a great deal more water in the Colorado than are indicated by streamflow records. While the compact assumed an average annual flow of 15,000,000 acre-feet, with 7,500,000 allotted to the Lower Basin, California claimed that the maximum safe annual yield was only 5,850,000 acre-feet for the Lower Basin.

Finally, California maintained that appropriative priorities have not been displaced by contracts with the Secretary of the Interior under the Boulder Canyon Project Act. California rejected Arizona's contention that its contracts with the Secretary of the Interior are valid, particularly if the contracts were assumed to have validity in the face of already established rights under the rule of prior appropriation. If Arizona's position were maintained, California argued, then all rights based on appropriation would be in jeopardy.

Apparently there remains no dispute over the definition of beneficial consumptive use and charges for evaporative losses. Arizona did not mention these in its revised bill of complaint.[42]

The position of the federal government, a party to the dispute, became critical, inasmuch as the federal agencies and their wards, the Indians, have claims on the Colorado River water which threaten to exclude the present or prospective users in either state from obtaining water from that source. The federal government asserted that the Indian tribes have prior claim on waters of the Colorado and that such claims are not subject to the Colorado River Compact. In an exchange reported between David E. Warner, representing the Justice Department and Simon Rifkind, the Special Master, the following was said: "That's like saying the Indians have a first lien on the water," remarked Judge Rifkind. "That's right," said Warner.[43] Each state, under these terms, would be required to share with the Indians whatever amounts the Indians claimed, these amounts being deducted from the amount alloted the state under the compact. California alleged that if all the claims of the federal government were honored they would amount to 12 million acre-feet, or 50 percent more than was originally given to the Lower Basin under the compact.[44] Such claims to priority would damage Arizona's position in particular since it has by far the largest Indian population capable of using Colorado River water.

In point of fact, the Colorado River Indians, located near Parker, began to take steps to increase their use of water by planning to lease a considerable portion of their land to a corporation for development by white settlers. This plan received the full backing of the Bureau of Indian Affairs. The Arizona Interstate Stream Commission protested this development prior to the settlement of water rights along the river, and received the support of California in a remarkable show of unity.[45] All such plans have now been suspended pending the decision on Arizona's title to the water.

The case of *Arizona* v. *California* came to trial before Special Master Rifkind on June 14, 1956. It ended on August 28, 1958, after 132 days of actual trial, 22,593 pages of testimony comprising 43 bound volumes, 4,000 exhibits, and testimony from 105 technical and scientific experts.[46] Judge Rifkind characterized the trial as "the greatest struggle over water rights in the latter day history of the West."[47]

On May 5, 1960, Judge Rifkind submitted his draft report on *Arizona* v. *California*.[48] His findings agreed essentially with the position taken by Arizona and the publication of them was hailed as a great victory for Arizona. Arizona's central contention that it had claim to 2,800,000 acre-feet of mainstream water under Article III(a) of the compact was upheld. California was limited to the 4,-400,000 acre-feet established by its own Self-Limitation Act. Also upheld was Arizona's claim that the tributaries of the Colorado River below Lee's Ferry were not included in the terms of the Colorado River Compact and the Boulder Canyon Project Act; therefore Arizona's appropriation of these waters is not chargeable under the act.

Thus, the decision granted Arizona 2,800,000 acre-feet annually, California 4,400,000, and Nevada 300,000. If there was a surplus above the 7,500,000 acre-feet to be released annually to the Lower Basin, then this surplus would be divided equally between Arizona and California, except that Nevada might contract for 4 percent of the surplus, which would be deducted from Arizona's share. If there was insufficient water to meet the annual allocation of 7,500,000 acre-feet, then the available water would be apportioned on the same percentage basis as allocated: Arizona 28/75ths; California, 44/75ths; and Nevada, 3/75ths. If there was inadequate water to satisfy perfected rights in the Lower Basin states, then the water would be allocated for use in accordance with the priority of present perfected rights without regard to state lines.

The draft report allocated certain quantities of water to certain Indian tribes, wildlife refuges, and the Lake Mead Recreation Area, notwithstanding state laws or lack of contract with the Secretary of the Interior. The Lower Basin states were made responsible for providing the 1,500,000 acre-feet required by the Mexican Treaty for Mexico when there is sufficient surplus, but when there is inadequate surplus, the two basins will share the burden equally.

Arizona's present uses of the water from the mainstream of the Colorado are limited to the Yuma area, the Wellton-Mohawk Project, and the Colorado River Indian Reservation near Parker. These projects have appropriation rights guaranteed by their early dates of diversion or by reservation of water rights in the creation of the Indian reservations. Arizona calculates, however, that there will remain sufficient water in the mainstream to warrant the construction of the necessary works of the Central Arizona Project to bring water into central Arizona. According to the original project plan, Arizona would divert one million acre-feet of mainstream water. Estimates of water legally available to Arizona for central Arizona run well over one million acre-feet.

For a considerable period while the trial was proceeding Arizonans active in the water struggle were optimistic that the state could finance its own Central Arizona Project by borrowing through long-term bonds. Arizona sought authorization from the Federal Power Commission to construct Bridge Canyon Dam, the key structure in the project. It now appears clear that Arizona will be unable to finance its own reclamation project and will have to return to Congress to obtain Congressional authorization and appropriations. California countered with its own application for a license to construct a dam at Bridge Canyon for the production of power. Originally Arizona pressed for an early decision on a license, feeling confident of the superiority of its application and its capacity to finance the works independently. It is now seeking to buy time by delaying a decision on a license in order to prepare its case for Congress.

As this book was going to press, the Supreme Court, on June 3, 1963, announced its opinion in *Arizona* vs. *California* in a decision that split the court 5-3. The court upheld, in its main outlines, the position taken by Arizona and adopted in general by Special Master Rifkind. Of primary importance, the court decided that the water from which California could calculate its share of Colorado River water was in the *mainstream* only. Tributary waters, the court said, were excluded from the terms of both the Colorado River Compact and the Boulder Canyon Project Act, being reserved to the Lower

Basin states in which they flowed. The waters of the Gila River were reserved to Arizona in the calculation of water allocated by the compact. This means that the total amount of water governed by the Boulder Canyon Project Act was sharply decreased, significantly reducing, therefore, the amount of water claimable by California as surplus. The decision also gave Arizona title to all its tributary water, to the 2.8 million acre-feet receivable under contract with the Secretary of the Interior, and to its share of available surplus.

To the consternation of those who believed the law of prior appropriation inviolable, including three justices who wrote biting dissents, the court declared that the allocations of water depended upon valid contracts with the Secretary of the Interior and the limitation of 4.4 million acre-feet of mainstream water to California under the terms of the California Self-Limitation Act. The long-cherished appropriation doctrine was replaced by the discretion of the Secretary of the Interior. This discretion would reach its maximum in times of shortages when he could reduce the amounts available to each state without the guidance of legislative standards.

The court also accepted the Special Master's view that the reservation of public lands for special uses implied a reservation of the water supply necessary to make the land useful. The amounts allocated to federal reservations, including Indian lands, would be charged to the states in which such reservations were located, and these amounts, estimated at one million acre-feet, would be based on potential rather than actual development.

The specific decree by the court will be made known during the court's 1963-64 term. Barring any significant change, the effect of the decision was to give Arizona legal title to 2.8 million acre-feet of mainstream water, some of which is already being used in existing irrigation developments, but much of which would still be available for the Central Arizona Project.

Arizona has already begun taking steps to press its case once again for the project. Undoubtedly the state will run into the same formidable opposition it faced in the period 1948-1951 when it was defeated because of the cost of the Central Arizona Project and the uncertainty of its title to water. California will most certainly press its claim that there is inadequate water in the mainstream to justify construction of the project, as well as its view that the project is economically unsound and designed to save a few exploiting landowners. The long fight over the Colorado is not yet over.

CHAPTER 7

Public Power and Politics

The policy questions involved in the generation and transmission of electrical energy have created political issues of transcendent importance in Arizona as they have in other parts of the nation. With a tremendous potentiality of power resources within or on its borders, the state has perennially faced power shortages. Proposals to alleviate these shortages through public power projects have always engendered the sternest opposition of the electrical utilities and their allies, thus making power policies of major political consequence.

As in the case of irrigation water, the federal government has played a principal role in the development of hydroelectric energy. The role of the federal government in this field antedates statehood, since the initial plans for the Salt River Project included provisions for the installation of generators to produce power for the pumping of additional water from underground reservoirs. Later a small amount of additional power was produced on the Gila River at Coolidge Dam by the Bureau of Indian Affairs. The modest amounts produced at these two installations did not stir up much controversy. But the proposals to introduce power produced on the main stem of the Colorado River brought the issue of public power to the front where it has remained ever since.

The availability of low-cost electrical energy has long been recognized as a primary factor in stimulating economic activity in an area. Lacking any other inexpensive domestic source of power at the present time, Arizona requires electrical energy for agriculture and industry as well as for the comfort and service to a public that needs lighting, heating, and perhaps most of all, cooling. The Federal Power Commission recognized this dependence in a survey report on Arizona's power situation in 1942, in which electrical energy was treated as a dynamic factor in the state's development. The report assumed that the state would "plan the utilization of its endowment in low-cost hydro-electric power as the foundation of the State's future economy."[1] With most students of Arizona's economy supporting industrialization as the basis for a strong economy, the importance of low-cost power cannot be overemphasized. The federal government has developed or is developing 2,495,000 kw along the Colorado but there yet remains a potential 2,000,000 kw to be developed.

The first public power developed in Arizona came as an adjunct of the Salt River Project. Since the dedication in 1909 the project has grown remarkably until by 1958 it was serving 52,302 accounts and had an installed capacity of 390,890 kw of hydro and steam power.[2] Alongside the Salt River Project the private power utilities grew also, providing by 1958 over 535,000 kw of power, primarily from steam generating plants.[3]

Regulation of the private utilities, as the primary producers of energy, was virtually nonexistent during the Territorial period.[4] Public dissatisfaction over the excessively high rates and low tax burden of corporations generally led to demands for strong regulatory authority over the corporations, and particularly the utilities.[5]

As a result the Arizona Constitution provided for a corporation commission with broad powers over corporations engaged in providing transportation, or furnishing heating, cooling, oil, gas, electricity, or water for public purposes, or the transmission of messages, and over all common carriers.[6] The elected commission however, has been anything but an effective regulatory agency. According to a preliminary study, during the first 12 years of its existence, from 1912 to 1924, the commission rejected only seven petitions of the power companies, initiated only one action against a power company, and settled that one out of court. The commission never investigated the rates charged by the mining companies, who were among the chief power producers, and never denied a petition to issue more securities in spite of evidence that the power companies were overcapitalized.[7]

The public utilities in Arizona were largely owned by out-of-state interests, including ownership by such companies as Cities Service Company and Electric Bond and Share Company.[8] Along with utilities interests throughout the country, the utilities promoted an effective campaign against the construction of Hoover Dam. The opposition was particularly effective in Arizona in view of Arizona's official opposition to the Boulder Canyon Project because of the state's unwillingness to accept the Colorado River Compact.[9] It is an interesting commentary on Arizona politics that the most vigorous opposition to the project came from the Salt River Valley Water Users' Association, one of the major producers of power in the state, a public district under the laws of Arizona, and one of the earliest beneficiaries of federal reclamation policy. The association maintained that it could adequately provide for any markets within the state.[10] According to Neal Houghton, a former member of the Arizona Power Authority, it was this early and sustained opposition by the power utilities which delayed until 1944 the creation of the Arizona Power Authority for the purpose of taking Bureau of Reclamation power.[11]

It is ironic that, while the power companies were resisting development of low-cost power and the sale of it within Arizona, electrical power rates in the state were among the highest in the United States.[12] The FPC reported in 1935 that residential rates were the second highest in the United States, following only New Mexico.[13] With the private companies preventing the state from utilizing the tremendous undeveloped capacity of the Colorado River, the power users of the state continued to pay high rates while Southern California power users contracted for Arizona's unused allotment of Hoover Dam power and bid fair to shut Arizona entirely out of Hoover power.

In spite of Arizona's opposition, the Boulder Canyon Project Act was passed in 1928 and Hoover Dam began to produce power for California purchasers on September 11, 1936. The act provided that Arizona should receive an allotment of 17.6 percent of the firm energy developed at Hoover Dam, but the regulations were interpreted to mean that Arizona could take such energy only in its "sovereign capacity," which therefore required that Arizona establish a state agency to bargain and contract with the federal government for the power.[14] Owing to public utility opposition and the general distaste for the Boulder Canyon development, Arizona refused to contract for the withdrawal of energy or to install generators at Hoover Dam.[15]

During the late 1930's, Arizona went through one of the most severe droughts in its history. The growth of the population and the increase in demand for electrical energy to pump water for irrigation combined with the drought to create a severe power shortage by 1939. To meet the emergency, the Bureau of Reclamation constructed a transmission line to the Phoenix area so that Hoover Dam power might be brought into Arizona by means of a connecting link at Parker Dam.[16] Later Arizona received its own power from Parker Dam. Even with this additional power, the FPC declared that only the high rainfall in 1941 saved the state from a critical power shortage in that year.[17]

Under these conditions the Federal Power Commission made a

power survey, the results of which were reported in 1942. The FPC
reported that power rates in the state were the third highest in the
nation, and that such rates precluded the "full application of the
State's water power to obtaining those benefits of industrial growth
and improvement in farm and home life which are being increasingly
enjoyed in the TVA area" and elsewhere where low-cost power was
available.[18]

ARIZONA POWER AUTHORITY

In the early 1940's public officials in Arizona, and particularly
Governor Osborn, began to reconsider the wisdom of Arizona's adamant opposition to the Colorado River Compact and the refusal to
participate in any way in the development of the Colorado River. This
change was initially expressed in his request in 1941 that the legislature establish a water and power authority to replace the existing
agencies in the field of water and power.[19] This agency would have
been responsible for the appropriation and administration of power
and water from the Colorado River, for the coordination of existing
irrigation projects, and the establishment of water and power projects.
This proposal met with almost no support in the legislature, dying
on the floors of both houses and receiving the approval of not a single
committee.

The next regular session of the legislature saw Governor Osborn
return to the matter of power again, this time armed with the recently
completed *Arizona Power Survey* which indicated the need for Arizona
to capitalize on its unused power resource, the Colorado River. The
governor lashed out at the opponents of Colorado power — mainly
the power companies — as "foreign holding companies, whose unjustified debt structures, wasteful duplications and inefficient and non-
integrated hodgepodge of facilities bleed the people of this state of
about three-fourths of a million dollars every year."[20] He pointed out
that these interests had argued in the past that they could handle any
markets in the area, that the transporting of power from the Colorado
was impractical, and that locally produced power would be cheaper,
while at the same time they connived to obtain Hoover Dam power
for themselves in order to sell this cheap power at their exorbitant
rates.[21] The following were the net incomes reported for private utilities in Arizona by the Federal Power Commission in 1943: Arizona
Edison: 6.6 percent; Arizona Power Corporation: 24.6 percent; Central Arizona Light and Power: 15.3 percent; Citizens Utilities Co.:
12.1 percent; Tucson Gas, Electric Light and Power Co.: 15.7 percent.[22]

The governor proposed, therefore, the establishment of a water and power authority to contract for water and power from the Colorado. It would bargain with and contract for power and water with the Bureau of Reclamation and provide these commodities on a reasonable basis to users in the state. Such an agency would operate on a statewide basis and would be empowered to construct, operate, and maintain facilities for the transportation and distribution of power and water.[23]

The governor's bill was radically revised in the Senate, which stripped the authority of the power to issue bonds, to build and maintain water and power projects, and to sell any of its power on a retail basis. The chief opposition again came from the Salt River Valley Water Users' Association which charged that the power brought into the state by the authority would be too expensive. At the same time the association argued that this power would compete with its power production in the central valley.[24] The small cooperatives supported the governor's bill through an organization known as the Arizona Citizens Power Protective Association.[25] Questions of constitutionality of the proposed legislation arose, involving the rate-making power of the proposed authority, inasmuch as the Arizona Corporation Commission is authorized by the state constitution to be the rate-making authority of the state.[26]

The bill, as amended, passed the Senate by a large margin but failed even to come to a vote in the House, owing to a vociferous minority which argued that the amendments made the power and water authority bill a give-away to the power companies.[27] This group charged that the bill guaranteed the existing contracts between the Bureau of Reclamation and the Salt River Valley Water Users' Association and other private utilities, and thus threatened to crystalize high power rates in spite of the cheaper power to be brought into the state by the power authority.

The governor broadened his attack on the private power companies in 1943 by asking that the legislature authorize a study of power rates charged by Arizona utility companies. He charged in his message to the legislature that the power companies and even the Corporation Commission opposed such a study and that this opposition indicated that there was something to hide in their rate structures.[28] The governor was equally unsuccessful in this measure.

The year 1944 proved to be the turning point in matters relating to the Colorado River. In a special session early in 1944 the legislature ratified the Colorado River Compact. Immediately thereafter Governor Osborn convoked another special session, one of the purposes of which was the creation of a power authority and authorization of a study of power rates.[29] Owing to the power shortages and the campaign by Governor Osborn, the private utilities had by this time accepted the fact that a power authority would be established and had compromised many of the differences that had separated them from the advocates of public power.[30] They accepted the power authority since it was clear that its establishment was necessary for Bureau of Reclamation power to enter the state under the terms of the Boulder Canyon Project Act, but they sought to amend the law to suit their own purposes.

Those in the legislature who favored the utilities offered amendments which would have made the authority the virtual handmaiden of the utilities, since the commissioners had to be experienced in the utility business, and the agency would be prevented from selling power to any organization except the private utilities.[31] These amendments were defeated.

The chief source of friction again arose over the amendment sought by the Salt River Valley Water Users' Association and the Central Arizona Light and Power Company which would confirm contracts entered into by them with the Bureau of Reclamation for power produced at Parker Dam. In spite of vigorous protests by the public power advocates, this amendment passed. Once these subsidiary issues were resolved the legislature passed the power authority bill by large majorities, each house vying to be the first.[32]

The bill providing for a power survey was rejected. It was rumored at the beginning of the session that if the power survey bill passed, the power authority bill would not get through.[33] The power companies again opposed the survey; others distrusted the Federal Power Commission, which would undertake the survey for the state, or expressed a desire that the Arizona Corporation Commission conduct its own survey.[34]

The Arizona Power Authority Act created an agency with the authority "to bargain for, take and receive . . . electric power developed from the waters of the main stream of the Colorado river by the state or the United States . . . ."[35] In addition, the APA was empowered to "acquire or construct and operate electric transmission systems, standby or auxiliary plants and facilities and generate, produce, sell at wholesale, transmit and deliver such electric power to qualified purchasers," and it could "enter into agreements for interconnection or pooling with projects, plants . . . of other distributors of electric power."[36] This broad grant of power gave the APA the necessary means to supplement Arizona's inadequate supply of energy. It could obtain power from the Bureau of Reclamation facilities at Hoover Dam, it could construct lines to transmit the power if necessary, it could establish an interconnecting grid with all other power producers in the state, and it could construct and operate its own power-producing facilities when needed. The most important restriction lay in the prohibition against selling power except at wholesale, unless no other sources of power were available.

The Arizona Power Authority is an independent agency governed by a commission of five, appointed by the governor with the consent of the Senate. Although there were attempts to write into the Power Authority Act requirements which virtually ensured utility control, the commissioners are required only to be "qualified by administrative and business experience," and not "associated with any public service corporation engaged in generating, distributing or selling power to the public generally in this state for profit."[37]

During the period between 1944 and 1952, when the first power from Hoover Dam came into Arizona, the APA was engaged in the negotiation of contracts and the construction of its generators at the dam. In 1945 a contract was signed between the APA and the Bureau of Reclamation permitting Arizona to withdraw its allotment of Hoover energy.[38] The legislature provided an original loan of $100,-000, which was later increased to $250,000, to enable the APA to engage in surveys, investigations, and negotiations, including the planning of a grid for interconnecting the various power centers of Arizona.[39] An agreement was consummated with the Bureau of Reclamation to connect the Davis Dam transmission system with the Hoover system in order to avoid duplication of transmitting facilities.

The APA continued to suffer from political troubles during this period. Governor Osborn again called for a power survey in the two legislative sessions held in 1945. There was no overt opposition to the bill during the second session, but in his special session message, Governor Osborn stated, "There is not opposition to it on the surface, but underneath, Oh, My! the opposition is violent and vicious. It can come from but one place and that is the power combine."[40] However, the bill passed, providing an appropriation of $50,000 for the Arizona Corporation Commission to be used to employ Federal Power Commission experts for a rate survey.[41] The bill directed the ACC to investigate property valuations of gas and electric utilities for the purpose of determining a reasonable rate-base upon which rates could be established.[42]

What appeared to be a final victory for the groups backing the power survey proved to be only a temporary setback to the private utilities. The survey law was attacked as unconstitutional in a suit brought against the Corporation Commission, the attorney general, and the state auditor for the purpose of denying funds to the ACC. The suit charged that the legislature had no power either to expand or contract the powers of the Corporation Commission or to direct the commission in what activities it should engage, since the commission was a creature of the constitution and therefore not subject to legislative authority. The suit went to the Arizona Supreme Court and the power survey law was invalidated by that court.[43] The court accepted the views of the utilities, stating that "it is not for the legislature to say whether the valuations of public service corporations ascertained by the Corporation Commission represent the fair value or that additional or different determinations as to their fair value should be made. Such matters are within the exclusive perogatives of the Commission, subject only to judicial review." Thus came to an inglorious end the last attempt to make a general inquiry into power rates and property valuations.

The private utilities made one more assault directly on the APA, an assault that excited what one observer described as one of the bitterest debates in Arizona's legislative history.[44] Supported by the Salt River Valley Water Users' Association, the Central Arizona Light and Power Company, and the San Carlos Irrigation District, a bill was introduced in the legislature in 1947 which provided for the transfer of many of the duties of the APA to the Corporation Commission. This bill would have denied the APA the power to construct the power grid advocated by the Federal Power Commission by which the private and public power producers would be linked together for maximum efficiency. It would have required the APA to use existing transmission facilities and to sell power only under certificates granted by the Corporation Commission. The bill would also have eliminated the preference provisions for public districts and municipalities while giving iron-clad protection to the utilities in the areas they served.[45]

This bill raised a storm of protest. The very idea of transferring duties to the Corporation Commission threw consternation into the ranks of public power supporters since the commission still retained its label of favoritism toward the utilities. The Power Authority director, who had originally led the fight for the creation of the APA in 1944, stated of the bill,

Having produced power shortage after power shortage in this state, by opposition to the creation of a state agency as required by federal laws, they now aim to convert the agency established into a power trust agency and to seize this power and put it to work for the profit of the people who have no interests in the state other than to extract the last possible dollar of tribute from our citizens.[46] The utilities attacked the APA on many grounds: that it was socialistic; that it would provide power no less expensive than steam power already being produced within the state; that it would compete unfairly with the private utilities; and that it would not provide firm contracts with the utilities. The backing for the APA came from the electric cooperatives and some municipalities.

In typical fashion, the crucial decisions were made outside the legislature by representatives of the interest groups involved. After numerous conferences, an agreement was arranged which actually enhanced the position of the APA and amounted to a resounding victory for the advocates of public power.[48] The APA retained its power to issue certificates, and to construct its own lines when necessary, but was directed to use the Bureau of Reclamation facilities when available; franchised territories were to be determined on the basis of the best service to an area. References to the Corporation Commission were deleted entirely and no fundamental changes were made in the system of preferences.[49] Accompanying this struggle over the APA was a fight between the cooperatives and the private utilities. The latter wished to subject the cooperatives to regulation by the Corporation Commision for the purpose of protecting their franchised territories. The utilities were again defeated; the cooperatives were not subjected to Corporation Commision control. A year later specific exemption from commission control was granted.

COLORADO RIVER POWER

As indicated above, Arizona received its first power from the Colorado River from Parker Dam in 1940. The Bureau of Reclamation sells the power allotted to Arizona from Parker directly to its customers rather than through the APA. The major recipients of the power are the Salt River Valley Water Users' Association, Arizona Public Service Company (formerly the Central Arizona Light and Power Company), and the Tucson Gas, Electric Light and Power Company, with relatively small amounts going to other distributors.[50] Out of an installed capacity of 120,000 kilowatts Arizona was receiving 95,250 kw before 1952; as contracts expire, however, this amount will gradually be reduced to 60,000 kw as California recaptures its allotted 60,000 kw.[51] This reduction will be compensated for by power produced at Davis Dam.[52]

While still awaiting completion of generating facilities at Hoover

Dam, Arizona contracted with the Bureau of Reclamation to receive its share of the power produced at Davis Dam, 67 miles south of Hoover.[53] Construction of Davis Dam began in 1942 but World War II delayed its completion until 1951. Arizona was allocated 90,000 kw of the total generating capacity of 225,000 kw. Of Arizona's share, 44,225 was required to be delivered by the bureau to federal agencies and former purchasers of power from Parker Dam.[54] The remainder was allocated by the APA among its customers. The energy is transmitted, along with Hoover energy, on the Bureau of Reclamation transmission system into the central valley area. Delivery of the first power from Davis Dam occurred in March 1951 in time to avert a serious shortage in power for use in operating pumps for irrigation.

In 1952, with the completion of the installation of the two 82,500 kw generators at Hoover Dam, the first of its power was delivered to Arizona. During the first full year of operation, ending June 30, 1953, the APA delivered to its purchasers over one billion kilowatt-hours.[55] The APA has continued to deliver approximately one billion kwh per year since that time, but it has had to supplement hydro power from the Colorado River with purchases of steam power in order to comply with its contractual obligations. Of the slightly over one billion kwh which APA delivered to its customers in fiscal 1960, a little less than three-quarters was hydro power from the Colorado, and the remainder was steam energy purchased from private utilities in the state.[56] Low water conditions on the Colorado have frequently caused severe reductions in the delivery of hydro power. In fiscal 1957, for example, only 62.8 percent of scheduled firm energy was produced at Hoover Dam and the APA had to reduce its deliveries of hydro power to 75 percent.[57]

While preference customers — municipalities, cooperatives, electrical districts and irrigation districts — take the largest share of power sold by the APA, the single biggest purchaser is the Arizona Public Service Company, the largest private utility in the state. Sales to non-preference customers total approximately one-third of the APA's total sales. In spite of the private utility opposition to public power, these utilities have come to rely heavily on it to supplement their own production.

In cooperation with the private utilities and preference customers, the APA has attempted to integrate and pool the power facilities of the state "so as to make such power available at the times and places and in the amounts needed . . . ."[58] At the present time the integrated transmission system is almost entirely owned and operated by the Bureau of Reclamation. As more sources of power are added

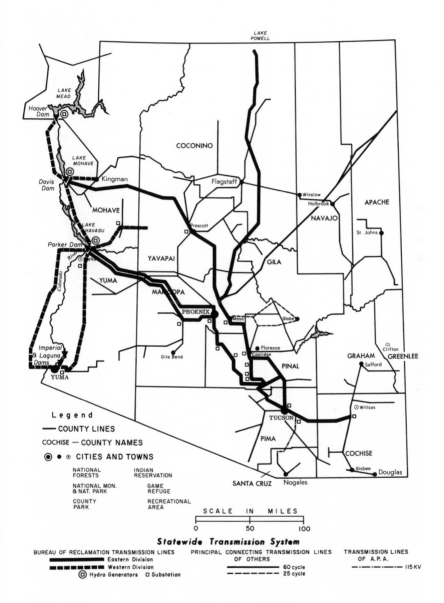

Power Transmission Lines

—*U.S. Bureau of Reclamation*

Most of Arizona's hydroelectric power comes from Colorado River dams like Hoover (above) and Parker (below). Lake Havasu at Parker is the water-diversion point for the proposed Central Arizona Project.

to the lines, expansion and improvement of these transmission facilities will be required. Unless the Bureau of Reclamation can accomplish this expansion, which would require appropriations from Congress, there is danger of expensive and uneconomical duplication of facilities. The APA has offered to purchase the bureau transmission lines but no agreement has yet been reached.

In Arizona, not only is power dependent in large part upon water, but the reverse is also true. The slogan of the Arizona Power Authority is very apropos in this regard: "More Power for Arizona: Where Electricity is Water." The agricultural economy as well as the municipalities are very much dependent on pumped water and this water is brought to the surface primarily by means of electricity. For example, of the 4,240 billion kwh generated for use in Arizona in 1953, 1,308 billion kwh, or 31 percent, were used for pumping water for irrigation.[59] Well over 50 percent of the power delivered by the APA is used for that purpose.[60] The largest single block of power used in the state is for irrigation purposes, far exceeding that used for residential, commercial, and large power purposes.[61] In spite of cotton-acreage limitations and restrictions on digging wells, the demand for power for pumping has remained high because of increased pump lifts. During the period 1944-1954 the average pump lift increased 106 feet, while the average amount of electricity required to pump an acre-foot of irrigation water increased from 109 to 405 kwh per acre-foot.[62]

The critical need for expansion of power production is indicated by the projected growth curve for energy consumption from 1950 to 1974. Energy consumption in 1950 was 3,285 billion kwh; by 1958 this figure had grown to 6,391 billion kwh. By 1965 energy consumption is expected to reach nearly seven billion kwh, and by 1974, with an estimated annual growth of 7 percent, it is expected to reach 16 billion kwh.[63] Private utilities are expanding their capacity rapidly through the construction of steam plants, but a considerable proportion of future growth must come from hydroelectric developments on the mainstream of the Colorado River. In the view expressed by the APA, "Failure on the part of the citizens of the State to secure and develop for Arizona these last available water and power resources of the Colorado River will not only seriously limit the potential economic growth of the State, but may even jeopardize continuation of its present economy."[64]

The APA attempted to increase its productive potential in 1956 by an alternative method: the construction of a stand-by or auxiliary steam plant near Benson, Arizona, to supplement the power from the

Colorado River. This proposal met with determined resistance from the private utilities which opposed such a scheme as "socialism" and denied the need for such facilities, arguing that the private utilities were providing for present needs.[65] The APA quickly gave up the proposal. It undoubtedly has the legal authority to construct and operate such plants, but politically it is unable to act.

The primary source of additional hydro power remains the Colorado River; the chief sites for the production of power are Glen Canyon, Marble Canyon, and Bridge Canyon. Glen Canyon Dam, now almost completed, will have an installed capacity of 800,000 kilowatts, capable of producing 4,337,000 kwh for the region it serves. The Arizona Power Authority has made application for a bloc of power from Glen Canyon Dam, but at this writing the Bureau of Reclamation has made no allocation among the applicants for Glen Canyon power. It is expected that all of the power allocated to Arizona from Glen Canyon will be sold to preference customers in Arizona. However, the role that the APA will play in the distribution of this power is not clear, for it is possible that the Bureau of Reclamation will itself sell the power directly to its preference customers. To serve these customers the Bureau of Reclamation is constructing a line from the dam to a delivery point near Phoenix where the preference customers must arrange to obtain the power. At that point, the Glen Canyon power transmission system will interconnect with the Parker-Davis system.

Before the bureau decided to construct its own transmission lines, the Arizona Public Service Company — the major private utility in the state — offered to transmit the power on a line which it is constructing from the Four Corners area into central Arizona. APSCO is building this line to transmit its own power produced at its newly constructed thermal plant in that region. The Bureau of Reclamation, however, rejected the proposal by APSCO on the grounds that it would have an adverse effect upon the repayment program for the Upper Colorado Storage Project, and that it might result in higher power costs to the bureau's customers.[66] Thus, two transmission lines are being constructed which virtually parallel each other, running from the Four Corners area into central Arizona.

It should be noted that the bureau accepted the proposals of several other utilities to transmit Glen Canyon power to other areas of the Upper Colorado Storage Project. The application by APSCO was the only one rejected. Many public power advocates were critical of the Secretary of the Interior for permitting private power companies to transmit any of the energy.

BRIDGE AND MARBLE CANYON PROJECTS

Arizona is putting its chief hopes for hydroelectric power development on the construction of dams at Bridge Canyon and Marble Canyon. In 1956 the legislature authorized the APA "to take such steps as may be necessary, convenient, or advisable to construct, operate and maintain at the Bridge Canyon Dam site or other sites, dams for the generation of electrical energy and in connection therewith facilities . . . for the storage and diversion of water."[67] The APA staff, in cooperation with the Interstate Stream Commission, began gathering data and information necessary in the formulation of plans for the development of the Colorado between Glen Canyon Dam and Lake Mead.[68] In August 1957 the APA employed the Harza Engineering Company to expand and expedite the investigations.[69] On the basis of these investigations, on June 24, 1958, Arizona submitted to the Federal Power Commission an application for a license to construct, operate, and maintain hydroelectric dams and power plants at both Bridge Canyon and Marble Canyon.[70] The dams would be financed through the sale of long-term bonds which would be paid off from power revenues.

The underdeveloped section of the Colorado River represents for Arizona two million kilowatts of power if fully developed. The actual plans for development by Arizona have varied from the time of the first application for a license, particularly in terms of an additional dam at the Prospect site and silt retention dams on the Little Colorado River. The discussion below is based on Arizona's amended application for a license of November 1959.

Arizona proposed to build a dam at Bridge Canyon impounding water approximately to the boundary of the Grand Canyon National Monument with provision for raising the height of the dam if such elevation is in the public interest and is approved by Congress. The dam would create a reservoir of 820,000 acre-feet. The proposed initial installation would include four generating units of 120,000 kw and provision would be made for four additional units, bringing installed capacity to 960,000 kw.

Construction of Marble Canyon Dam would begin four years after the start of construction at Bridge Canyon. This dam would create a reservoir of 480,000 acre-feet which would back up water to the Glen Canyon Dam. Initially, four units rated at 85,000 kw would be installed with provision made for two additional units.

The power transmission at the two power dams would be integrated with the transmission system at Glen Canyon to bring the power to the major population centers at Phoenix and Tucson.

To protect these two dams against sedimentation, the APA proposed to construct two dams on the Little Colorado River system — Tolchico and Moenkopi dams. These dams would have no power-production capacity, but they would ensure a 119-year life for Bridge Canyon Dam and a 105-year life for Marble Canyon Dam. The following represents the proposed schedule of construction:[71]

	Initial Installation	1980 *Installation*	*Ultimate Installation*
Installed Capacities	MW (megawatts)	MW	MW
Bridge Canyon	480	720	960
Marble Canyon	340	510	510
Total	820	1230	1470

A major conflict appears to be developing over further construction at these sites among the state of Arizona, the city of Los Angeles, and the conservationists. Los Angeles has also filed an application for construction of a dam at the Marble Canyon site, with a proposal for power development guaranteed to arouse the opposition of the conservationists. Los Angeles proposes to divert 92 percent of the water of the river by tunnel to Kanab Creek where generators would be installed. The Colorado River would be reduced to a small stream as it passed through Grand Canyon National Park, but would be restored to its full flow below Kanab Creek. The conservationists, as suggested earlier, look upon such a proposal as desecration of a national treasure and for this reason they question all of the developmental projects in the area because of the threats these constitute to the Grand Canyon National Park and Monument.[72]

At this writing Arizona has won the recommendation of the FPC trial examiner for a license to construct Marble Canyon Dam as a power project. But the APA has not pressed for a decision on the Bridge Canyon project in view of the decision by leading Arizona water experts that the Ceneral Arizona Project cannot be financed as a reclamation project without federal assistance. The APA agreed reluctantly to such a delay since its main concern is with the production of power to satisfy Arizona's growing power needs. It feels able to proceed with the construction of a power dam without federal assistance, and is less interested in utilizing the power and its revenues to justify the reclamation features of the Central Arizona Project. The latter, of course, requires Congressional authorization.

As chief power agency for the state, the Arizona Power Authority has the responsibility for defending Arizona's interests in power
matters. The APA has been critical of those managing the power
operations at Hoover and Davis dams. It charges that the Integrating
Committee, dominated by Southern California interests, is managing
the flow of the Colorado River to benefit Southern California and to
injure Arizona. It claims that Hoover Dam is being operated in such
a fashion as to maximize power production in the winter months when
California experiences its peak of demand, causing Arizona to experience shortages during the summer owing to the fact that the
managers of the dams are storing rather than releasing water. Arizona
experiences its peak demand during the summer growing season.[73]

On another matter, however, California and Arizona were
united. Both states protested against Bureau of Reclamation plans for
filling Glen Canyon reservoir (Lake Powell), charging that these
plans would seriously curtail power output at Hoover Dam. They
asked that the Secretary of the Interior give written guarantees for
their water and power rights, and that Powell Lake be filled only
during those years when surplus water is available.[74] After a protracted struggle, during which time the Western states introduced
into Congress the Water Rights Settlement Act (see Chapter 2), the
Secretary of the Interior in April 1962 announced the filling criteria,
which should quiet both states' fears. Under these criteria, the United
States agreed, among other things, to: (1) satisfy all uses of water,
other than for power, below Glen Canyon while filling of the reservoir takes place; and (2) make an allowance for recompensing either
through money or power for any necessary reduction in firm power at
Hoover Dam as determined by a formula.[75]

Conflict over the role of the APA has racked the agency with
internal dissension, some charging that it has become the "captive"
of the power utilities. In the latter part of 1954 and early 1955 an
open breach developed among the members of the APA Commission
over the advisability of a proposed contract with Arizona Public Service Company for 30,000 kw of power. Some members of the commission demanded outside review of the contract by a competent engineering firm before consummating the contract. Others, including
the chairman, urged speedy action without further consultation.[76]
With apparent reluctance, the APA engaged an engineering firm
which filed a preliminary report on the contract with the commission
in early 1955. The report generally favored the contract with some
modifications. A confused dispute developed over alleged complicity
by the chairman of the commission and the administrator with the

Arizona Public Service Company owing to the fact that the preliminary report was shown to the company and discussed with its officials in violation of a commission order.[77] The result of this alleged breach of trust, which later involved severe attacks by the attorney general on the commission, was a mild censure of the offending officials and the resignation of the APA's chief engineer.[78]

Other evidence cited to demonstrate commission subservience to power utilities concerns recent contracts between power districts and private companies. Contracts have been written in the last few years which permit the preference customers, such as electrical districts, to turn over the administration of the power business to a private utility which in turn sells the power to the district's customers. This practice has raised the specter in the minds of some that the private utilities might claim exemption from the Corporation Commission regulatory power which now governs their rates since they are operating as agents for exempt electrical districts. It is alleged that this practice would allow the utilities to obtain cheap hydro power from the Colorado River without having to pass on the benefits to the customer, as is required by the Arizona Power Authority Act.[79]

The outstanding advocates of public power on the APA Commission were replaced when their terms of office were completed. Their replacements, however, have not demonstrated any pronounced subservience to the utilities. The proposal for any construction of stand-by steam plants discussed previously was highly unpalatable to the utilities.

In its 1946 *Report* the Arizona Power Authority stated:

> Nowhere in the nation does electric power mean more to present life and future development than in Arizona. Lacking developed coal, oil or gas fields as sources of motive power, the State must depend largely upon hydro-electric power for its growth and welfare.
>
> Electric power constitutes our greatest present accessible natural wealth and in the future development and utilization of its potentially low priced hydro-electric power lie the sound foundation of our future economy, social betterment and progress.[80]

Much has been done to realize this economic progress through the development of the hydroelectric power in Arizona. The future demands made by increased population, industrialization, and a higher standard of living will require even greater efforts if Arizona's potentialities are to be fulfilled.

CHAPTER 8

State Management of Water Resources

The administration of water resources in Arizona illustrates the problems of adjusting the management of such a vital commodity to the social, economic, political, and technological conditions of a time and place. The early arrangements for managing the water supply originated from the necessity of communal sharing of the burden for providing water where and when needed. These early practices have placed an indelible stamp on present-day methods. As improved means were found for conserving water supplies through engineering, and increased demands were made on the water supplies, new arrangements were required to provide a broader base for financial support and to regulate more adequately the relationships among water users. And finally, means were required to make the adjustments necessary to guarantee the maximum beneficial use of water at a particular time and in a given political and economic condition.

The agencies responsible for the administration of water resources, therefore, are involved in a complex set of relationships and an allocation of responsibilities determined less by human forethought than by force of circumstances. While rationality has not been absent, its impact has been less than might be expected or desired where water supply is so critical.

The formal division of powers between the states and the federal government envisaged primary responsibility for the management of natural resources in the states. In fact, however, federal responsibility for natural resources has been the dominant theme in resource management during long periods of our history. In the public lands states, of which Arizona is one, the responsibility for the conservation and development of natural resources has always been that of the federal government. In Arizona, at the time of the Mexican Cession in 1848, all land that had not been vested in individuals became the property of the United States and therefore its charge. As a result of this historical fact, and the reinforcement for federal control provided by the conservation movement, the United States continues to bear a heavy burden of responsibility. The federal government owns or holds in trust more than 70 percent of the land area of the state, and provides assistance to farmers and ranchers on their privately owned lands[1].

This reliance on the federal government helps to explain the reluctance on the part of the state to bear a very significant share of the financial or administrative burden. The role of the state in the resource field has been primarily that of facilitating federal operations through the passage of legislation and the publication of data relating to federal programs. In the field of water resources, in particular, the state has never had an action program except in the accumulation of technical information. The state has had to rely on the federal government for funds for its reclamation projects, for a large part of the money used in making technical studies of water problems, and for much of the money used in improving the utilization of water in agriculture.

The state has certainly facilitated the conservation of water by authorizing the establishment of irrigation and water conservation districts. Because these have been given taxing powers and public-district status, they have been able to encourage maximum development of the water supply and the most efficient management of it. Among the most important districts are the Roosevelt Water Conservation District, Maricopa County Water Conservation District No. 1, Roosevelt Irrigation District, Buckeye Irrigation District in Maricopa County, and the San Carlos Irrigation and Drainage District in Pinal County.[2] In similar manner, the authorization of soil conservation districts on farm and ranch lands has allowed the farmers and ranchers to engage cooperatively in programs which have resulted in better utilization of the water. Nonetheless, except in the game and wildlife field, the state has done little to develop or improve the water supplies of the state through action programs. Whether it will do so if the Colorado River waters are made available to Arizona is, of course, problematical.

The backwardness of the state in managing its water supply is matched by its unwillingness to undertake administrative reforms. While other states have been creating departments of natural resources or conservation, Arizona has steadfastly refused to permit the integration of its resources agencies. Indeed, it might be said with some justification that the entire reorganization movement so popular during the last decade has almost passed Arizona by.

An explanation of this reluctance to undertake administrative reform is to be found in Arizona's constitutional history. Arizona became a state during the height of the Progressive movement in 1912. The movement's influence is indicated by the constitutional emphasis on popular government and division of power and responsibility. Election of heads of administrative agencies, commission organization, and limitations on the power of the governor typified Arizona administration. When new responsibilities were assumed by the state, the tendency was to create an independent agency in order to shield it from "politics." The most recent example of this was the creation of the State Parks Board by the 1957 legislature. It is an independent agency headed by a six-member board appointed by the governor. Although opposition was expressed to the creation of another separate agency, the livestock and other agencies pushed the bill through.[3] In the study made by Griffenhagen and Associates in 1949, it was revealed that there were no less than 115 separate agencies of administration with a wide variety of forms of administration.[4]

This situation of course applies to the agencies in the natural resources field. The State Land Department, the Game and Fish Commission, the State Parks Board, the Department of Mineral Resources, the state engineer, the Arizona Power Authority, and the various agencies active in the agricultural and livestock industries, such as the Agricultural Extension Service, all go their separate ways, depending upon whatever means of cooperation they can establish to create harmony in the state administration. In theory the governor may exercise some authority over these agencies, particularly through the appointing power, but in fact these agencies often become the handmaidens for those interests which are most directly concerned with their activities. In the natural resource and agricultural fields, one finds that interest groups directly affected by the operation of an agency often control the activities of the agency, either by formal requirements as to membership on the commission or through informal consultation with the administrators. For example, by law two of the members of the Parks Board must be representatives of the livestock industry in the state. Under such circumstances, it is not difficult to understand the resistance to integration of organizations and functions into a single major resource department.

Added to these conditions is the tradition of political spoils in Arizona. The merit system, except for those agencies supported by federal funds, is virtually nonexistent, other than in the Game and Fish Department. There is no centralized personnel system or agency.

Each agency is therefore responsible for recruitment, promotions, and removals, with the attendant possibilities for partisan removals. For example, as a result of the elections of November, 1956, there was created a new majority on the Corporation Commission which proceeded to remove, according to reports, 41 of 62 employees. Included was one expert on railroad rate-regulation who was in the middle of highly technical proceedings. The removal of Roger Ernst, the former state land commissioner and until 1960 Assistant Secretary of the Department of Interior, was accomplished to provide a position for a "deserving" Democrat in spite of Mr. Ernst's outstanding record. There is no quarrel with partisanship here, but merely an indication of the importance of partisan ties in Arizona. To add to the difficulties of administration, the salaries are relatively low, causing a high turnover of technical employees who can command better salaries either with the federal government or with private industry.

In examining the administrative agencies concerned with water resources, it should be noted that in many instances these agencies have other important responsibilities which may or may not be related to the management of water. These other interests often tend to push the interest in water to the background, resulting in neglect of this most critical of all matters.

ARIZONA STATE LAND DEPARTMENT

The State Land Department undoubtedly ranks first in potentiality as a central resource agency in Arizona. It forms the nucleus around which a department of natural resources, as suggested by the 1949 report on administrative reform, might be established. At the present time, however, the department is limited almost entirely to housekeeping and record-keeping functions, unable and rather uninterested in advancing its work toward developmental activities. While its fields of formal responsibility are broad — land, water, soil conservation — its powers to manage and conserve are severely limited.[5]

The reasons for these limitations are complex. It is difficult to obtain funds in very significant amounts for an agency whose responsibilities are housekeeping and regulatory in nature. Conservation, furthermore, has generally involved the regulation of the use of some resource, and this regulation naturally offends groups that are weighty in influence in the state legislature. The department has not been able to enlist the support of powerful interest groups which could back its proposals to introduce conservation programs in the state. In fact, the department has often been under attack for excessive responsive-

ness to the cattle growers and the water users. Finally, the concept of
the public interest is much more difficult to define and popularize
among a generally apathetic public, whereas the concept of the private
interest is readily identifiable by the private interest groups. Where
there are some signs of growing support for the State Land Department as a resource-development and conservation agency, there are
also continuing signs of resistance to proposals to broaden the department's powers.

For much of the period between statehood and 1942, the responsibility for management of water resources was lodged in an independent state water commissioner. Between 1923 and 1925 the state
land commissioner doubled as water commissioner but lost this added
responsibility to a newly-appointed water commission who retained
the job until 1942 when his functions were again transferred to the
Land Department, where they have since remained.[6] Invariably the
excuse for organizational changes was "economy and efficiency," but
economy usually took precedence over efficiency. It should be noted
that administrative rationalization may be only a facade for the actual
destruction of a program. The elimination of the Underground Water
Commission by transferring its functions to the Land Department in
1954 was clearly a case in point.[7]

Whatever the motives behind these changes, it seems certain that
water and land problems are so highly interrelated that complete
separation is not desirable. Land development, even for urban and
industrial purposes, is dependent on the quantity and reliability of the
supply of water. Erosion control is a necessary facet of water conservation and dam protection. In innumerable ways the two resources
interlock and demand coordination. As the Griffenhagen Report
pointed out in 1949, there may be exceptional circumstances wherein
an agency, such as the Interstate Stream Commission, should be excluded from such a multi-purpose department, but this arrangement
should be the exception rather than the rule.[8]

The state land commissioner is appointed by the governor with
the consent of the Senate for a term of six years. He can be removed
only for cause, and then only after a public hearing, written charges,
and an opportunity to confront witnesses.[9] In effect, the state land
commissioner is independent of the governor during the period of his
incumbency except in cases of malfeasance in office. (Two such occurrences are recorded involving a financial shortage in the commissioner's
office in the period 1930-33, and more recently in the resignation of
a commissioner for allegedly making a personal profit from a land
sale he negotiated.) The State Land Department, and therefore the

land commissioner, has "general control and supervision of the waters of the state and of the appropriation and distribution thereof, except distribution of water reserved to special officers appointed by courts under existing judgments or decrees."[10] Besides the general responsibility for the management of water, the land commissioner

> may make surveys, investigations and compilations of the water resources in the state, and their potential development, and may cooperate for such purposes with the United States; he shall maintain a permanent public depository for existing and future records of stream flow, and other data relating to the water resources of the state; he may formulate and prescribe rules and regulations governing the appropriation and distribution of water.[11]

The Land Department is thus the chief state research agency for water, according to the statutes, although in fact its potentialities are far from being realized.

In the field of research, the most important long-term program has been in cooperation with the United States Geological Survey. By an act of 1921 the state water commissioner was authorized to enter into an agreement with the USGS for stream gauging, and a small amount of money was appropriated for that purpose.[12] The federal government matched the money appropriated by the state. The actual operation of stream gauging was put under the exclusive management of the USGS although the decisions concerning the streams to be gauged were to be made after consultation between the USGS and the water commissioner.[13]

Stream gauging by the USGS has been continuous since 1921 except for two years — July, 1933 to June, 1935 — when it was discontinued for reasons of economy. (It is ironic that these were relatively wet years and would have contributed important information on streamflow.) This program has always been of the greatest importance to the water users of the state. When the program was initiated the chief concern was the determination of rights on the Gila River and the mainstream of the Colorado River. The determination of rights on the Gila was necessary in order to settle rights prior to the construction of Coolidge Dam. Also it was very important for Arizona to determine how much water was contributed by its rivers and streams to the mainstream of the Colorado, since the Colorado River Compact did not apportion water among the Lower Basin states.[14]

The cooperative arrangement has suffered over the years from an almost continuous unwillingness on the part of the legislature to appropriate sufficient funds for comprehensive gauging, although complaints concerning the paucity of funds have been less frequent in recent years.[15] In addition to the gauging stations maintained exclusively by the federal government on interstate streams, the federal-state agreement in 1956 sustained 76 intrastate gauging stations. Irrigation districts also lend financial support "in order to broaden the scope of the investigation."[16] The importance of stream gauging is described succinctly in the State Land Department annual report for 1955-56:

> In addition to national interests relating to the interstate and international waters originating in or traversing Arizona, to irrigation, flood control, and public lands, and to a sound and stable development of the country's resources and to national prosperity, the State's interest relates to administration of water and its distribution among many users, to local regional developments, domestic and public water supply, to benefits to its mining industry, to watershed management, and to providing facts upon which to base decisions resulting from litigation.[17]

In fiscal year 1957 $128,500 was expended in stream gauging, including the matching funds and other federal funds from cooperating agencies such as the Corps of Engineers and Bureau of Reclamation. The state land commissioner recommended that appropriations continue at that level, although it was suggested that "additional stations are required to meet growing demands for records providing general information, for determining the source of the water in the main streams, for the irrigation economy, power development, ground-water-replenishment studies, and the development of data on hydrologic relationships."[18]

In the biennial report for 1937-38, the state water commissioner recommended

> an appropriation and suitable legislation for an underground water survey. When this survey is made, and a report is made showing the extent of underground water in different sections of the state, and if the report so warrants, I recommend irrigation projects in commensurate [sic] with the amount of water shown in the report that might be developed. In the past investors have expended vast sums of money in an endeavor to develop underground water to find within a few years the supply wholly exhausted, and their investment a total loss.[19]

As a result of this recommendation money was appropriated by the legislature to allow the state water commissioner to enter into a cooperative agreement with the USGS for ground-water investigations. These investigations began on August 1, 1939, and have been continued since that time. Although the land commissioner has periodically complained of inadequate funds for ground-water studies, the legislature has been willing to provide more for these investigations than for stream gauging. In fiscal year 1959 the Arizona State Land Department spent $63,600 in matching funds for ground-water surveys, an amount which was then duplicated by the USGS.[20]

The objectives of this program are "to continue the collection of basic hydrologic data throughout the State, with emphasis in areas of extensive irrigation development having large withdrawals and depletion of ground water in storage."[21] This part of the work results in an annual water-level report for the state.

A second objective is "to continue reconnaissance investigations of specific areas, groundwater basins, and known aquifers to provide the necessary geologic and hydrologic data to evaluate the groundwater resources of the State."[22] This phase is directly related to the long-term research necessary for comprehensive knowledge of the water resources of the state and the geologic structures in which underground water is found.

Finally, studies are designed "to intensify the continued collection of hydrologic and geologic data, as related to the solution of specific problems of occurrence, movement, recharge, storage, discharge and fluctuation of groundwater."[23] Such information contributes to an understanding of the relationship of precipitation and runoff to recharge, the productivity of aquifers, rates of depletion of ground water, and other unsolved problems. Although the land commissioner in 1956 asserted that these last two phases of the program should receive increased emphasis, for in these phases lie the answers to many of the problems which cannot be resolved by the mere collection of data, the lamentable fact is that little is being done on this basic research.[24]

The State Land Department is responsible for the registration of appropriations of surface water and the registration of wells. It is also responsible for conducting initial hearings in litigation over water rights. The department has been hard put, however, to carry out these responsibilities because of a lack of money and staff. As early as 1926 the state water commissioner complained that he spent "the greater portion of his time acting in his quasi-judicial capacity, determining rights as between applicants on various waterways."[25] Later, the commissioner reported that the department was running at times as much as two years behind in the determination of rights among disputing parties.[26] In many instances, as a result, appropriators of water were taking the water for their use without following the procedure prescribed by the State Water Code. The water commissioner recommended that the legislature amend the code to provide that anyone who appropriated water in any way other than that prescribed by the code had not established a legal right to the use of the water, but the legislature showed no willingness to accept this suggestion.[27]

A controlled burn of chaparral on a ranch in Yavapai County (above).
The range is in poor condition because the chaparral has choked out
the forage grasses. Another area (below) on the same ranch after
controlled burning and reseeding with weeping lovegrass. The range
condition has been improved and opportunities for erosion decreased.

—Robert R. Humphrey

The situation in the department improved immeasurably during the tenure of office of Roger Ernst, who became land commissioner in 1953. Litigation was brought nearly up to date, and business-like procedures were established to eliminate the backlog of work. In spite of the fact that applications for appropriation of water have remained at a steady figure, decisions on them have been expedited so that a petitioner receives an answer within months instead of years.

The problems surrounding registration of wells have not improved the administrative processes in the Land Department. Already finding it difficult to enforce the provisions of the surface-water code, the department was also made responsible for the enforcement of the Ground Water Code as well. During the first five years of the operation of the code (1948-1953), there were numerous violations of it. The department was unable to prevent the occurrence of violations or to punish those who had perpetrated the violations. The land commissioner reported in 1952 that owing to the shortage of funds all the Ground Water Division could do was to review applications made for the drilling of wells. "It has had no funds to carry on the field-checking operations as to what may be happening in the way of illegal wells, although much information has been gathered by the division with respect to many such operations. Also due to lack of funds no prosecution could be started upon such violators."[28] Attention has been given to the problem by the department, and in 1954 field men began township surveys to determine whether all wells were registered and whether they had been drilled legally.[29]

The violators of the Ground Water Code are finding it somewhat more difficult to avoid the restrictions of the code; nonetheless, according to one member of the Land Department staff, the enforcement requirements of the code are all but impossible to administer. To use his own phrase, they would require "a policeman to stand by each well." Others contend that the problem is not one of incapability of enforcement but of lack of will to enforce. The code stipulates that in critical areas no land may be brought into cultivation by ground water which was not irrigated during the five-year period prior to 1948. An investigation to determine whether a particular piece of land falls within this category involves lengthy hearings, field investigations, and checking of records. The Land Department is ill-equipped to handle such proceedings and interest appears to be lacking.

The adjudication of disputes over water rights has always occupied a central place in the operations of the department. The introduction of informal hearings sped up the process considerably but it still consumed a considerable amount of the time devoted to water-resource matters. With the 1957 change in administration in the Land Department, however, there appeared to be a new point of view concerning the authority of the department to adjudicate conflicts between water-rights disputants. In spite of what seems to be the clear intent of the surface-water code, the administrators began contending that the department is little more than a registration agency for those seeking to appropriate water, and that the department was not set up to adjudicate rights or determine priorities, but only to determine whether a vested right exists when another application is filed. In the view expressed by a department official, the administrators were thinking of eliminating all water hearings since they felt they had no authority to bestow a water right or to determine who has a valid water right among many users. It is pointed out that hearings are time-consuming and costly and the courts always available. However, the department has continued to hold hearings and to make judgments on disputed applications, indicating continued concern that some administrative machinery operate to protect the water supply.

Increased pressure has been exerted on the department to provide other services to the community, particularly to landholding or lending companies. "Loan companies, banks, realtors, both in and out of state, are now requesting an abstract of all water rights held by their applicants before making a loan or selling, and the same procedure is being followed by some escrow companies."[30] As more and more of the agricultural land in the central valley, particularly around Phoenix, is taken for residential or industrial purposes, the State Land Department can look forward to a steady increase in these requests for services. It is somewhat ironic that the Land Department is beginning to discount the importance of registered water rights at the very time that landholding companies and mortgage companies are demanding such evidence before making sales and loans.

The depository character or housekeeping nature of the department is indicated by the placement in the department of responsibility for licensing of those who engage in weather control or cloud modification.[31] Each licensee is required to report the results of any given project. At the present time, four licenses have been given but only one project has been reported as completed. Except for the work of the Institute of Atmospheric Physics, described in a later chapter, there is no activity in "rain-making."[32]

The newest, and most promising activity of the Land Department concerns the management of the state's watersheds. A Division of Water Resources was created in 1956 to cooperate with the Salt River Valley Water Users' Association and the University of Arizona in the study of Arizona's watershed areas. This study was undertaken for the purpose of improving conditions on the watershed areas "so as to better utilize a greater percentage of the annual rainfall; to determine the reasons and causes of the increasing loss of annual runoff of rainfall on the watershed areas," and to coordinate the efforts of all agencies concerned with these problems.[33] Results of the initial study were incorporated in the now-famous Barr Report, which recommended significant changes in the management of the watersheds.[34]

The work of this division has involved promotion of research on watershed-management practices as they affect water yield, and the dissemination of information to policy-makers and the public generally on the possibilities of increasing the usable water supply. The division works closely with the Arizona Water Resources Committee, which is made up of leading figures in the state's economy in "planning a program of watershed management aimed at recovery of a larger percentage of rain and snowfall than is being realized under present watershed management practices."[35] The Division of Water Resources and the Water Resources Committee have taken credit for Congressional appropriations of approximately one million dollars for additional watershed research initiated since the Barr Report was published.

Much of the effort of the two organizations has been devoted to public information. Each year they have sponsored conferences on watershed management, in which there have been progress reports on research activities in watershed management.[36] They have put out brochures and reports and have presented a slide film show not only throughout Arizona but also in Washington, D. C. for the Conference of Western Senators.[37] By an intensive campaign designed to demonstrate the value of watersheds as water-producing areas, and the technical feasibility of managing the watersheds without destructive results, they have sought to counter public suspicion of anyone who attempts to manipulate the forested areas, asserting at the same time that such manipulation is not antithetical to the multiple-use principle.

Since the management of watersheds in Arizona is for the most part the responsibility of the federal government, the Water Resources Division will probably remain relatively small, particularly since the Water Resources Committee plays such a vital leadership role. If, however, the state assumes some responsibility for watershed-development work, in cooperation with the federal government, the division will undoubtedly expand. If the state does assume such responsibility it will be virtually the first action program in the water-resource field in Arizona history.

The State Land Department, as the land management agency of Arizona, has many informal relationships with other state and federal agencies related to water and conservation problems. It cooperates with the Bureau of Reclamation and the Corps of Engineers on irrigation and flood-control projects. It works very closely with the Soil Conservation Service through its Division of Soil Conservation, even putting its own lands into soil conservation districts and bearing its share of the expense for conservation practices in the districts. It cooperates with the Bureau of Land Management and the Forest Service in the protection of its grazing and forested land, in part to protect the soil against erosion and flooding. Its efforts along these lines, however, are extremely limited.

The department cooperates with the Arizona Game and Fish Department by making land available for the latter's projects and in authorizing appropriations of water for lakes. It works closely with the Arizona Interstate Stream Commission in its battle over the Colorado River as well as in other activities concerning the use of water from Arizona's streams. The state land commissioner is an *ex officio* member of the Interstate Stream Commission, and no little of his time has been taken up with deliberations of this commission.

In summary, it is fair to say that the State Land Department, in its water-resource responsibilities, shows signs of becoming an important instrument for conservation of water. It is responsible for a wide variety of activities — soil conservation, administration of surface and underground water law, research, watershed management, and weather modification — and in some areas, notably watershed management and at times in ground-water regulation, it shows signs of taking the lead. Under favorable circumstances it might become something more than a depository for these separate but interrelated functions and assume responsibility for leadership in the entire field of soil and water conservation. At the same time, the department personnel indicate a very real unwillingness to undertake leadership, a desire to avoid controversy, and a distrust of the scientific research necessary to advance the cause of conservation in the state.

ARIZONA INTERSTATE STREAM COMMISSION

The most prodigious efforts exerted by the state of Arizona in the field of water resources have concerned the Colorado River. Since 1922 the state has struggled with the federal government and the other Colorado River basin states in an attempt to obtain what it has considered to be a fair share of the Colorado River. Except for the

Wellton-Mohawk and Yuma projects, and some development of Indian land near Parker, Arizona has been singularly unsuccessful. It is now near the culmination of what appears to be its final effort in this long struggle to determine its rights to the river's water.

The Arizona Interstate Stream Commission was created in 1948 to prosecute the state's claims to the Colorado River water in Congress and before the courts.[38] It is the heir of the Colorado River Commission which had endeavored to protect Arizona's interests on the Colorado during the late 1920's and the 1930's when the Boulder Canyon Project Act was passed and Arizona sought to prevent development of the river along lines which were objectionable to it. With the ratification of the Colorado River Compact by Arizona in 1944 the Colorado River Commission went out of existence. The state land commissioner took the responsibility of getting Bureau of Reclamation approval of the Central Arizona Project. It was felt, however, that there was need for a separate agency which could give its undivided attention to the project and also serve as a focal point for the various groups and interests that were concerned with bringing additional water from the Colorado River into the state.

The responsibilities of the AISC were broadly defined in the enabling legislation: to prosecute and defend all rights, formulate plans and the development program, represent the state in conferences and negotiations, obtain licenses for reservoirs, dam sites, and numerous other requirements related to Arizona's network of streams of an international or interstate character.[39]

The commission consists of seven members coming from no fewer than six different counties of the state. These members are appointed by the governor with the consent of the Senate for terms of six years. The state land commissioner and the chairman of the Arizona Power Authority also serve in *ex officio* capacities.[40] In the creation of the commission it was deemed essential that it be a "nonpartisan group, since the matter involved is for the common good of the entire state," that it be statewide so that all interests might be represented, and "one in which the Executive and Legislative branches of the state have confidence . . . ."[41] Judging from the statewide editorial support given the commission, and the backing of both political parties, the AISC has fulfilled the role expected of it. The single major source of opposition has come from groups along the lower reaches of the Colorado River who fear that the development of the Central Arizona Project will result in a diminution of the water supply presently serving their projects. In 1957 residents of Yuma County, particularly those around Parker, called on the governor to "fire" all the present members of the commission for attempting to delay the development of 65,000 acres of land near Parker until after the resolution of the dispute over the Colorado River.[42]

During the years 1948-1951 most of the commission's efforts were directed toward gaining the approval of Congress for the Central Arizona Project. The Bureau of Reclamation had already made preliminary studies of the feasibility of the project and had recommended its authorization; this recommendation had received the approval of the Secretary of the Interior and was forwarded to Congress in 1948.[43] The job before the commission was to convince Congress of the worth of the project and to overcome the bitter resistance of California.

In spite of intensive efforts by the commission in presenting evidence and testimony before the committees of Congress, its efforts came to naught. The authorization bill passed the Senate twice — in 1949 and 1951 — at least in part because of the influence of Arizona's senators, Hayden and McFarland. But each time the House Committee on Interior and Insular Affairs bottled the bill up and effectively thwarted further consideration of the matter.[44]

After receiving assurance that Congress would no longer consider the merits of the Central Arizona Project until the dispute between California and Arizona over water rights was resolved, Arizona shifted its attention to the legal aspects of the dispute and to the courts. When California began withdrawing water from the Colorado in excess of that which Arizona claimed she was entitled to, the AISC, as the agent for Arizona, filed suit against California. From 1952 to 1956 the commission was occupied with the aggregation of material and evidence and the filing of motions before the United States Supreme Court. With the completion of briefs by the two parties, the trial before the Special Master began on June 14, 1956.[45]

During the presentation of evidence and legal argument, Arizona's position shifted materially, and considerable dissatisfaction was manifested within the state toward its presentation. Several of the original attorneys pleading Arizona's case departed and were replaced. These internal problems did not receive much public attention, however, and did not noticeably disturb public confidence.

As part of its struggle with California, Arizona resisted all attempts by California to make the Upper Basin states parties to the dispute. Arizona, through the AISC, successfully encouraged the Upper Basin to oppose inclusion, and its position was upheld by the Special Master. Only Utah and New Mexico are parties to the dispute and only to the extent that they have claims on water in the Lower Basin. Both states have tended to support Arizona's legal position.

Arizona has supported the Upper Colorado Storage Project, at least in part "to cement further the union of interests between Arizona and the Upper Basin States as opposed to California's efforts to pre-empt all of the Colorado River."[46] Arizona is party to the Upper Colorado River Basin Compact and is allotted 50,000 acre-feet annually to meet its potential Upper Basin needs,[47] although estimates indicate that Arizona will need only 30,000 acre-feet.[48]

The interests of the AISC include other streams which criss-cross the state, some of which are interstate and/or international in character. The commission is responsible for conducting whatever negotiations are necessary in connection with those streams. It has in recent years taken part in negotiations dealing with such rivers as the Santa Cruz, the Virgin, the Gila, and the Little Colorado.

In addition, the AISC has been active in Washington in behalf of other projects on Arizona's lesser streams. The commission retains a representative in Washington whose efforts have been directed toward authorization of a number of watershed projects and flood-control structures along the Gila River system. Two of these projects, the Painted Rock Reservoir and Whitlow Ranch Reservoir, are completed, and construction on another, the Tucson diversion channel, began in 1963. The Alamo Reservoir on the Bill Williams River has been authorized but no money has been appropriated for construction.

When Arizona failed to obtain Congressional authorization for the Central Arizona Project the state began to make plans for the construction of the project "without federal funds, or with partial federal aid." Considerable research had already been done by the Central Arizona Project Association when in 1956 the legislature gave authority to the AISC to pursue these investigations in association with the Arizona Power Authority. The legislature appropriated funds to be used for the purpose of "investigating works, plans or proposals pertaining to interstate streams, and . . . formulating plans and development programs for the practical and economical development, control and use of the water of interstate streams."[49] In order to placate those in the Yuma Valley, the legislature provided that no action taken under the act should affect the existing contracts and vested rights of all lawful users of Colorado River and other water.[50] To coordinate the efforts of the AISC and the APA, the state land commissioner was appointed state water and power coordinator.[51]

On May 5, 1960, Judge Rifkind made public his *Draft Report* on *Arizona* v. *California*. His recommendations followed in general outline the position taken by Arizona, and were viewed in both Arizona and California as a major victory for Arizona. His report must still be considered by the Supreme Court but Arizona is hopeful that his recommendations will be upheld. The state is therefore pressing forward with its plans for water and power development. It is clear, however, that the Central Arizona Project cannot be constructed except under Congressional authorization and federal financing. The economic studies undertaken by the state and by the Central Arizona Project Association have demonstrated the impossibility of the state financing a project that would require a considerable subsidy from power revenues for the water-delivery features of the project. When it was still hoped that private financing could be obtained, Arizona had been pressing the Federal Power Commission for a decision on a license to construct Bridge Canyon Dam, the key structure in the Central Arizona Project. Now that the state-financed plan has been ruled out, the AISC will again begin to prepare the necessary material to present to Congress in justification of the Central Arizona Project while the state seeks to delay a decision on the Bridge Canyon Dam which might prove favorable to California's application.

The AISC works closely with other public and private groups throughout the state. It has been actively associated with the many irrigation districts, the Arizona State Reclamation Association, the Association of Soil Conservation Districts, the Agricultural Extension Service, the Arizona Farm Bureau Federation, the Game and Fish Commission, the Arizona Game Protective Association, the University of Arizona, and the Ceneral Arizona Project Association.[52]

The most important of these is the Central Arizona Project Association which describes itself as "an Association of agricultural, business, professional, and industrial people whose purpose is to obtain supplemental water for Arizona's agricultural economy."[53] This organization came into being in 1946 to give support to the effort to obtain Congressional authorization for the Central Arizona Project. It has worked very closely with the AISC in the suit against California, sustaining the commission on expensive matters when there were either legal or economic reasons for the commission's inability to attend to them. This involved such things as "providing research assistance, paying for extra stenographic help, setting up a petty cash fund, and establishing a sizable revolving fund which remains the asset of this Association but is of great help to the Commission . . . ."[54] At one time the same person served as chief counsel both to the association and the commission, and the present chairman of the commission was formerly the president of the association.

The CAPA is supported by interests in the state that will benefit directly or indirectly from the importation of Colorado River water into the central valley. Irrigation districts, industries, professional, agricultural, financial, and business interests all contribute according to the nature and the amount of interest involved.[55]

The future role of the Interstate Stream Commission depends entirely on the final decision of the United States Supreme Court in *Arizona* v. *California*. If the court rules in Arizona's favor, the AISC will undoubtedly play an important role in mobilizing political support, economic justification, and detailed plans for the Central Arizona Project. Owing to the favorable situation the commission has always occupied within Arizona there is little doubt that it could obtain the support of the legislature if that were all that was required. Since the commission will have to seek Congressional authorization, its prospects for success are not so favorable.

In summary, the state of Arizona has facilitated the effective management of the water supply through passage of laws permitting the establishment of irrigation, drainage, and water-delivery districts. It has supported conservation research (discussed below), and in some instances has cooperated with the federal government in conservation programs. Because of the crucial role of the federal government in a state so largely in public ownership or trust, the state has not felt called upon to engage in action programs to protect or develop its own resources. It has failed to participate fully and in a meaningful way in increasing the water supplies or in conserving the limited amounts presently available. In this regard Arizona is hardly unique, however. Vincent Ostrom has observed that

> in most states some resource agencies are doing an effective job on some phases of resources administration, such as the development of its parks, the management of state lands, forest fire prevention, or conservation of wildlife, but no state is carrying on a well coordinated multiple-purpose program of resource administration. It is questionable whether the states have either a legislative or administrative arrangement competent to define or to satisfy the public interest in the effective control and utilization of all natural resources. Rather, the states seem best organized to serve the purposes of groups of users relating to a particular resource utilization.
>
> If the western states are to assume a greater role in the conservation and development of natural resources, they must first put their own houses in order . . . . Otherwise, so far as resource development is concerned, the states are apt to become anachronisms like counties, to be tolerated as they muddle through.[56]

The state shows signs of awakening in regard to watershed management, but it is doubtful that its role will change significantly until there is a greater awareness of the need and the benefits resulting from effective state action.

CHAPTER 9

Federal Activities in Water Management

The federal government has played an extremely important role in the development of water supplies in Arizona, dwarfing the role of the state government. Beginning with the Salt River Project, completed in 1911, the Bureau of Reclamation in particular has bulwarked the reclamation efforts necessary to turn the desert into a productive community and establish one of the most efficient agricultural centers in the United States. Furthermore, the people of Arizona look to the federal government for guidance and assistance on many enterprises which are as yet only dimly seen.

The two major water resource agencies of the federal government, the Bureau of Reclamation and the Army Corps of Engineers, have both performed important functions in the state. But there has been little of the interagency rivalry which has so plagued other river basins. The activities of the Corps of Engineers have been limited almost entirely to flood control, and completed projects of this type are few. The Bureau of Indian Affairs has undertaken some major projects which serve both the Indians and white men. The Bureau of Reclamation, however, continues to play the dominant role in river development in Arizona and the entire Colorado River basin.

BUREAU OF RECLAMATION

The Salt River Project is nationally famous as the first project authorized under the Reclamation Act of 1902. Agricultural development by means of the normal flow of the Salt River had nearly reached its maximum around the turn of the twentieth century. It was apparent that further development and stability of existing developments depended on the construction of storage facilities. To negotiate with the federal government, guarantee repayment of costs, and assume responsibility for management of the project the farmers in the Salt River Valley formed the Salt River Valley Water Users' Association.[1] Pursuant to an agreement signed June 25, 1904, the Reclamation Service undertook construction of Roosevelt Dam and other project works which were finally completed in 1911. The cost of the project was $10,166,021 or $60 per acre, far in excess of the original estimates.[2]

The Salt River Project provided water for an approximate 240,000 acres.[3] The project also provided water for 90,000 acres of so-called Warren Act lands and electricity for the pumping of water on many project and adjacent lands.[4] In 1917 the project was turned over to the Water Users' Association for operation and maintenance and has been so managed since that date.

The project has expanded greatly since 1911, both through Bureau of Reclamation assistance and bond issues floated independently by the Salt River Valley Water Users' Association. Between 1922 and 1930 the association constructed Mormon Flat, Horse Mesa, and Stewart Mountain dams on the Salt River to increase storage and to produce larger quantities of power. Between 1936 and 1939 the Bureau of Reclamation constructed Bartlett Dam on the Verde River and made spillway improvements on the four Salt River dams. The final structure, Horseshoe Dam on the Verde River, was built by the Defense Plant Corporation and completed in 1945. The total storage of these dams is 2,076,713 acre-feet, and the generating plants at these dams have a capacity to produce 70,900 kilowatts.[5] Although the project is fully managed and operated by the association, title to the irrigation works constructed by the bureau remains in the federal government.

The original debt of better than $10,000,000 was repaid in 1955. Currently a rehabilitation and betterment program is underway, under contract with the Bureau of Reclamation, which will ultimately cost $8,000,000. The money will be used for extensive improvements in the distribution system, including such features as lining irrigation canals and laterals.[6]

In 1937 the project was given the position of a political subdivision of the state.[7] The purpose of this was to allow the project to levy assessments in order to expand its facilities for storing water and producing power, and also to exempt its works from taxation. It was further permitted to subsidize its irrigation-water users by allowing it to sell surplus water and power. This favorable economic and political situation has caused some consternation in other sections of the state not receiving the benefits of the project. The *Arizona Daily Star*, an influential Tucson newspaper, has continually editorialized to the effect that the power operations of the project should be taxed. Bills have been introduced in recent sessions of the legislature to strip the association of its preferred status but have met with no success. However, in 1963 the legislature passed a bill which provided that the association could make "voluntary contributions" in lieu of taxes.

The Salt River Project has paid off handsomely in terms of the economic development of the central region of Arizona. The Bureau of Reclamation, with characteristic modesty, describes it thusly: "Horatio Alger's Theme on a magnificent scale — not the rise of an individual, but of a community of over one-third of a million people — is found in the story of the irrigation development of the Salt River Valley in south-central Arizona." The gross value of crops in the project in 1960 was $69,049,902. With the production of power by the project and retailing of power developed by the Bureau of Reclamation on the Colorado the Phoenix area is becoming a center for light industry.

The reclamation projects near Yuma on the Colorado River rank second only to the Salt River Project in economic importance to the state. The Yuma and Yuma Auxiliary Projects were authorized early in the century — 1904 and 1917, respectively — and have long since reached their full development. The Yuma Project serves an area of 53,000 acres and has the highest gross crop value per acre irrigated of any reclamation project in the state, with a 1960 average of $543 per acre. In 1955, when only 45,539 acres were planted, the project had a gross crop value of $14,617,590, a figure that rose to $24,793,175 in 1960.[8] This project cost the United States a total of $5,806,743; the Yuma County Water Users' Association has contracted to pay for its share of what amounts to all but $200,000 of the cost of construction. Since 1959 the project has been managed and operated by the association, taking over from the Bureau of Reclamation in that year.[9]

The Yuma Auxiliary Project was originally designed for 45,000 acres, but the irrigation works constructed were unable to serve the entire area. In 1949 the project was restricted to 3,305 acres with the remaining acreage being served as part of the Yuma Mesa division of the Gila Project. The Auxiliary Project cost $902,651 of which the water users were required to pay and already have paid a total of $724,033.[10] The gross value of the crops was $1,100,492 in 1960, with a gross crop value per acre irrigated of $374.[11]

The Gila Project was under consideration as a reclamation project from the early 1920's. Work began on the project in 1935 but World War II substantially retarded its completion. In 1947 the project was reauthorized and considerably reduced from its original 150,000 acres to the 115,000 acres of the present project.[12] The Gila Project consists of two main divisions: the Yuma Mesa division and the Wellton-Mohawk division. In addition, there are two units, the North and South Gila Valley Units, consisting of around 15,000 acres, which have been under cultivation for many years and are now served by the project. The Gila Project is still in the process of development but experience indicates that it will prove financially sound. By the end of 1955 there were 53,627 acres under cultivation, and in 1960 these acres produced $15,130,194 in gross crop value, with a gross crop value per acre irrigated of $191.68. This compares favorably with many other reclamation projects in the West, although it does not reach the standards of some projects in Arizona.[13]

These reclamation projects are the basis of the economy in southwestern Arizona. A recent survey by the Stanford Research Institute found agriculture to be the primary source of income and employment and the most important factor in economic development in the future.[14]

The Bureau of Reclamation is much more renowned for other developments in Arizona, but these have provided water-supply benefits primarily to water users outside Arizona. Hoover Dam, for example, with its capacity of nearly 30 million acre-feet of storage provides no water for irrigation in Arizona. This was one of the facts about the construction at Boulder Canyon that caused opposition to the project in Arizona. The importance of Hoover Dam for Arizona lies primarily in its flood control features, its production of electrical energy, and the facilities for recreation provided at Lake Mead.[15] Hoover Dam and its companion structures lower down on the river have tamed the once raging Colorado River, have provided protection for downstream developments and made possible diversions for irrigation from downstream works. Since 1935, when Hoover Dam was completed, floods have been virtually nonexistent.

Davis Dam, 67 miles farther down on the Colorado River, provides river regulation and maximum development of power in conjunction with operations at Hoover Dam.[16] With a storage capacity of 1,820,000 acre-feet the dam, operated by the Bureau of Reclamation, is chiefly beneficial to Arizona in providing power and recreational facilities, and refuges for wildlife. It went into operation in 1951 after a long delay owing to World War II.

Parker Dam, like the other units on the Colorado River, is chiefly important to Arizona in its capacity to control floods, produce power, and provide recreational facilities. It is located 80 miles below

Davis Dam and has a capacity of 717,000 acre-feet. Completed in 1938, its chief purpose is to provide a diversion point for water transported by the Colorado River Aqueduct to Southern California.[17] Lake Havasu, the reservoir created by Parker Dam, may prove to be an important structure relative to Arizona's future water supply if plans for the Central Arizona Project, discussed in detail below, are approved.

The Bureau of Reclamation may yet play an even more illustrious role in the development of Arizona in the future than it has in the past. Evidence for this is found in the monumental study of the Colorado River by the bureau, published in 1946, in which the bureau found 34 potential projects in the Lower Basin, all of which would be of interest to Arizona.[18] Although not all of them would provide water for reclamation projects, all would be important in flood control and desiltation. Not all of these projects could be built, even though they are feasible both from economic and engineering viewpoints, since there is not enough water in the river to supply the needs of all the projects. The problem, therefore, is to choose the projects yielding the highest benefits in relation to costs, and those most needed in the region.

Glen Canyon Dam, the key structure in the Upper Colorado Storage Project and now under construction, is primarily of interest to Arizona for its power production (which will be approximately 900,000 kilowatts), silt retention, hold-over storage, flood control capacities, as well as recreation opportunities.[19] In spite of an active storage capacity of 20,000,000 acre-feet, the dam will not provide water for Arizona since there are no lands in the area susceptible of cultivation.

There are possible areas for development on the Little Colorado and the Virgin rivers and Kanab Creek but they are limited. The Little Colorado projects, if fully developed, would bring 32,250 acres into production, provide storage of 284,000 acre-feet, and deplete the stream by 48,700 acre-feet annually. They would also provide flood control and silt retention. Proposed projects on Kanab Creek and the Virgin River would bring only 21,500 acres into production. The importance of works on these streams, besides irrigation, would be in power production, silt retention, and flood control.

Central Arizona Project

The project of most importance and which is under the most serious consideration as a means of relieving the water shortage in Arizona is the Central Arizona Project. This project has become a center of controversy, not only between the states of Arizona and California but also between those who look with sympathy on reclamation projects and those who consider these projects to be unnecessary subsidies to farmers. Requiring as it does authorization by Congress, the project has become a national issue evoking considerable attention in the national press.[20] Because of the importance attached to this project in Arizona, and because it is one of the few ways in which the water supply of the state can be supplemented, it is necessary to discuss it at some length here.

Under the proposal approved by the Bureau of Reclamation[21] and supported by the state of Arizona, 1,200,000 acre-feet of water would be diverted from Lake Havasu behind Parker Dam by means of pumps which would lift the water 985 feet. The power for this pumping would come from Bridge Canyon Dam. The water would be carried 241 miles in an aqueduct by gravity flow to central Arizona. A portion of this water would be made available for the present Salt River Valley development. This supply would free 484,000 acre-feet of Salt River water which would be carried by canal to the Gila River and thus to the lands in Pinal County, a distance of 74 miles. The remainder of the water would be stored in McDowell Reservoir above Granite Reef Dam near Phoenix. Other structures would be built to provide for capturing all currently wasted flood water, for river regulation, silt control, flood control, and the production of power.

The need for the water in Arizona was summarized in the Bureau of Reclamation report which stated that the Central Arizona Project would

(1) permit reduction of pumping and thus limit withdrawals from the ground water basin to its safe annual yield, (2) permit delivery of a supplementary supply to lands now inadequately irrigated, (3) permit delivery of an adequate supply to developed lands now idle for lack of water, (4) permit delivery of an adequate municipal supply to the city of Tucson, and (5) permit carrying of excess salts out of the basin.[22]

It was claimed by officials of the Bureau of Reclamation that "without Colorado River water about one-third of the productive capacity of the agriculture development will be lost, which will result in large-scale abandonment and migration."[23] Without this supplemental water only 414,000 acres of the 672,000 acres in production in 1945 would continue in cultivation, causing distress not only to the agricultural industry but to all the other industries related to agriculture and the entire economy of Arizona.

Bills authorizing the Central Arizona Project were introduced

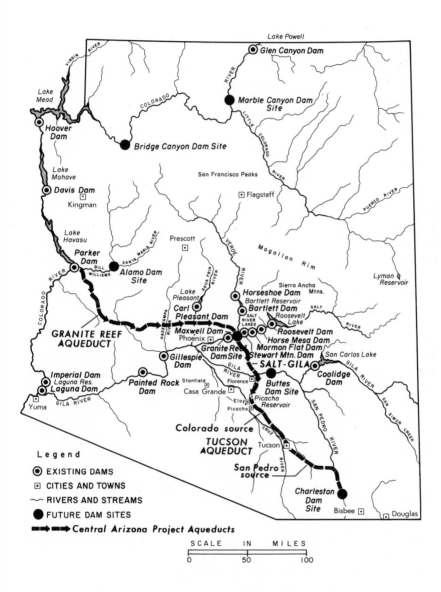

Lake Powell

⦿ Glen Canyon Dam

Lake Mead

Marble Canyon Dam Site

COLORADO RIVER

VIRGIN RIVER

LITTLE COLORADO RIVER

⦿ Hoover Dam

● Bridge Canyon Dam Site

Lake Mohave

San Francisco Peaks

⦿ Davis Dam
⊡ Kingman

⊡ Flagstaff

PUERCO RIVER

Lake Havasu

SANTA MARIA RIVER

⊡ Prescott

VERDE RIVER

Mogollon Rim

Lyman Reservoir

Parker Dam

BILL WILLIAMS RIVER

Alamo Dam Site

AGUA FRIA RIVER

HASSAYAMPA RIVER

COLORADO RIVER

Sierra Ancha Mtns.

Lake Pleasant

⦿ Horseshoe Dam
Bartlett Reservoir

SALT

Carl ⦿
Pleasant Dam

⦿ Bartlett Dam
SALT RIVER LAKES
Roosevelt Lake

GRANITE REEF AQUEDUCT

Maxwell Dam⦿
⊡ Phoenix

⦿⦿ Roosevelt Dam
Horse Mesa Dam

RIVER

⦿ Gillespie
Dam

Granite Reef
Dam Site

Mormon Flat Dam
Stewart Mtn. Dam

San Carlos Lake

SALT-GILA⦿

GILA RIVER

GILA RIVER

Coolidge Dam

Imperial Dam
⦿ Laguna Res.
⦿ Laguna Dam
⊡ Yuma

GILA RIVER

Painted Rock
Dam

⊡ Stanfield
Casa Grande

⊡ Florence

Buttes
Dam Site

Picacho Reservoir

⊡ Eloy
⊡ Picacho
Picacho

SAN PEDRO RIVER

SAN SIMON CREEK

Colorado source
TUCSON AQUEDUCT

⊡ Tucson

SANTA CRUZ RIVER

Legend

⦿ EXISTING DAMS

⊡ CITIES AND TOWNS

〰 RIVERS AND STREAMS

● FUTURE DAM SITES

➡➡ Central Arizona Project Aqueducts

San Pedro
source

SAN PEDRO RIVER

Charleston
Dam
Site

⊡ Bisbee
⊡ Douglas

SCALE IN MILES
0   50   100

Central Arizona Project and Proposed Dams

in Congress in 1949 and for the next three years were the subject of strenuous efforts by Arizonans to get them passed. Twice such bills passed the Senate only to be defeated in the House Interior Committee. Chief opposition came from California interests who argued that Arizona did not have title to the water it claimed under the Colorado River Compact and which it proposed to divert through the project. Others, both in and out of California, disputed the economics of the project, portraying it as a gigantic raid on the public treasury in behalf of agriculturists who had overextended themselves and now wanted the taxpayers to come to their rescue. Confronted by Congressional unwillingness to pass authorizing legislation until the legal title to the water was settled, Arizona then sought to quiet its title in the Supreme Court.

In the late 1950's serious consideration was given by the leaders of Arizona's water industry to the possibility of constructing the Central Arizona Project under a license from the Federal Power Commission, with financing through the private bond market. To this end, in August 1957 Arizona Power Authority employed the Harza Engineering Company to make studies and investigations of potential hydroelectric projects between the headwaters of Lake Mead and the Glen Canyon Dam site. Harza reported the initial results of its studies in December 1958.[24] The original recommendations called for construction of dams at the Bridge Canyon, Prospect, and Marble Canyon sites on the Colorado, and the Coconino site on the Little Colorado. While the report was conceived primarily in terms of power production, it was noted that with little additional expense works could be later constructed which would provide for water diversions.[25] The company suggested several alternatives: diversion from Marble Canyon to a reservoir on the Little Colorado from which the water would be pumped over the Coconino Plateau to the Verde River; diversion from both Prospect and Bridge Canyon through a 77-mile gravity tunnel to Big Sandy Creek and then by a 250-mile canal to McDowell Dam near Phoenix; or diversion along the same lines from Bridge Canyon alone. The Parker pump-lift diversion also remained a possibility.

After the original Harza report, the plans for Colorado River development were revised twice, once in an application by the Arizona Power Authority for license to construct Bridge Canyon and Marble Canyon dams before the Federal Power Commission and again in an amended application for license to construct at Bridge Canyon, Marble Canyon, and two sites on the Little Colorado.[26] The amended application for license eliminated Prospect Dam and changed the elevation of Bridge Canyon Dam to a height which would back water only up to the lower boundary of Grand Canyon National Monument, thus eliminating any controversy with conservationists who would protest an invasion of the monument itself. Instead of a single dam on the Little Colorado two dams would be constructed — one near the Grand Falls of the Little Colorado, and one on Moenkopi Wash. These two would be for sediment control purposes. No provision was made for water diversion works on the amended application, but it was noted that either Bridge Canyon or Marble Canyon provided possible locations for diversion of desilted water to various parts of Arizona for irrigation, and for industrial and municipal water supply.[27]

In June 1959 the Central Arizona Project Association ordered a complete re-examination of the project in view of the changed conditions in Arizona. The president of the association was quoted as saying:

> It is apparent that although the project was originally conceived almost entirely as an agricultural irrigation development, a substantial portion of Arizona's share of the river may find its ultimate usage in direct support of expanding urban and industrial segments.[28]

By this time it was becoming apparent that private financing would place such a heavy financial burden on the primary beneficiaries, the farmers, that it would not prove feasible. At this point the Central Arizona Project Association and the state, realizing the need for Congressional approval of construction and financing under the Reclamation Act, requested the Bureau of Reclamation to undertake a reappraisal of the project. At the same time they persuaded the Arizona Power Authority not to press for a decision on its license to construct a dam at Bridge Canyon.

The focus of the bureau's concern shifted toward municipal and industrial water needs, while maintaining an active interest in sustaining the agricultural economy.

> On a quantitative basis, the water need is, and probably always will be, for irrigation, since agricultural water demands will exceed urban demands under all foreseeable conditions. However, from the standpoint of preserving the economic health and well-being of the area and providing for future growth and development, the municipal and industrial water requirements may be of equal or greater importance . . . much of the urban growth has been, and will continue in the future to be, at the expense of agricultural acreage for which a local water supply is available . . . . Without import water future municipal and industrial demands could be met only by the purchase and retirement of irrigated lands outside the urban area. Such forced reduction of the agricultural economy undoubtedly would be reflected in reduced urban growth.[29]

With the expected growth in urban population during the projected 50-year period, it is expected that urban water use will increase from the present 300,000 acre-feet per year to about 1,400,000 acre-feet per year, while there will be a decrease in annual irrigation water use of more than 1,000,000 acre-feet.

Under the 1962 plan, as under the original plan, Bridge Canyon Dam would provide the power to pump 1,200,000 acre-feet of water from Lake Havasu into central Arizona. This dam would have a powerplant with a capacity of 1,500,000 kilowatts, which would generate on the average of 5,800,000 kwh per year. Approximately one third of this power would be required for pumping the water the 985 feet necessary for it to flow by gravity to central Arizona. The remaining power would be sold to commercial power users in a market which is rapidly expanding in Arizona as well as in southern California and southern Nevada.[30]

In addition to Bridge Canyon Dam, the revised proposal envisions the construction of Granite Reef Aqueduct to transport the water from Lake Havasu to the Salt River system. Additional canals would convey the water to Pinal County where the water would be integrated with the San Carlos Project, and also to the Tucson metropolitan area. Several other structures would be added to conserve water presently being wasted, including the Buttes Dam and Charleston Dam on the San Pedro, the Tucson Aqueduct to bring water from the Charleston Dam to Tucson, and Hooker Dam on the Gila River. Of the total 1,200,000 acre-feet diverted from the Colorado, 814,000 acre-feet would be devoted to agriculture and 256,000 acre-feet to municipal and industrial uses. Most of the remainder would be lost to evaporation and seepage.

The Bureau of Reclamation estimated the cost of the government-constructed facilities at $971,329,000, with the water users paying an additional $100,000,000 for a distribution system. (Estimated cost of the entire priject in 1948 was $738,408,000; at that time the project plan also included construction of major silt retention works on the Little Colorado, now made unnecessary by the construction of Glen Canyon Dam.) Cost of operation, maintenance, and replacement would be approximately $7,352,000. In the bureau's economic analysis, benefits would exceed costs by a ratio of two to one, calculated on the basis of a 50-year pay-out period, and including both direct and indirect benefits in the analysis. Using only direct benefits the ratio would be 1.4 to 1. Commercial power users would repay the preponderance of capital costs — over $465,000,000 — with the power being sold at six mills. Irrigation users, paying $10.04

per acre-foot at canalside, would repay $317,173,300, and the remainder, $203,889,000, would be charged to municipal and industrial users, calculated on the basis of $33.52 per acre-foot. Nonreimbursable costs would total $26,385,000.

The decision of the Special Master in Arizona's suit with California has yet to be passed upon by the Supreme Court itself. However, unless the court chooses to reverse its Special Master it would appear that the argument concerning legal title has no validity. Perhaps more serious are the arguments raised by those who feel that the regional project would be a gift to land "boomers" who profligately wasted the precious water resources by overexpanding and now wish to be bailed out by the federal government. It has also been argued that not all of the ultimate costs to the taxpayer were included in calculating the cost-benefit ratio, that the allocation of the interest component on commercial power to defray irrigation costs is a gigantic raid on the public treasury amounting to billions of dollars, that the power is wastefully used in pumping, and that the ability of farmers to pay the costs of the water has been overestimated. Arguments of a similar character will undoubtedly be heard with regard to the present plan of development.

Another factor which may play an important part in Congressional considerations is the effect of construction of Bridge Canyon Dam on Grand Canyon National Monument and Grand Canyon National Park. The project proposal calls for the dam to be built to create a reservoir which will back water up through the monument and into Grand Canyon National Park. In the lower reaches of the park the water level would be raised as much as 89 feet during normal flow. Conservationists, who oppose tampering with natural conditions in national parks — as they successfully did with Dinosaur National Monument in Colorado — are vigorously opposing such a proposal. As the Bureau of Reclamation points out, if a decision is made to reduce the level of the reservoir to prevent the invasion of the monument and park, the loss in power revenues would have to be made up by charges on other users.[31]

Still another factor complicating the proposals for additional construction on the Colorado comes from the results of U.S. Geological Survey studies which indicate that present projects and those under construction will bring reservoir capacity to an optimum — "the minimum capacity necessary to provide the maximum obtainable regulation." Because of corresponding increases in evaporation, increased storage is unlikely to increase the total water supply.[32]

The Project Development Office of the Bureau of Reclamation in Phoenix has proceeded with investigations of other projects in Arizona. It has completed a survey of Marble Canyon, and, in cooperation with the Army Corps of Engineers, it has made studies of Buttes Dam and other possible developments on the Gila River.

U. S. ARMY CORPS OF ENGINEERS

The Corps of Engineers has completed four projects in Arizona, one of which is a flood-control project near Holbrook, to protect the town from the rampages of the Little Colorado River. In 1959 it completed the construction of Painted Rock Dam on the Gila River below Gillespie Dam. This flood-control project, capable of impounding nearly 2,500,000 acre-feet of water, was designed to prevent the destruction wrought by periodic flooding of the Gila. The undependable water supply and high evaporation rate make it unsuited for irrigation storage. In 1960 the Corps finished the Whitlow Ranch Reservoir which regulates runoff from Queen Creek into the Gila River, and Trilby Wash Detention Basin which controls flows from that tributary to the Agua Fria River at a point 20 miles west of Phoenix.[33] The Corps is presently engaged in a project of improving the channel of the Gila River from the Upper Gila Valley to the Buttes Reservoir site. This project will create a cleared floodway 94 miles long, by removing phreatophytes from an area of 14,300 acres. This project is designed to provide partial flood protection in the Safford Valley and also to increase the water supply for agriculture by as much as 19,800 acre-feet. These channel improvements are part of a more general plan for development of this area which would involve construction of dams in the Safford Valley and at the Buttes site. Projects authorized but not financed include the construction of a dam at the Alamo site on the Bill Williams River for flood control protection in the Lower Colorado River. The Tucson Diversion Channel has been authorized for the purpose of protecting the city from floods. Construction on this project began in 1963.[34] The Corps also has obtained authorization for levees and channel improvements on the Gila and Salt rivers but no work has yet been undertaken.

Controversy over the construction of a dam at the Buttes site prevents the Corps of Engineers from further expanding its work.[35] The Corps and the Bureau of Reclamation have jointly studied the feasibility of the Buttes Dam in conjunction with a dam which would be constructed at the head of Safford Valley on the Gila River. The water users in the San Carlos Irrigation District, desperate for water, conceive of the Buttes Dam as a means for controlling the periodic floods on the San Pedro River and providing storage for their water-short area. The water users in the Safford area, however, oppose the construction of the dam without reservations to protect their water rights. Both groups depend on the Gila River Decree of 1930 under which the water rights of the Safford water users are related to the quantity of water behind Coolidge Dam. The Safford irrigationists fear that the construction of Buttes Dam would allow the San Carlos Project to be operated in a fashion to damage them materially. Water impounded at Coolidge Dam could be stored at Buttes, reducing the storage at Coolidge, and therefore reducing the amount of water which Safford-area farmers could use. The Safford water users would consent to the construction of the Buttes Dam only if it were built in such a way that it could not be used as a storage dam, and if the Gila River Decree could be modified to protect more adequately their interests. The Safford water users want a dam at the head of their valley to control floods and provide a normal flow of water in the Gila over a longer period of the year. More water would be made available for the Safford water users and for storage at Coolidge Dam, and less would be lost to phreatophytes during floods.[36] It is apparently assumed that the Corps would be the construction agency for the Buttes Dam, while the Bureau of Reclamation would construct the Gila River Dam in the Safford Valley.

The Corps of Engineers has made numerous surveys in Arizona for future flood control projects. In the recent past it has made studies on the Little Colorado River and its tributaries, on the Gila River, the Colorado River above Lee's Ferry, Kanab Creek, tributaries to the Gila near Tucson, the San Pedro River, and the Florence area. The Corps and the Bureau of Reclamation have cooperated in investigations of channel improvements along the Salt and Gila rivers where salt cedars and other useless growth have so ·clogged the channels that surrounding communities are seriously threatened at flood stage. The building of levees and the removal of the growth, now under way, will relieve the flood threat and conserve water now lost to the useless plants.

BUREAU OF INDIAN AFFAIRS

The Bureau of Indian Affairs also has a role to play among the public agencies concerned with the management of water in Arizona. Its role is not restricted to the Indians alone for in at least two areas the white man depends on the Indian reclamation projects. The Colorado River Indian Reservation near Parker has witnessed rapid development in recent years. Served by Headrock Dam, built by the Indian Service on the Colorado River, this reservation had under cultivation 42,484 acres in 1956.[37] The Indians contemplate further development eventually reaching over 100,000 acres, but all development work has been suspended until the Supreme Court renders its decision in the *Arizona* v. *California* case. The prospects for this reclamation project are indicated by the statement of the Arizona Commission on Indian Affairs that this reservation is "one of the wealthiest, for its size, of any Indian reservation in the country."[38]

A potentially prosperous project is the San Carlos Project on the Gila River consisting of lands owned both by white men and the Gila River Indians. Designed to serve 50,000 acres of land in the heart of Arizona's agricultural community, it has suffered seriously from the drought conditions prevailing in the state for decades. Coolidge Dam, completed in 1929, and designed to store 1,200,000 arce-feet of water, has never been filled and at times is completely dry. The dearth of surface water has forced the farmers in the project to pump increased amounts of underground water and seriously deplete the ground-water reservoirs. Other factors complicate the picture for the Indians: inefficient management of their water supply, uneconomic units of land, and ignorance of the intricacies of irrigation farming.[39] Agriculture in this section is seriously threatened by the shortage of water and many acres have already gone out of production. The Bureau of Indian Affairs, which manages and operates the San Carlos Project and Coolidge Dam, has come under criticism for inefficient administration but it appears that the basic problem is a shortage of water rather than mismanagement.

It would be inaccurate to say that the federal government is responsible for the economic development in Arizona, but it has certainly been one of the most active agents in developing and conserving the resources of the state. Individual initiative has of course been important, but without the major contribution made by the federal government, since repaid in large part, it is doubtful that Arizona would find itself one of the most efficient agricultural areas in the nation and a rapidly growing center for other industries. It appears, furthermore, that the future progress of the state in managing its water resources will depend largely on the cooperative effort of the people of Arizona and the government of the United States.

CHAPTER 10

Land Management

The intimate relationship between soil and plant management and water supply is a widely recognized fact. While the total amount of water in a particular region depends on precipitation, the quantity of usable water depends in large part on the way in which man cares for the land surface on which the precipitation falls. Soil, denuded of its plant cover, loses its capacity to retain water through infiltration, resulting in floods in certain periods of the year and drought in others. The soil itself is removed, destroying the productive capacities of the land. Silt-laden streams replace clear streams, depositing quantities of the silt behind dams, in irrigation ditches, and on fertile farm land. The quantity of usable water is dissipated through torrential floods during parts of the year and through evapotranspiration during the remainder of the year.

In the desert region the need for careful management of the land is even more pronounced than in other areas. It is obvious that certain geologic processes, such as those that produced the Grand Canyon, cannot be controlled in a substantial way. But man himself has contributed heavily to the acceleration of erosion in areas where at least a state of equilibrium had been reached. Some date the acceleration of erosion in Arizona from the time of the arrival in large numbers of settlers who made their livelihood from grazing and farming.[1] With thin soils, low precipitation, and high rates of evaporation, the lands of Arizona were particularly susceptible to erosion under mismanagement, and eroded they have been.

The pattern of land management in Arizona involves a sharing of responsibility among a number of federal agencies, the Arizona State Land Department, and the many private landowners. As a result of this divided responsibility, and because 32 percent of the state has never been surveyed, it is impossible to get accurate figures on land ownership within the state, for some federal agencies have overlapping jurisdictions, and records are kept on several different bases.

However, all figures show that the amount of land subject to state control is small. Figures for land owned by the State Land Department range from 8,728,536 acres (12.01 percent) to 9,943,557 acres (13.68 percent). The figures for private land vary even more, ranging from 8,810,895 acres (12.12 percent) to 12,566,799 acres (17.28 percent). The remainder of the land — somewhere between 50,725,328 and 54, 015,101 acres, or more than 70 percent — is under federal ownership or administration. Of this, the Bureau of Indian Affairs holds somewhere between 19,472,906 and 21,514,516 acres (around 27-29 percent) in trust for the Indians.[2]

As a result of this landownership pattern, the federal government must bear the primary burden of protecting these lands against mismanagement, and of restoring the lands already suffering from erosion. The lands under federal administration range from the very best forested land in the state — the state owns virtually no forested land — to the most depleted land. All of this land, however, requires conservation management or treatment, since the condition of the lands in one part of a water system seriously affects the water and land situation elsewhere. As Higbee has put it, "The use to which mountaintop timber lands are put in Colorado is of vital importance to irrigators in Yuma, Arizona."[3]

The land-management picture is complicated still further by the division of responsibility among several federal agencies. The management of federally owned land is the responsibility of the Forest Service, the Bureau of Land Management, the National Park Service, the Fish and Wildlife Service, and the military departments. In most instances, each of these agencies manages these lands for many purposes other than that for which they were originally set aside. The Park Service manages land valuable for its grazing capacities; the military departments manage lands that are valuable for fishing and wildlife. Whether these secondary values are always given the recognition due them is often questionable. Such division of responsibility also raises the problem of coordination of efforts where conservation on contiguous or related land is required.

The final complicating factor is the checkerboard ownership of land in the state. Federal, state, and private lands in many parts of the state are scattered among each other, making it oftentimes difficult even to ascertain who owns particular parcels of land. Obtaining overall management plans and programs in such a situation becomes a very thorny problem.

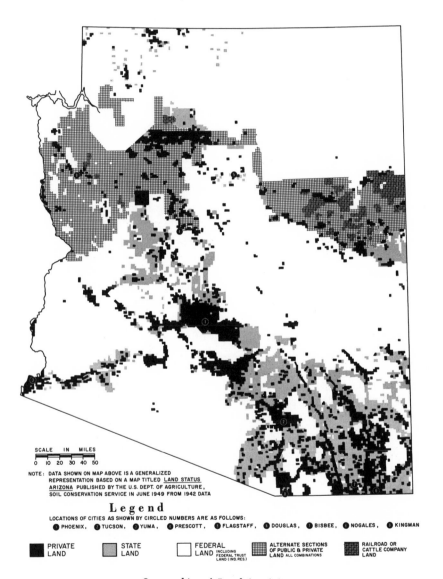

SCALE   IN   MILES

0   10   20   30   40   50

NOTE : DATA SHOWN ON MAP ABOVE IS A GENERALIZED
        REPRESENTATION BASED ON A MAP TITLED LAND STATUS
        ARIZONA PUBLISHED BY THE U.S. DEPT. OF AGRICULTURE,
        SOIL CONSERVATION SERVICE IN JUNE 1949 FROM 1942 DATA

Legend

LOCATIONS OF CITIES AS SHOWN BY CIRCLED NUMBERS ARE AS FOLLOWS:

● PHOENIX, ● TUCSON, ● YUMA, ● PRESCOTT, ● FLAGSTAFF, ● DOUGLAS, ● BISBEE, ● NOGALES, ● KINGMAN

PRIVATE LAND

STATE LAND

FEDERAL LAND INCLUDING FEDERAL TRUST LAND (IND. RES.)

ALTERNATE SECTIONS OF PUBLIC & PRIVATE LAND ALL COMBINATIONS

RAILROAD OR CATTLE COMPANY LAND

Ownership of Land in Arizona

ARIZONA STATE LAND DEPARTMENT

The State Land Department administers about one-seventh of the land area of Arizona. These lands have great economic importance inasmuch as they are sources of support for the educational system. In spite of this, the state has done very little to take advantage of these resources either by protection or development. For many years the lands were managed for the benefit of the users of the land, with the view that such policies contributed to the economic development of the state. Only in recent years has the state attempted to realize its full value from the lands, a change in attitude resulting in part from scandal in the administration of the State Land Department.

This department is given statutory authority to manage or dispose of state lands with considerable discretion. It can sell, lease, classify, and appraise lands. It can include state lands in irrigation projects or districts, and can pay the assessments chargeable to that land. It can investigate reclamation projects and report recommendations to the legislature. It may put its grazing lands in Taylor grazing districts. The department is charged with conserving or otherwise administering the timber, stone, or other products found on state lands. With the approval of the department, lessees may make improvements on state lands and such improvements may be sold to a new lessee. The state restricts grazing on its lands to 110 percent of rated capacity unless written consent for a variance is obtained from the department.[4]

It appears that there is no lack of statutory authority to manage efficiently the state lands and to perform conservation work on them. The inadequacy lies in the unwillingness of the legislature to provide money and personnel to perform the work. There is virtually no action program to improve state lands. Individual lessees are performing some conservation work on leased lands such as clearing of noxious growth. The Land Department suggests that such work should be encouraged, but does little to encourage such conservation activity,[5] other than to promote cooperation with the federal government in its cost-sharing programs for farm and range improvements.

In addition to doing little to improve state lands the department has in the past failed to exercise protective custody of these lands, allowing them to deteriorate, and doing little to protect these lands against misuse. The Arizona Education Association, naturally concerned about the revenue from the state leased lands, investigated the policies of the Land Department in 1948 and discovered that all land was being rented for grazing purposes at the flat rate of 3 cents per acre per year.[6] The capacity of the land was ignored in setting the rate, as was the price of beef, which had soared during the years of World War II. It was also discovered that land was sold or leased in such fashion that cattlemen could obtain virtual monopolies of certain sections of the range. "Kinfolk" purchases and leases were permitted with a minimum of competitive bidding. Similar favors, it was charged, were granted the copper companies which were interested in potential deposits on state lands. The land commissioner, strongly supported by the cattlemen of the state, fought to retain office after failing to be reappointed by the governor in 1949, but ultimately lost, and was later convicted of illegal activities in land management.

Prior to 1950 there were no regulations regarding the number of cattle or sheep which could be grazed on state lands. The stockmen ran as many head of livestock on the land as they saw fit. No appraisals of the land were made by state officials. Consequently these lands deteriorated badly, and were supporting considerably fewer cattle than they had in previous years. As a result of the attack on the State Land Department by the Arizona Education Association, which was joined in by the state American Federation of Labor organization and the *Arizona Daily Star* of Tucson, amending legislation was passed in 1950 which provided for appraisals of all state land, graduated fees based on the carrying capacity of the land and the price of beef, and the limitation to 110 percent of rated grazing capacity.[7] Many cattlemen persisted in the view that they should have unrestricted right to determine the number of cattle on state land but their views were not sustained.[8]

This legislation resulted in increased revenue to the state, but was of limited value in terms of conservation. Little of the revenue is utilized for improved conservation practices. The limitation on livestock numbers has been relatively ineffective owing to the inability of the State Land Department to police the restrictions. The livestock owners protest that they are conservation-minded since their livelihood depends on the capacity of the land,[9] but there are indications that they have not fully lived up to their obligations.

The State Land Department has been responsible for the administration of the State Water Code since 1943, and yet it appears that there has been failure on the part of the officers there to recognize the interrelationship of this responsibility with that of land management. Critical areas were established in depleted ground-water basins in order to limit the amount of land in production and thus protect the remaining water supply. At the very time that such critical areas were being established, however, the department was granting changes in leases from grazing to agricultural after lessees had developed water on their land by means of pumps which were so large that they could have been designed for no other purpose than irrigation. As early as 1941 State Land Department officials were warned that they should restrict sales and leasing of land, particularly around Eloy in Pinal County where the most serious declines in the water table have occurred,[10] but the department ignored these warnings, with the resulting rapid expansion of agricultural lands in those areas and depletion of the underground basins.

The State Land Department, in summary, is primarily a land-holding and leasing organization, with some sales occurring when they are deemed beneficial from an economic standpoint. Almost nothing is done to restore these lands or to prevent further destruction of them. Some commissioners have recognized the need for investing money and effort in these lands but their views have not been taken seriously. It is not unwarranted to predict that the state-owned lands will continue to decline unless there is a radical departure from present views on conservation activity, and such a departure is not in the foreseeable future.

THE FOREST SERVICE

The forested lands in Arizona are by far the most important in terms of water supply. The national forests in the state occupy 11,-381,541 acres or approximately 16 percent of the area of the state, receive 25 percent of the total precipitation, and furnish 37 percent of the runoff by volume.[11] Compared to an average annual precipitation outside the forests of 10.7 inches, the average for the forests is 18.4 inches. It is not only in quantity of water that the forests contribute so heavily to the water supply, but also in the timing of the runoff. The forests retain the water during the winter and early spring, and make it available during the late spring and early summer.

The management of the forested lands is mainly a responsibility of the federal government, with primary administration shared by two of its agencies. The Forest Service manages 2,250,000 acres of commercial forest land, the Bureau of Indian Affairs manages 815,000 acres, with a small remainder shared by the Bureau of Land Management, or in private ownership,[12] making a total of 3,180,000 acres. These forested areas also constitute some of the finest range lands in the state.

Federal administration of the forest lands in Arizona is a fairly well accepted fact today.[13] Periodically there are movements, usually associated with nationwide efforts, to divest the federal government of this responsibility, but such movements have not gained much

—*Courtesy James R. Hastings*

In 1884, the hills around Tombstone (above) were part of a luxuriant desert grassland covering much of southeastern Arizona. In 1961, the same area (below) supports a cover of creosote bush and other useless vegetation. Several factors combined to accelerate erosion which permitted the invasion of noxious plants into the former grassland.

—*James R. Hastings*

momentum except among the cattlemen who always chafe under what they consider too restrictive management by the Forest Service. The cattlemen opposed the creation of the National Forests originally and were supported in their stand by the mining interests. In 1931, 1947, and 1953 there were revivals of interest in state ownership of these lands, each time with approval of the cattlemen.[14] It was stated by one observer in 1948 that "Arizona is a crown colony — not a state."[15] The stockmen are perennially critical of the Forest Service because of its reductions of stock numbers on depleted ranges.

Water is unquestionably the most valuable product of the forests.[16] However, under the principle of multiple use, the forests are managed to provide for the realization of other values such as the grazing of livestock, the growing of timber, and the leisure-time activities of the general public. While these values are often complementary and therefore protected by the same management practices, there are frequently conflicts over the application of the multiple-use principle owing to the claim of one or another interest that its use is being neglected. As a result, the Forest Service frequently finds itself in a cross-fire while trying to maintain a balance in its management practices.

The importance of the lands under Forest Service administration can be indicated by reference to timber and grazing statistics. The Arizona forests are primarily ponderosa pine with some spruce-pine at the higher elevations. Demand for this timber has increased greatly and has resulted in more timber being cut at higher prices. In 1932 some 57,577,000 board feet were cut at an average price of $2.31 per 1,000 feet. By 1956 this had risen to 207,624,000 board feet at $9.10 per 1,000 feet. The total volume cut from national forest lands in Arizona had a value of $1,524,814 in fiscal 1961.[17] While some deprecate the importance of managing forest lands for maximum timber production, Forest Service studies indicate that demand for timber products will increase tremendously in the next few decades. It is predicted that there will be an increase of anywhere from 67 percent to 105 percent in the demand in the year 2000 as compared with 1952.[18] The importance of the timber industry in Arizona is indicated by recent construction of pulp and timber-treating plants at Snowflake and Prescott which will provide jobs for many people, and will result in a beneficial thinning of the forests.[19]

The forest lands also provide some of the finest grazing land in the state. The large bulk of lands under Forest Service administration are in fact not timber lands at all, but are chiefly valuable for their forage. In 1960 the Forest Service issued 950 cattle and horse permits

—*Robert R. Humphrey*

Clearing a chaparral-infested area by chaining (left) permits the recovery of grasses important for forage and for retarding erosion.

under which 152,529 animals grazed, and 26 sheep and goat permits which represented 67,297 animals.[20] There was no significant reduction of grazing units during the years 1953 through 1957 in spite of the drought, but the 1960 figure for sheep and goats represents a reduction of 14,350 animals.

The recreational significance of the forests in Arizona is of the highest order. With 4,999,800 visitors in 1960, the demands for enhancement of the recreational values of the forests will become increasingly strong. Visitors to campgrounds and picnic areas alone accounted for 2,255,900 visits in 1960. The sportsmen, who speak with the loudest voice on the recreational use of the forests, almost doubled their visits to the national forests in the years from 1953 to 1956. In that latter year, 423,000 sportsmen visited the forests and spent 894,000 man-days there.[21]

The Forest Service has recognized the necessity of taking positive steps to slow down and ultimately reverse the erosive trends on Arizona's watersheds. While its work is hampered by a lack of sufficient funds, it undertakes conservation work which contributes signally to erosion control. According to Salmond and Croft, Forest Service policy has followed three fundamental approaches in dealing with watersheds: multiple-use management when several values are involved; restriction or closure of areas to land uses when water-supply values are paramount; and experimental management in the interest of increasing water yield.[22]

Multiple-use management is as easy to state as it is difficult to define and practice. Each one who benefits from the forest feels that present management practices are discriminatory against his interest. Reconciliation of these conflicting interests continues to be one of the important responsibilities of the Forest Service.

For some time the Forest Service paid little attention to the methods used in logging and cutting timber except to ensure sustained yield. But it has come to recognize that these operations must be conducted in such a fashion as to minimize erosion. The service imposed layout plans and construction standards for roads and skid trails to reduce runoff and ground disturbance; erosion control structures and reseeding or replanting have been required. Much effort has been devoted to the control of wildfires which bare the soil and leave it without protection against the intense summer storms. These fires further damage the capacity of the soil to hold the water through the process of infiltration, and therefore reduce the steadier and more prolonged stream flow.

Overgrazing constitutes a serious cause of erosion and much attention has been directed toward limiting the numbers of stock permitted licensees on the ranges, and requiring seasonal distribution on the range to prevent excessive concentrations of the stock in limited areas. The Forest Service cooperates with the Arizona Game and Fish Department in setting game kills on the national forests.

The Forest Service has recognized the need to improve the forest and range lands under its jurisdiction as well as to regulate their use. The service recently developed a 12-year program outlining forest protection and management practices, forest research, and trail construction needed during that period. It is notable that the largest cost items in this plan concern soil and water conservation, including water rehabilitation on 281,000 acres, and 1,500 miles of gullies, roads, and streams, 70 flood prevention projects, and soil surveys for 2.4 million acres.[23] The plan calls also for revegetation of 776,000 acres of range land and construction of 1,600 water developments. Significant amounts are included for improvement of wildlife habitat on 130 miles of streams, 1,800 acres of lakes, and other small watering facilities. Indicative of the rate of progress in recent years is the fact that in fiscal 1957 the Forest Service revegetated 81,740 acres of range land, planted 1,670 acres of trees, and constructed 2,935 range water developments.[24] The long-range plan also includes significant amounts for recreation facilities development in view of the anticipated increase in the number of visitors.

In some areas, because of the critical nature of the water supply, the Forest Service is permitted to close the forest to entry except by forestry and municipal officers. One such closure has occurred in

Arizona forests in a watershed near Flagstaff where the water shortages have become acute in recent years.

Because of shortages in the water supply for agricultural areas of the state, a great deal of interest has centered on experimental efforts to increase water yield. Foresters have been accused, both by interest groups and by members of the forestry profession itself, of having given too little attention to forest management in the interest of water supplies. Municipal officers, cattlemen, and agriculturists have all criticized the Forest Service for overprotection of trees with resulting loss of downstream water supply and crowding out of forage.[25] Foresters have charged that their own training has neglected forest management for water supply, and that as a result foresters have failed in living up to their creed of multiple use.[26]

THE BARR REPORT

These criticisms, along with experimental work done elsewhere, led to proposals for modifications of forestry practices in Arizona in order to increase the supply of water reaching the streams. In 1956 the legislature appropriated $25,000 to be used by the State Land Department in cooperation with the Agricultural Experiment Station personnel at the University of Arizona, and the Salt River Valley Water Users' Association for the purpose of examining these new methods of watershed management. The result was the publication of the Arizona Watershed Program's *Recovering Rainfall: More Water for Irrigation,* popularly known as the Barr Report after the program's chairman, George W. Barr of the University of Arizona.

After examining the Salt River watershed intensively, the members of the Watershed Program severely criticized the Forest Service for favoring timber in its management policies. It predicted that:

> The day will come when the forest lands of this watershed are managed primarily for water and recreation and in minor degree for timber. This assumes that water will increase in value as rapidly as timber increases in value. Adjustment in management practices is politically difficult, because the public has become indoctrinated with the idea that the forest by serving one purpose well also serves other purposes well. It is also difficult because the watershed "landowner" as represented by public agencies is inclined to favor those products that bring direct financial return. Timber production provides the largest source of revenue; grazing, some revenue; but recreation and water, none.[27]

Relying on experimental data derived from small watersheds, the Barr group suggested an immediate program of vegetation modification. They admitted, however, that "predictions of the effects of

watershed modification were based largely on generalization from
the results of plot experiments and on consideration of hydrologic
processes."[28] It is this reliance on plot data and knowledge of general
hydrologic processes that caused some disquietude among forest managers. One researcher, however, advocated the program for the very
purpose of stimulating research. He felt that only with dramatic
measures could public support be gained for further investigation,
and he was willing to run the risk of error. There had been no experiments of such a broad nature as suggested by the Barr Report that
could be used as the basis for prediction regarding increases in water
supply; neither had previous experiments been conducted in areas
having the conditions found in Arizona. But, said the committee:

> The need for additional water is so urgent and the chances of obtain-
> ing it from the watershed so promising, that an extensive coordinated
> program should be started as soon as possible to increase the yield of
> irrigation water from the watersheds of the state. The day has passed
> when water can be considered merely a by-product, unchangeable and
> inexhaustible, of a watershed devoted chiefly to growth of timber and
> forage. Water production is now the most important use of this state's
> land surface above 3,000 foot elevation. Timber, grazing, even recrea-
> tion in some areas must be subordinated to the demands of water produc-
> tion.[29]

In strict economic terms, the Barr Report charged, the present
forestry practices are shortsighted. Theoretically, clear-cutting of the
forests — which was definitely *not* proposed by the Barr Report —
would bring a greater return to the federal government if the money
derived therefrom were invested at 2 percent. For other reasons, of
course, such a measure was not advocated. However, with better management more water could be obtained which, if calculated to replace
high-cost pumped water, would bring a greatly increased return to the
federal government. With heavier cuts, it was asserted, long-term
timber growth would be stimulated also, owing to the removal of
dense and unproductive stands. Forage, which brings a return from
the grazer, would be increased. And with the dense stands of timber
and other growth removed, the forests would be more desirable
places for the recreationist.

In physical terms, the Barr Report advocated manipulation of
the forest and range lands by replacing plant life using relatively
large amounts of water with plant life using less water. By thinning
dense stands of timber, clear-cutting some areas, block-and-strip cutting in others, by removal of streamside vegetation, and elimination
of uneconomic growth — such as pinyon and juniper — it was expected that the state could save hundreds of thousands of acre-feet

of water annually. The Barr Report estimated that 285,000 acre-feet of water could be saved from the Salt River watershed alone by adopting the practices suggested by the study.

The largest amount of water would be derived from thinning the ponderosa pine, and converting the poorer sites entirely to grass. In the sub-alpine forests some areas would be converted to grass by strip-and-block cutting, reducing transpiration and evaporation losses. At the lower elevations, the pinyon and juniper would be removed and grass planted. Some treatment might be given the chaparral and desert shrub areas but the costs would be extremely high. In these areas the primary effort would be to remove trees and shrubs along the banks of streams.

Technical consultants on management of watersheds differed widely in their views on the conclusions of the Barr study. Some felt that no widespread treatment of vegetation should be undertaken until much more experimental data was available; others felt that a vegetation modification program should be initiated immediately. For example, A. R. Croft, forester in charge of watershed management of the Intermountain Region of the Forest Service, stated:

> Lack of guiding knowledge is one of the major handicaps at the present time in undertaking a program of vegetal manipulation to increase water yield. . . . Determining the effects of wholesale treatment of the vegetation over the watershed lands. . . would not be an easy matter, and it would certainly be unwise to treat large areas without first providing a means of determining, on small control areas, the effects on stream flow of any treatments that may be given.[30]

At the other extreme H. G. Wilm, one of the pioneers in the field of watershed manipulation in the Forest Service, concluded that

> substantial increases in water yield may be obtained through the intensive management of the various watershed vegetation types on the Salt River Watershed, for the special purpose of water production. If soundly executed and maintained through wise range and timber management practices, these operations should also increase range values and provide adequate protection of watershed stability.[31]

The supporters of the Barr Report argued that all other values would not be subordinated to the production of water. Reckoning with the expected opposition from the recreationists, the report maintained:

> Watershed modifications should be planned with the intention of preserving and improving recreational values. At the same time, it is through the *use* of resources that man satisfies his wants. Locking up and preserving the watershed as it is today in the name of recreational needs will not best serve the interest of the developing commonwealth, nor even the majority of the recreation users.[32]

In fact, the report continued, improvements in the recreational capacity of the forests might result; the pine forest, for example, would be "open and park-like, a more attractive forest than the brush-choked stands of today."[33] Obviously the stockmen would stand to benefit from replacement of trees with forage. With increased cutting of timber the lumber industry would be stimulated for a number of years although it would later decline. The resulting slack, it was argued, might be taken up elsewhere in areas where water production was not so important, or by the increasing importance of the tourist industry. However, by 1959 the vice-chairman of the Arizona Water Resources Committee stated, "At some point in this application of priorities on water use, the principle of multiple use of watershed lands in the public domain will die as a matter of necessary policy in the public interest."[34]

The Barr Report had a decided impact on the viewpoints of public and private groups both within and outside of Arizona. In the state, it was the subject of numerous editorials, a central topic of discussion of conventions of groups interested in the watersheds, an arguing point for scientists, and a source of some discomfort for the landholding agencies. The Forest Service had to justify its existing management practices under frontal assaults made by the cattle growers and the irrigationists. As a practical matter, the Barr Report stimulated research as indicated in the chapter on U. S. Department of Agriculture research.

The Forest Service took a cautious approach to the suggestions in the Barr Report. It was the view of the service that too little evidence was available to substantiate the claims made by the proponents of vegetative manipulation, or to justify the undertaking of the radical departures in forestry management suggested by the report. The regional forester, in speaking before the Arizona Cattle Growers Association, was unwilling to dismiss the timber value of the forest since, he asserted, values change from time to time. Furthermore, it was his view that removal of vegetation, particularly from streambanks, would be detrimental to recreational values. He advocated an increase in research activity to determine the validity of the claims made in the Barr Report. Only the sportsmen came to the defense of the Forest Service in an organized way. The Arizona Water Resources Committee and the Water Resources Division of the State Land Department applied public presure to gain more rapid acceptance of their views which more or less parallel those found in the Barr Report.

Since the publication of the report, a remarkable amount of effort has been devoted by federal and state agencies to determine the validity of the claims of the report. A large number of significant projects have been undertaken on Forest Service land in various parts of the state, representing a broad range of land and plant conditions, to determine whether the manipulation of the plant cover will result in significant increases in usable water, and whether the manipulation can be accomplished without permanent damage to the water producing resource. Moreover, there has developed a general recognition among those who criticized the service that no widespread manipulation practices can be undertaken without careful evaluation of experimental results from tests conducted on the watersheds. Such tests require many years for their completion. Some resource users are naturally impatient with the slow processes of science, but most are willing to take a careful approach. The exhortations to action found in the Barr Report have been replaced with careful assessments of experimental data and limited actions based on them. It is perhaps noteworthy that the annual proceedings of the meetings of the Arizona Watershed Program are devoted almost exclusively to reports of technical work on the watersheds.

BUREAU OF LAND MANAGEMENT

The Bureau of Land Management rivals the Forest Service in the extent of the land area of Arizona which is subject to its administration. As of 1960 the Bureau administered 13,096,108 acres of public land.[35] While the Forest Service administers what is generally conceded to be the most valuable public land in the state, the Bureau of Land Management is responsible for some of the poorest and least productive land in a state where a good share of its land area tends to fit the description of the public domain as "the driest, rockiest, steepest, least accessible and generally least desirable tracts in each locality."[36]

Much of Arizona has not been suitable for any economic pursuit other than livestock raising. This land, lying generally in the western section of the state, lacks sufficient water to provide for agriculture of any kind. Precipitation is generally less than 10 inches annually, and may be as little as 3 to 4 inches in the southwestern section. Lacking the essential element of water, this land has remained unclaimed by the public, although at least in theory still subject to public entry.

Owing to the lack of precipitation and high rates of evaporation, these lands are, for the most part, unimportant in the production of water for use in any place other than on the land where the precipitation falls. Floods at times occur as the result of torrential downpours, but these waters are generally lost through evapotranspiration. These lands are highly subject to erosion as a consequence of the thin soil and sparseness of vegetation. For this reason, it is argued by some competent observers that, instead of trying to manage these lands for increased water, "it is logical to manage them for the control of summer flood peaks and the reduction of sediment, and for other products which will not jeopardize watershed potentials," and "to supply the needs for a protective plant cover to control erosion."[37]

Mismanagement in the past and the inability presently to apply available knowledge to management practices has resulted in erosion of a very serious nature. Cutting and gullying continue on many watersheds, destroying valleys known in the past to have provided excellent forage for large numbers of livestock. Some, such as the San Pedro and the Santa Cruz valleys, are considered virtually beyond restoration. One management official estimated that 500,000 acres of the best grassland of the state have been destroyed during the past half century through the elimination of the grass cover due to erosion.[38]

Until the passage of the Taylor Grazing Act in 1934 this range land was unregulated and unmanaged, and had been the object of intense competition among those who sought to exploit its meager resources. By the time of the passage of the act, these lands had become "overgrazed, wind-eroded expanses, interspersed with rocky peaks and barren slopes."[39] The last quarter of the nineteenth century was a boom period in cattle and sheep raising in Arizona Territory. Cattlemen, who had apparently learned little from their experience in Texas, brought their herds to Arizona under the influence of land promoters who described Arizona as a land with " 'unsurpassed facilities for grazing,' " where " 'expense was reduced to a minimum,' " and where " 'not an acre of grassland . . . is unavailable for raising stock because of lack of water.' "[40] The number of stock increased enormously, riding on a wave of high prices. Competing for the grasslands under the open range policy of the federal government, the cattlemen gained control of watering places and then claimed the right to utilize the surrounding territory to feed their stock. Frequently they fenced off sections of the range and considered it "theirs."[41]

Unrestricted grazing and competition for the range led to rapid depletion of the range and its resources. Lush and productive valleys were virtually destroyed for grazing purposes. The water that had previously spread out on the flood plain began to concentrate in channels and cut deeper and deeper into the valley floor.

The formation of the arroyos in southern Arizona valleys is spread

over considerable time, and details of the process have seldom been recorded. But the change from aggradation and the building of flood plains
to channel-trenching can be placed in the 1880's in most of the important
valleys, though many of the tributaries were not affected until the nine-
ties.[42]

In many places the unpalatable grasses and the noxious plants were
the only growth to remain after excessive grazing. These were then
free to spread and choke out and prevent the return of the nutritious
grasses that previously had grown in abundance.

This condition, begun in the latter part of the nineteenth century,
continued down through the period when the Grazing Service was
established in the Department of the Interior in 1934. During much
of this period there was argument among the stockmen over the form
of management of these lands. Some maintained that only those
whose livelihood depended on these lands should exercise control over
them, while others argued for public management and control. By
the time of the passage of the Taylor Grazing Act, however, this argument had been resolved in favor of federal management although
there remained, and still remain, those who demand private management for these lands.[43] Movements for disposal of public lands either
to the states or to private individuals for nominal fees still meet with
support among stockraisers.[44] However, the Bureau of Land Management, in contrast to the Forest Service, has not been the subject of
severe criticism by the stockmen.

In gaining control of the deteriorated range, the Grazing Service faced the enormous job of appraising the lands to determine the
carrying capacity of the land in order to ascertain the numbers of stock
to be allowed on the range. After such surveys were made, grazing
permits were issued to stockmen under conditions stipulated by the
service and upon the payment of grazing fees. In many instances
these initial surveys were inadequate owing to lack of funds and personnel. It became increasingly evident that the original assessments
of carrying capacity were much too high. It was necessary, therefore,
to reduce stock numbers in many areas. Stockmen have been reluctant to do this, but considerable headway has been made.

The number of animal-unit months of grazing permitted on
BLM lands has varied widely, from as high as 925,460 AUM's in 1952
to only 570,318 in 1957. These variations have resulted in large part
from the rapid changes in the conditions of the range. But range
managers continue to report instances of trespass operations, even
among stockraisers who serve on BLM advisory boards. However, in
fiscal 1960 there were only 22 new trespass cases, most of them apparently resulting from failure to file applications in the allotted time.

One of the major accomplishments of the BLM has been the settlement of disputes over rights to the range. The settlement of grazing privileges generally followed the practice of conferring the privilege of using the public domain for grazing on the owner of the water rights in an area and the owner of a ranch who had used the land before the establishment of federal regulation. With the adjudication process completed the state director stated in 1956 that the "principal work to be done will involve grazing capacity rechecks in some districts for the issuance of term permits and preparation of management plans for each allotment."[45] By 1961 virtually all range adjudications had been completed and individual allotments established. Range surveys of one type or another have been completed on all district lands, and further rechecks have been undertaken in cooperation with the Soil Conservation Service.[46] The problem is to obtain surveys that are sufficiently refined to be useful in the regulation of land utilization.

The management of the public domain in Arizona is complicated by the fact that land ownership is of the checkerboard variety. Besides the grants made to the state by the federal government, grants totalling 7,789,798 acres were made to railroads in Arizona.[47] Other private operators have acquired tracts of land throughout the grazing lands as bases for livestock operations. The complexity of ownership of these lands makes management programs difficult since all agencies are not equally concerned or advised regarding conservation practices. The Bureau of Land Management is not authorized to do conservation work on state or private lands even though these are interspersed throughout the public domain. The federal grazing lands may, however, be put in soil conservation districts. The Soil Conservation Service and the BLM are now cooperating in drawing up management plans for many of these lands. The management problem is complicated even further by small tracts — sometimes no larger than a section — of land which are in public domain but are so isolated that they cannot be included in grazing districts. These lands total over two million acres, nearly all of which are leased to private stockraisers.[48]

The BLM is responsible for classifying the public domain for its "highest" use. The principal demands on this land, other than grazing, are for homesteading, mining, or business purposes. During fiscal 1956 the bureau disposed of 601,861 acres in various ways.[49] Pressure is brought to bear on the bureau to dispose of the lands for purposes that raise serious questions about the effect of such disposition on the public interest. This has been particularly true regarding ag-

—Courtesy Robert R. Humphrey

The grassland which once covered this Cochise County area along the U.S.-Mexican boundary had, by 1895 (above), completely disappeared. In 1956 (below), the condition of the range had improved to the point where it could be called "fair." Grazing limitations, more rain, and decreased erosion were the main factors causing this improvement.

—Robert R. Humphrey

ricultural entry. The state director stated in 1956, "If no control were exercised over the granting of such agricultural entries, attempts to irrigate the additional lands might exhaust the water supply and destroy the existing agricultural economy."[50] The Solicitor of the Department of the Interior has ruled that no valid entries can be made under the Desert Land Act since that act requires a demonstration of a right to appropriate water.[51] Since the Arizona Supreme Court in *Bristor* v. *Cheatham* declared that ground water — the only available source of water in the desert — is not subject to appropriation, no such entries have been permitted. Only limited interest is found in mineral entry on the public domain in Arizona, but such claims do cause concern to the land manager. Location work is often done with a bulldozer, raising havoc with the land surface and the forage resource.[52]

Types of conservation work engaged in by the BLM include revegetation of overgrazed ranges, control of eroded surfaces, brush and weed control, water diversion and spreading, development of stock water and small storage reservoirs, and fencing to ensure better forage utilization.[53] While the officials of the bureau in Arizona recognize the necessity of fully implementing the accepted and desirable conservation practices, they find that their capacity to do so is restricted through lack of funds. Congress has not yet comprehended the increased revenues which would come from greater investment in these lands.[54] Much work has been done but the problem lies in the enormity of the task of restoration.

Activity has been concentrated largely on the construction of costly soil and moisture projects where the benefits are too general and long-run to be reimbursable, such as construction of detention dams and dikes for the reduction of flood peaks and prevention of erosion. Money put up by private operators on public land is matched with federal government funds for projects undertaken by the private operators, with maintenance underwritten by these operators.[55]

The BLM has divided the land it administers in Arizona into four major sub-basins which are considered work units for its long-range conservation plans. It has been calculated that the program, which would run for 20 years, will cost as much as $19,240,480, with the BLM bearing nearly two-thirds of the cost.[56] This program involves the treatment and intensive management of virtually all land owned by the United States or under its jurisdiction and subject to administration by the BLM. The success of such a program depends, of course, in large part on the willingness of Congress to appropriate funds.

Some doubt the possibility of restoring Arizona's range lands.

The BLM has undertaken this effort but warns that the destruction cannot be stopped either easily or cheaply. As one range management specialist put it, "We can't just fence the area off and keep stock out for 20 years. This will not bring the [valley] bottoms back. We must first get control of the flood waters."[57] In order to control the flood waters it is necessary to establish detention and control structures — dams, spillways, and waterspreading devices. With accurate analyses of hydrologic data to guide construction of the installations and good management practices to maintain them, it is felt that these depleted ranges can be restored. The controlled water moves slowly over the bottom of valleys, depositing silt instead of causing additional erosion through cutting, and storing moisture for grass production. "Once such conditions are developed the native grass comes in, and the whole grazing pattern changes back to that existing before the destruction of the natural grass flood bottoms." By this method "highly productive grassland bottoms have been developed from eroding waste areas."[58]

The approach being used in the rehabilitation of these ranges is the community watershed plan. Since there is not sufficient money to devise detailed plans for entire sub-basins, an attack on the erosion problem is being based on the smaller watersheds which feed the major streams. An example of this is the Railroad Wash Community Watershed above Safford Valley and a tributary of the Gila River.[59] One-third of the lower drainage area was denuded by erosion caused by railroad construction, excessive cattle numbers, and floods. This watershed was described in 1944 in the following way:

> Intensified sheet erosion and the more spectacular gullying and arroyo-cutting have followed the removal of vegetational cover through heavy utilization by livestock and the recurrent effects of drought. Notable in showing the effect of overuse by livestock is the sheet-eroded and entrenched San Simon Wash — now nearly barren of palatable grasses and forage plants but once characterized by dense growths of sacaton and tobosa in its bottom lands.[60]

To rectify this situation, the BLM plan calls for the erection of detention and diversion structures, brush eradication, weed control, and fencing. The railroad tracks will be protected from washouts, the highways from flooding, and the lands from flooding and siltation. The BLM will pay the major portion of the construction costs, with others who will benefit paying one-fifth. Maintenance will be the primary responsibility of the local interests.

These measures to spread or to slow down the flow of the flood waters are frequently opposed by downstream irrigation interests. This has been particularly true regarding the work of the BLM on the

San Simon Wash. These interests contend that the conservation measures deny them water otherwise available. The BLM maintains, however, that this flood water would never reach the lower valleys, that it is lost to evaporation, transpiration, and siltation. The only reasonable use, the bureau asserts, is on the watershed to prevent erosion, restore the forage, and protect against destructive floods. In the opinion of the state soil and water conservation officer of the BLM, silt is the major "thief" of water. Those who say that they will be glad to take the silt if they can get the water are wrong, he asserts, since the silt deposited in the stream channel causes to be lost "two acre feet of water . . . for each acre foot of silt deposited."[61] For this reason, he contends, all sources of erosion on the forest and range lands should be checked.

In common with the Forest Service range lands, the public domain is also faced with the invasion of noxious or unpalatable weeds and plants. These plants use up the soil moisture over a broad surrounding area, replace the more valuable forage, and contribute to the acceleration of erosion.[62] Much effort is being devoted to removal of this growth, but it is doubtful that eradication is keeping ahead of the spread. Over large grazing areas

> many of the grasses have been thinned out or stands have been entirely killed. In many places low-value shrubs such as creosote bush, shadscale and cacti have come in as the grasses have gone out. This has changed the aspect from one of grassland, or grassland with scattered shrubs to one of shrubs with or without grass.[63]

Various techniques have been used to eradicate these shrubs — burning, chaining and cabling, diesel oil, and chemicals. There is considerable disagreement concerning the causes of the spread of these shrubs, with Robert R. Humphrey attributing the spread to the control of fires and others finding other factors more important.[64] The disagreement concerning causes results in inability to agree completely on methods of control.

BLM officers are not convinced about the advisability of extensive removals of pinyon and juniper which constitute the most important woodland species on the public domain in Arizona. These trees are in some demand for posts and poles. These officials are doubtful of the effects of removals of this kind of growth on water production. They fear that the net results of broad removal programs might be a reduction in the recovery of water since the danger of erosion is so great and the replacement of woody species so difficult. The BLM requires reseeding of lands from which the natural growth has been removed.

In the management of its land, the BLM cooperates with other federal and state agencies. It works with the Soil Conservation Service in ranch planning where there is private land and public domain interspersed. It manages grazing land on the Lake Mead Recreational Area for the National Park Service, and withdrawn lands for the Bureau of Reclamation and the Federal Power Commission. As the result of recent legislation, the BLM is responsible for managing the surface of lands entered for mineral purposes.[65] It assists in the management of the Fish and Wildlife Service game ranges and cooperates with the Forest Service in fire control. The one area of federal-state cooperation concerns management of wildlife. The BLM and the Arizona Game and Fish Commission have agreed to study the effects of range rehabilitation programs on wildlife species.[66] The sizable increase in the number of both big and small game has increased the need for joint attention to the problems of access to the public lands, particularly for the sportsmen.[67]

The stockmen in Arizona are well organized and are a potent force in range management. Operating through advisory boards, and organized in every county and often allied with the mining interests, the livestock industry is in a formidable position to influence range operations. (The stockmen have long sought similar influential positions through advisory board arrangements with the Forest Service but without success.) There are indications, however, that the BLM is beginning to step up its enforcement activities and to make more accurate surveys of range capacities.[68] Such efforts may result in the same criticisms which the stockmen now direct at the Forest Service, criticism which is now almost entirely lacking toward the Bureau of Land Management. The present Secretary of the Interior, Stewart Udall, has indicated an affirmative interest in grazing-district management, particularly along the lines of multiple-purpose management.

BUREAU OF INDIAN AFFAIRS

The Bureau of Indian Affairs of the Department of the Interior is responsible for the largest total land area in Arizona. This land, for the most part held in trust by the federal government for the Indians, constitutes 27-29 percent of the state's area. The quality of this land varies widely, ranging from forest land of high value, excellent grazing land, and land suitable to high quality agricultural production, to some of the poorest, most arid, eroded land to be found in the United States. In general, the Indians live on lands of very poor quality and barely eke out a miserable existence.

Arizona has the largest Indian population in the United States, numbering 83,387 persons in 1960.[69] The majority live on 19 separate reservations, although many have moved off the reservation completely, or live on it for only part of the year. The largest reservation, the Navajo, has in Arizona alone over 48,000 inhabitants, while the Yavapai has only a few families. Although reservation Indians receive a few services from the state, they are almost entirely the charges of the federal government. Any attempt to impose the burden of their welfare on the state meets with determined resistance not only from the state but from the majority of the Indians. The state recognizes the enormity of the task of caring for and raising the living standards of these people, while the Indians fear the loss of their lands should the reservations be transferred to state control.

The problems of achieving conservative use of the land in order to protect the soil and water resources is exacerbated on Indian lands because of the cultural differences of the Indians. They have not easily adapted themselves to modern techniques. Their systems of landholding often make management extremely difficult. They are at times motivated differently than the white man. And those who seek to teach them frequently fail to understand or respect these cultural differences and therefore fail to transmit their ideas and methods of conservation.

The fundamental problem on most of the Indian lands is an excessive number of people on an extremely limited resource base. The only long-range solution to the problem is to find employment off the reservations unless there is some unexpected development of resources where they are now living. On the Navajo Reservation this is now a distinct possibility for the majority of the population, since natural gas fields have already come into commercial production, and other valuable mineral deposits have also been discovered. Other reservations have reclamation possibilities but for the most part there are few resources to maintain the expanding population. The Bureau of Indian Affairs has made efforts to relocate Indians off the reservations with some success, but it has failed to reduce in a significant way the pressure on the limited resource base.

Because of the seriousness of the depletion of Indian lands and the extent of the area, the soil and water conservation program of the Branch of Land Operations of the Bureau of Indian Affairs is the largest of any public agency in the state. In fiscal 1957, $3,521,085 was budgeted for land management operations alone, which included forest and range management, development of stock water supplies, agricultural extension, soil conservation, operation and maintenance

Most of Arizona's rivers and streams are intermittent, running only after a storm or during the spring thaw. Many of them suddenly disappear underground, only to emerge again a few miles downstream. Especially in the high plateau country (above) and in the desert areas (below), runoff is unpredictable and difficult to capture. It is not all wasted, however, for much of it finds its way into the underground table.

—Chuck Abbott

—Chuck Abbott

of irrigation systems, irrigation-project construction, and power operations.[70]

A projected 20-year program envisions total expenditures of $107,674,353. The Indians themselves will pay more than one half the cost while the remainder will be provided by the BIA. Farm and range plans have been put into effect on nearly 2,500,000 acres of farm and ranch land and satisfactory compliance has been obtained. However, 15 times that number of acres have not received any planning whatsoever.

A discussion of the conservation efforts on each reservation is impossible here, but a survey of several important reservations will serve to illustrate existing conditions and what can be done to improve them.[71] The Papago Reservation — which actually consists of the large Papago and the much smaller San Xavier and Gila Bend reservations — in southern Arizona is an extreme example of a reservation with a dearth of resources, particularly water and forage, in relation to a large and growing population. There are no live streams and no natural waterholes. Moreover the lands are suffering from serious erosion. In 1954 the erosion damage on the total 2,855,921 acres was classified as follows: 197,814, slight; 1,230,622, moderate; 1,190,000, severe, and 237,405, critical.[72]

It is estimated that of approximately 11,000 Papagos some 5,000 live most of the time on the main Papago Reservation. Only 14 families are engaged in permanent agriculture on the reservation's single irrigation project of 1,640 acres. Other farms, called flash-flood farms, are utilized during favorable years, but can be regarded only as marginal farms, providing only a part of the subsistence of any one family.[73] The most productive use of water, except for isolated irrigation development, is for watering stock and the production of forage. The land is much more suitable for grazing than for agriculture, and, through the development of waterspreading structures and the eradication of brush, the amount of forage could be increased considerably. It is believed that irrigation projects could be expanded to cultivate no more than 14,200 acres under optimum conditions. In 1960 the Papagos added three million acres to soil conservation districts as a step toward improving their soil and water conditions, but even optimum conditions are not very favorable for farming since the evaporation rate is very high and erosion causes the irrigation structures to fill quickly with silt.

Even with complete development and maximum control of the range and farming resources, the reservation cannot hope to provide a standard of living comparable with the average Arizona white farm family; it is estimated that at least one-third of the Papagos will have to find employment off the reservation. The costs for development and protection of these lands will run to more than $20 million, with the BIA paying $9,628,700 and the Papagos paying $10,653,-150, chiefly for maintenance.[74]

In contrast to the Papago Reservation, the Colorado River Indian Reservation is described as "one of the wealthiest, for its size, of any Indian reservation in the country."[75] It is located on the banks of the Colorado River near Parker, Arizona, and consists of 2∠5,914 acres of land, nearly half of which is suitable for cultivation. The Indians now have under cultivation more than 42,000 acres, and have planned the development of an added 65,000 acres. All such development has been suspended, however, awaiting the decision of the United States Supreme Court in the suit between Arizona and California over the allocation of the waters of the Colorado. Like other Colorado River bottomland soils, these lands are very productive. They are presently devoted primarily to cotton, alfalfa, milo maize, and grain, but there is effort to establish a more diversified economy.[76]

Conservation activity has been greatly expanded in recent years on this reservation. By 1954, 153,674 acres had been analyzed by soil scientists. An adequate distribution system supplied water to 37,334 acres. Although few of the ditches were lined, future plans called for their lining under the Agricultural Conservation Program. Perhaps of greatest importance is the recognition by the Indians of the need for soil conservation work. In 1954 they formed the Colorado River Indian Soil Conservation Association in cooperation with many of the business interests near and in Parker.[77] In 1957 this association became a soil conservation district.[78] According to BIA officials, the Colorado River Indians are taking full advantage of the BIA's soil and moisture and agricultural extension programs. The 20-year conservation program is expected to cost the BIA $884,400, while the Indians will spend over $23 million.

Another potentially prosperous Indian settlement is the Gila River Indian Reservation in the heart of Arizona's agricultural community south of Phoenix. The Gila River runs through the reservation and provides irrigation agriculture through the San Carlos Project. The project was designed to supply water from Coolidge Dam for 50,000 acres of Indian land and an equal amount of land owned by white men.[79]

Although this land is similar to agricultural land which has been successfully developed in the same area, the Gila River Indians have never prospered agriculturally. The lack of success has been related only in part to poor land usage. One of the problems has been that the land for which San Carlos Project water was to be allocated was never precisely designated. It has been stated that "no Indian can be secure because any 'irrigable allotment' may be at any time excluded from the San Carlos Project water supply by the same arbitrary means through which the land was included. . . ."[80] The program of consolidation of lands in order to best utilize the distribution system of the project was in some areas a complete failure and in others only a qualified success.

A second problem that continues to plague the Indians is that of inheritance. Much of the land on the reservation has been individually allotted, and under the Pima inheritance system the property of a deceased person is divided among the direct descendants. This has resulted in the division of land into such small parcels that none of the landowners holding these small parcels can make a living. As a consequence, the land usually remains idle. While the solution to this problem might be a system of leasing, such a system when tried has not been very successful.

A third problem results from the inadequate preparation of the Indians for irrigation agriculture of the modern type. For many centuries the Pimas were prosperous irrigation farmers, but their water supply disappeared in the 1870's and 1880's as the white settlers upstream on the Gila diverted the water for their own use, and the Pimas were forced to seek other means of subsistence. By the time plans for the San Carlos Project were drawn up the Pimas had lost interest in and knowledge of farming. Furthermore, the Indians were generally not consulted by Indian Service officials concerning the plans, nor the crops they were going to plant. In many instances the agents failed to plan adequately by means of soil surveys what land would most benefit from the application of water. The Indians were given insufficient credit. They were not fully educated concerning the intricacies of the irrigation system and the techniques of irrigation agriculture.[81] Even today "they are inefficient in the use of their water supply, requiring much more water for the same purposes than surrounding non-Indian farmers."[82]

A final factor, over which no one had any control, was the drought which set in about the time of completion of Coolidge Dam in 1929. The dam was designed to store 1,200,000 acre-feet of water but has never been completely filled and at times has been completely dry. (Writing in the *Arizona Republic* for August 29, 1956, columnist Don Dedera said, "At last look, Coolidge Dam held so little water, its storage figures were printed with a minus sign.") With water supplies from the Gila River inadequate, the Indians, as well as the white farmers, have had to depend on wells. The Indians have embarked on a program of leasing some of their land to white men who are developing the land by drilling wells. This has been opposed by the state of Arizona because of the depletion of the underground water. However, the Pima Tribal Council has negotiated to lease out the tribal farm which consists of 10,000 unallotted acres. Inasmuch as the farm is right next to the river, depletion of the water table should not be serious.

Located on the watersheds east of the central valley of Arizona are the San Carlos Apache and Fort Apache reservations. These offer relatively few agricultural opportunities but they contain some of the finest timber and grazing land in the state.

In spite of the small acreage and high costs of development of agricultural land on the San Carlos Reservation, the Stanford Research Institute considers the cultivation of this land extremely important to the reservation economy.[83] If 3,400 acres — the minimum estimate in 1941 of what could be farmed on the reservation — were irrigated, they would provide support for 55 families; 6,000 acres would support 100 families. Such development would relieve the heavy burden which the grazing economy must bear. The Indians, however, are not yet educated in soil and water management. Arroyos have caused extreme damage when they have gone uncontrolled. Much of the land suitable for cultivation has grown up to bushes and perennial weeds. Pipes have cracked and the wastage of water is prohibitive.[84] Erosion has taken a heavy toll as the following estimates indicate: critical acreage, 400,000; severe, 938,919; moderate, 300,-000; and slight; 5,000.[83] It is doubtful under the present crop prices that such land will be developed for irrigation agriculture.

The chief use of the San Carlos Indian lands is for grazing. The reservation contains 900,000 acres of largely undeveloped timber lands. This timber constitutes an important economic resource for the future, but without adequate protective measures it is deteriorating. The Stanford Research Institute suggested in 1954 the removal of around 100 million board feet within a ten-year period — as heavy a cut as possible — "in order to clean up the forest and reduce the heavy beetle infestation."[85] The Indians are perhaps the leading advocates of prescribed burning and use this technique extensively. They have also been leaders in eradicating the undesirable growth such as pinyon and juniper.

Owing to its economic importance the Indians find it necessary to concentrate their efforts on improving their range resources. Some

450 of the 775 self-supporting families on the reservation make their principal living from farming and livestock raising, and 90 percent of the agricultural production comes from livestock operations. They have had serious problems in managing their range owing to over-stocking and failure to move their herds to make full utilization of the entire range. Fencing has been inadequate and the Indians have had only a minimum of education in conservation methods. There has also been inadequate water development on most of the ranges. The BIA has exerted considerable effort along these lines and has succeeded in reducing the cattle numbers to a point near carrying capacity, and improving range usage.

The Navajo and Hopi Indian reservations lie in the northeastern part of the state, with the Hopi Reservation being entirely surrounded by the Navajo, which is the largest of all the reservations. Together they contain 16,881 square miles. This region is described as "one of the most isolated and thinly settled sections of the Southwestern United States," with the lands being "among the poorest in terms of resources in the State."[86] In discussing the agricultural capabilities, Kluckhohn and Leighton stated that

> none of these situations is really favorable to agricultural production save where irrigation water is available. . . ; in general flowing water is rare. Rainfall is scanty in most parts of the Reservation. . . . High temperatures during summer and sub-zero weather during winter, high winds, frequent sand storms, and high evaporation rates are characteristic.[87]

These lands are generally of relatively poor quality for grazing also, suffering from the results of overgrazing.

> . . . the wasting of grasslands by gully-cutting has been catastrophic. The loss of much of the vegetative cover has likewise resulted in a great wast-age of the scanty rainfall, for there are not enough plant roots to hold it, and it runs rapidly down the slopes, carrying valuable topsoil with it. . . . the total productivity of the Navajo lands has probably been reduced by at least half since 1868.[88]

The Secretary of the Interior reported that "erosion has reached a stage where it can be stopped by nothing short of an all-out effort by both the Navajos and the Federal Government."[89] The condition of the land is indicated by the fact that it delivers only 2.5 percent of the water which flows into Lake Mead, but supplies no less than 37.5 percent of the silt.[90]

As a result of the public awakening in the late 1940's concerning the dire plight of the Navajos, Congress in 1950 authorized the Navajo-Hopi Long Range Rehabilitation Program. This program authorized the expenditure of $88,570,000 over a 10-year period, of which $10 million was devoted to soil and moisture conservation work.[91] This work has included "construction of dikes, jetties, diversion dams, water spreaders, ponds and charcos; land levelling, contour plowing, tree planting, brush eradication, seeding, soil and other surveys, land classification, [and] educational activities."[92] The federal appropriations were supplemented from funds contributed by the tribal councils, the Agricultural Conservation and Stabilization Service, and by labor donated by the Indians.

In 1958 irrigated acreage totaled 31,000 acres and dry-farm land totaled 36,219 acres. It was expected that irrigated acreage on the Navajo and Hopi reservations would reach 45,886 acres. Present estimates indicate that an over-all total of 175,000 acres might be farmed on the Navajo Reservation if all potential agricultural land were developed.[93] Ambitious plans for increasing irrigated agriculture are becoming reality with the completion of the Hogback Project on the San Juan River and the San Juan Project. The latter will provide water for the subjugation of 110,000 acres of land near Shiprock, New Mexico, and will create 1,200 farm units ranging from 90 to 105 acres each.

The grazing ranges constitute the most important economic resource now and for the foreseeable future. Much needs to be done, however, to bring these lands to a more productive level. In order to get a better distribution of stock, a range-water development program was begun which was to provide 246 new wells during the fiscal years 1950 through 1956. In order to increase the forage supply the Indians and the BIA technicians have begun a sagebrush-clearing operation which promises to increase forage production by 160 pounds per acre. The BIA is actively interested in range management to assure its use on a sustained yield basis. This phase of range protection is extremely difficult since the Indians consider the number of stock they own a sign of prestige and thus resist reductions in numbers.

In spite of the efforts made by the Indians and the Bureau of Indian Affairs to develop Indian resources, there will still be an excessive number of people on the reservations. Relocation of these Indians has been attempted with some success, but there is some fear, based on past experiences, that irrigation development will cause some of those relocated to return to the reservation. One attempt to relocate some Hopis and Navajos on the Colorado River Indian Reservation has had extremely limited success. Only 44 of the 113 Navajo who were relocated remained in 1958. The Navajos failed to adjust to the climate and new techniques of agriculture.[93]

The disastrous effects of mismanagement and the basic inadequacy of the lands for the number of people living on those lands undoubtedly are the most serious conditions facing those who strive to manage these lands according to conservation standards. Much effort is being made along these lines, but it is doubtful whether much can be accomplished unless the Indian himself is motivated to adopt the proper practices. Agricultural extension workers have tried to use the same techniques on Indians that they used on white farmers and they have often failed utterly, largely due to cultural differences and to lack of comprehension on the part of the Indians of what the extension workers were trying to accomplish.

The problems associated with erosion remain critical in Arizona in spite of the efforts of many of the land-management agencies. Great progress has been made in controlling the factors which led to erosion, such as overgrazing or improper methods of cultivation. But such controls, in themselves, will not reverse the downward trend. Much greater effort must be exerted in restoring the land by positive conservation practices — revegetation, reforestation, flood-control structures, and waterspreading. Such practices and structures are expensive but pay long-range dividends in restoring valleys, preventing ruinous floods, and protecting major storage structures against siltation.

CHAPTER 11

Soil and Water Conservation

The cost of farming in Arizona is high. The limited water supply and the cost of bringing the water to the land require that the utmost be done to conserve the supply. For this reason soil and water conservation agencies play an important role in Arizona agriculture, making it among the most efficient farming systems in the nation.

The enormity of the task imposed on the soil and water conservation agencies is indicated by a report of the regional office of the Soil Conservation Service in 1947:

> Soil and water wastage is the No. 1 problem of Arizona agriculture.
>
> Reconnaissance erosion surveys . . . show that over 90 percent of Arizona's total land area of almost 72½ million acres is affected by soil erosion varying from slight to very severe.
>
> So far as its present usefulness is concerned, a large part of Arizona's 775,000 acres of irrigated land is being destroyed slowly but surely by causes which are often far removed from the lands which are in jeopardy. Deterioration of the headwaters areas miles away, because of overgrazing, incorrect farming methods, or poor woodland management, robs the farm lands of irrigation water upon which they are dependent for crop production.[1]

The possibilities inherent in the technical approach to soil and water conservation are indicated by estimates that crop production could be doubled if all the technical information now available were actually made part of agricultural practice. To put it another way, the present head of the Arizona State Office of the Soil Conservation Service asserted that the application of present knowledge regarding soil and water conservation could result in the saving of 100 percent more water than is now utilized by plants, an amount "more than equal that expected from all the storage reservoirs now being planned."[2]

Both the state and the federal government take part in soil conservation activities. The state's role, like its role in many other fields, is limited to encouragement to the farmer to adopt the programs made available by the federal government, or to general agricultural education from the extension service, of which soil and water conservation is only a part.

AGRICULTURAL EXTENSION SERVICE

Extension work dates back to 1915 when Arizona accepted the terms of the Smith-Lever Act, by which the federal government made grants of money to the states for professional assistance to the farmer.[3] Under the terms of the act, the state was required to make appropriations matching the federal funds. The Board of Regents of the University of Arizona was instructed to accept the monies and to organize and conduct agricultural extension work in connection with the College of Agriculture at the university.[4] From very modest beginnings in which there were only four staff technicians, the Extension Service expanded rapidly until during the 1930's the service was able to place men in nearly every county as county agents as well as to maintain home demonstration agents and staff specialists at the university.[5]

In 1921 the legislature authorized the creation of county farm bureaus which were to be "the official body within the county for carrying on extension work in agriculture and home economics within the county in cooperation with the University."[6] Each farm bureau draws up a plan of work for the ensuing year and a budget which is submitted to the county Board of Supervisors for inclusion on the tax rolls.[7] The university, within limits, matches the funds appropriated by the counties.[8]

The Arizona statutes state that "cooperative agricultural extension work shall consist of giving practical demonstrations in agriculture and home economics, and imparting information on those subjects through field demonstrations, publications and otherwise."[9] The broad nature of extension work, dealing with all phases of agriculture, is clearly beyond the scope of this study. The Extension Service, however, as part of its practical demonstration work, does educate the farmer in the management of his water supply, both for the purpose of conservation and for the purpose of increasing the return to the farmer on his investment.

Interest in improving irrigation practices has existed since this type of agriculture was introduced in the region. Extension Service participation in irrigation education began only in 1923, however, when the farmers became concerned about the increasing depths to water. Later, additional instruction was given in such matters as reclaiming alkali lands and protecting the soil against floods by terracing and other forms of control.[10] Since that time the work has expanded greatly to include proper land use, contour and strip cropping, terracing, grassing waterways, water supply, storage and distribution, drainage, irrigation, and other practices more or less related to water management and conservation.[11] In the year 1955 the county agents or specialists assisted 828 farmers in the categories of water supply, storage, and distribution, drainage, and irrigation alone in connection with soil conservation. In accomplishing this the Extension Service had the assistance of 111 volunteer local leaders who contributed their services.[12]

It should be pointed out that these figures represent only a small part of the work of the Extension Service. Less than 5 percent of the total man-days of work was devoted to adult educational work in the field of soil and water conservation in 1955.[13] The very breadth of extension work prevents the agents from concentrating on this one agricultural problem area, even though it is central to all other agricultural activity. The publications of the Extension Service are also indicative of the inability to devote more than a small fraction of its time to water and soil conservation. From 1944 to 1960 only three articles on water management were published by the service in its circulars.[14] Much more attention was devoted to other matters such as crops, homemaking, and 4-H Club work.

The Extension Service cooperates with many of the federal and state agencies in carrying out programs of mutual interest, although there are no formal cooperative agreements binding them together. During the year 1955 the service devoted a significant part of its time to cooperation with the Bureau of Land Management, the Bureau of Reclamation, the Fish and Wildlife Service, the Forest Service, and other federal agencies, and to such state agencies as the Agricultural and Horticultural Commission and soil conservation districts.[15]

SOIL CONSERVATION SERVICE

The Soil Conservation Service operates essentially in the same field as the Agricultural Extension Service, creating a certain amount of tension between the two agencies.[16] Both agencies are responsible for farm planning and for providing professional and technical advice on farm betterment. One official of the Extension Service observed that it had failed to take the initiative in farm planning and technical assistance which thus became the specialty of the Soil Conservation Service, but he felt that Extension Service personnel are better qualified to advise the farmer, and said that these workers mildly resent the intrusion of the SCS. In spite of this undercurrent of conflict, it is clear that positive efforts are being made to create a cooperative climate between the two agencies in Arizona. The agents of the two services do cooperate in the field and assist each other in helping the farmers in drawing up farm plans and in suggesting improved soil and water conservation practices. Indicative of improved relations was the fact that by February, 1960, six soil conservation districts had completed agreements with county agents with regard to educational activities in soil and water conservation, and many more were in process.

The Soil Conservation Service was established in the United States Department of Agriculture in 1937, but the state of Arizona was reluctant to pass enabling legislation. It did so only in 1941 under heavy pressure from the irrigation farmers, but with limitations which severely hampered effective work by the soil conservation districts authorized by the act.[17] The act created in the State Land Department the Division of Soil Conservation headed by an administrative officer appointed by the state land commissioner who was himself designated the state soil conservation commissioner.

The duties of this official are to: (1) assist the soil conservation district supervisors; (2) transmit information among the districts and coordinate their efforts; (3) require annual reports of the district supervisors; (4) secure the cooperation and assistance of the federal and state agencies; and (5) disseminate information to the public.[18] Both legally and in practice, the state soil conservationist is concerned primarily with facilitating the operations of the federal Soil Conservation Service, by giving advice, information, and encouragement in the work that is supervised and conducted by the SCS. The state soil conservationist has no money to spend from state coffers, and has no regulatory powers other than that of ensuring the observance of soil conservation district laws.

His role should not be minimized, however. While he is not in a position to give technical advice to the farmer and rancher, he promotes the work of the federal SCS and encourages the creation of new soil conservation districts and the inclusion of new lands in older districts. He works closely with the supervisors of the districts and is one of the principal advisers of the Arizona Association of Soil Conservation Districts. He is the official coordinator of public and private agencies, federal and state, with the State Land Department for programs of soil and water conservation.[19]

The work of soil conservation has grown enormously since 1941. As of 1960 there were in Arizona 47 districts containing 56 million acres, or 78 percent of the land area of the state.[20] Until 1954 only farm lands were eligible for participation in soil conservation districts. By an amendment passed in that year, after nearly a decade of struggle, range lands were also made eligible.[21] Within a very short period of time, over 37 million acres of range land were added, taxing severely the resources of the SCS. Over 5 million acres within the districts are owned by the state of Arizona; approximately 9,500 farms and ranches are within the districts as members.[22] Although the area of responsibility has increased greatly, there has been no commensurate growth in the state Soil Conservation Division of the Land Department, leaving this work undermanned. It is an interesting fact that some of the opposition to this work was found within the State Land Department itself, there being those who opposed it as wasteful spending of money. At the present time, the state of Arizona pays less than half of the division chief's salary, the greater share coming from the federal government.

The role of the state Soil Conservation Division in the State Land Department is an anomalous one administratively, and yet an extremely significant and correct one politically. Lacking the capacity to assist farmers and ranchers in planning farms and conserving natural resources, its importance lies in its ability to give the state point of view on conservation matters, and to make "respectable" federal programs that might otherwise meet with resistance. The Soil Conservation Division chief is a state officer, and therefore is more knowledgeable, it is assumed, about local needs and problems. This perspective is, of course, not technical but related to political and economic questions which recur in administrative situations. The testimony of all concerned supports the advisability of a procedure by which a local officer encourages the adoption of conservation techniques and the utilization of federal programs.

The responsibility of the federal SCS is "to assist farmers and ranchers to develop and apply basic soil conservation plans fitted specifically to the soil and water resources involved."[23] It is not an "action" agency in the sense that it applies practices to agricultural or range lands. Its task is limited to giving technical advice and assistance to the supervisors of soil conservation districts and the individual farmers who choose to utilize the services. Operating entirely through the medium of districts, it is dependent on the willingness and interest shown by the supervisors and cooperators to make use of its information.

The soil conservation districts are empowered to engage in a wide variety of activities, including the conduct of surveys, investiga-

tions, and research relating to the soil, its character, methods of cultivation, erosion, seeding, and eradication of weeds. They may conduct
demonstration projects. They are authorized to enter into cooperative
agreements with any agency of the state or federal government in
carrying on various improvement practices. They may acquire property and equipment which they can make available to landowners,
and may develop plans for conservation of soil resources.[24] In 1959
the districts owned assets worth $500,000.[25]

The state enabling act declared it the policy of the state "to provide for the restoration and conservation of lands and soil resources
of the state and the control and prevention of soil erosion, and thereby
to preserve natural resources, preserve wildlife, protect the tax base,
protect public lands and in such manner to protect and promote the
public health, safety and general welfare of the people."[26] To accomplish these ends a voluntary cooperative program was authorized
through the soil conservation districts. No person is required to
become a member of a district or to include his land in the soil conservation program of an organized district. Even members of the
soil conservation districts undertake soil conservation work at their
own discretion, regardless of the suggestions and advice and planning
provided by the district officers and the technicians of the SCS.[27]

Perhaps the key to the voluntary nature of the district system in
Arizona lies in the fact that the districts are not given the power to
impose taxes or assessments against their members. The district responsibilities, as such, are supported by contributions of members and
nominal membership fees based, in part, upon the acreage to be included within the district. The lack of power to obtain funds through
assessments or taxation presents a real problem in the effective prosecution of soil conservation work, but there appears to be overwhelming sentiment against the granting of taxing power to these districts.

Districts are created by a petition of 25 or more landowners
owning not less than 20 percent of the land in the proposed district.
This petition is directed to the soil conservation commissioner who
makes a judgment based on such factors as topography, soils, patterns
of erosion, and land practices.[28] With his approval a referendum is
held which requires the affirmative vote of 65 percent of the landowners voting who own 50 percent of the land. At the same time
supervisors are elected to manage the affairs of the district.[29]

The Soil Conservation Service has responsibility for a wide
variety of water conservation and utilization measures and programs
— not all of which are utilized at present in Arizona — including
soil-erosion protection, flood-control programs, and small watershed

projects.[30] Its chief purpose is supplying technical assistance to farmers and ranchers in developing and applying basic conservation practices to soil and water resources.

The SCS disclaims any responsibility for the creation of soil conservation districts or for the establishment of work plans and programs for the districts. It is a technical agency established for the purpose of giving information and technical advice on farm techniques and methods. The responsibility for the creation and operation of the district lies with the individual and officers of his district. The administrator of the SCS asserted in 1954 that

> each soil conservation district has the legal responsibility for developing a district-wide soil and water conservation program aimed at solving local soil and water problems. The district is responsible for carrying forward that program by helping, upon request, land owners and operators, individually and in groups, to plan, to apply, and to maintain technically and economically sound soil and water conservation measures.[31]

While the district officers and the individual farmers are responsible for the work done in the district and on the farms, they may call on the assistance of federal agencies in their enterprises. For this purpose the Department of Agriculture enters into memoranda of understanding by the terms of which the department will, through its agencies "cooperate with and assist the district in carrying on erosion control and soil conservation work."[32] Supplemental agreements are then made by the districts with the SCS, by the terms of which the soil conservation technicians provide the actual assistance to the district or the farmers.

The basic feature of the work of the SCS is the development of the farm plan. This involves the analysis of soil and water problems of the individual farm in its own setting and the determination of the methods and techniques suitable to solve these problems in order to make maximum use of these vital resources. The SCS advises on the application of these techniques and reviews the plan with the farmer in light of progress made with a view toward maintaining the program. Each year the soil conservation district establishes a schedule providing for SCS assistance "after a joint review of the district's needs and Soil Conservation Service resources."[33] The district determines the work priorities on particular farms and ranches for which agreements have been signed with the district, and the SCS provides the technical assistance in accordance with those priorities.

The SCS occupies a position of considerable influence because of its technical expertise and its membership on many committees, some of which have a direct bearing on the availability of loans for conservation. Besides advising on programs and practices, the field conser-

Prolonged drought and overgrazing result in the elimination of the vegetative ground cover, leaving the soil vulnerable to the effects of gully and sheet erosion, and resulting in desolation like this (above). A similar area (below) has received conservation treatment consisting of erosion-control structures and reseeding of grasses.

—U.S. Bureau of Land Management

An area badly damaged by erosion (above), and the same area a few
years later (below) after the installation of erosion-control dikes
and reseeding with native and exotic grasses. The reestablished
vegetation cover maintains soil stability, captures moisture — thus
decreasing erosion — and provides forage for livestock and wildlife.

—U.S. Bureau of Land Management

vationist works with representatives of other state, federal, and local agencies and associations, handles the technical aspects of planning the annual Agricultural Conservation Program, facilitates cooperation among various persons and groups in the adoption of conservation practices, conducts an information program, and sets standards for state policy.[34] In playing such critical roles on the technical side of soil conservation, the conservationist is in a position to make a decided impact on the decisions made by the farmer.

The water users, while desiring the technical aid available through the conservation districts, were careful to write into the enabling act limitations on the diversion or use of water. Nothing in the act was "to affect existing water rights or in any manner contravene the provisions" of the water code. A soil conservation district may not perform any work resulting in the consumptive use of water, such as the construction of dikes or channels or dams for storage, spreading, diversion, or conveyance of water on any watersheds which contribute water to irrigation districts without first obtaining the consent of the districts involved. In cases where consumptive use of water is made by the soil conservation district, such "diversion, application or use of water by means of any improvement, constructed, maintained or operated . . . shall not be construed to be an appropriation of or vest any right to the use of public water."[35]

The great concern over consumptive use of water on the watersheds to the detriment of the irrigationists in the valleys resulted in the restriction of soil conservation districts to farming land until 1954 when the soil conservation law was amended to allow soil conservation on the range lands. This same concern was manifested in the opposition to authorization of Public Law 566 projects in Arizona. The SCS long favored such projects, asserting that

> there is recognized need for watershed protection projects in Arizona. Physical characteristics of prospective projects are such that mechanical or structural control measures are adaptable to a high degree. Land management or vegetative measures are also applicable; however, because of the extremes in climate and rainfall, management and vegetative measures will have little immediate effect on surface run off. It is expected that a large percentage of the applications received can be justified from an economic standpoint and funds for local sponsors should be available for the better projects.[36]

Water users in the lower valleys resisted these projects, believing that conservation measures on the watersheds would result in less water in the storage dams, a contention over which there was hot dispute. The state conservationist of the SCS has assured downstream users that the quantity of water would not be diminished.[37] Technicians argue that very little — between 1 and 5 percent — of precipitation on semiarid range land ever reaches a point of downstream use as surface flow, although they are interested in means of increasing stream flow by vegetation changes.[38] The irrigation districts and others until 1959 successfully resisted the pressure of upland farmers and ranchers for amendment of the soil conservation act to authorize P.L. 566 projects. The Salt River Valley Water Users' Association led the opposition to amending the act; its general manager stated at the annual convention of the Arizona Association of Soil Conservation Districts in 1956, "We are unalterably opposed to anything that will increase consumptive use of water on the watersheds, and therefore to amendments of state law to permit construction on the watersheds under Public Law 566."[39] Others opposed amending the law on the grounds that it would require the districts to assume taxing authority and thus destroy their voluntary nature.

In spite of the opposition, the state legislature passed legislation in 1959 authorizing the creation of special flood-control districts.[40] This authorization, in addition to soil conservation district legislation, is sufficient authority for P.L. 566 projects. As a result, by August 1961 two projects had been authorized for construction in Graham and Pinal counties, involving over 200,000 acres of land. A total of 22 applications had been received, a number of which were found to be ineligible, lacking in local support, or impractical.

The work of the farmers within the soil conservation districts with or without aid from the federal government has been extensive, involving a tremendous investment on their part. Because of the high costs of agriculture in Arizona, conservation practices are a necessity which most of them have long since recognized. The SCS has developed long-range plans for the work needed in the state as well as time tables for annual work required under the long-range program. It was estimated in 1961 that the job of farm-land conservation is about one-third completed in terms of these long-range needs.[41] These long-range plans contemplate some hard facts that few public officials or farmers are willing to face, with or without the adoption of water-saving farm practices.

> Reservoir storage capacities in central Arizona total 3,445,300 acre feet and is considered adequate. The trend of amount of gravity water yield is downward. This situation along with continuous lowering of the water table in several of the major irrigated areas indicates the necessity finally of a reduction of irrigated acreages.[42]

Unless new water supplies are tapped, the only way to prevent a reduction in acreage is through better use of water.

The role of the SCS in the process of granting loans for farm practices was noted above. One of the most important programs is the Agricultural Conservation Program administered by the Agricultural Conservation and Stabilization committees. The Agricultural Conservation Program makes money available on a cost-sharing basis "for conservation measures that establish or improve a protective cover of grasses, legumes, or trees, conserve or dispose of water, and other measures that protect, improve, and make better use of farm and range land."[43] Under this program, the federal government provides around 50 percent of the costs for practices "which it is believed farmers or ranchers would not carry out to the needed extent without program assistance," and "which have not become a part of regular farming operations on their farms or ranches."[44]

The role of the SCS is that of technical adviser and participant in the committees which formulate the programs at the state and county levels. The SCS determines the feasibility and necessity of proposed practices in order that they might qualify for cost-sharing and then supervises the installation of them and certifies their completion.[45]

Work under the Agricultural Conservation Program has been widely distributed, it being estimated that 90 percent of Arizona's range and farm land has received assistance. Most of the practices for which money was given related directly to better use of or increased production of water, including development of springs, construction of livestock reservoirs, land leveling, lining of ditches, and construction of irrigation dams.[46] These practices and others permissible under the program represented an outlay of $1,450,000 by the federal government for the fiscal year 1957 alone. The Agricultural Stabilization and Conservation committees are formally responsibile for giving an independent judgment on the requests made for money, but such independence is mitigated by the interlocking membership on these committees and the boards of supervisors of the soil conservation districts.

The SCS performs a similar function for the Farmers Home Administration which insures private loans to farmers for improved practices such as construction and repair of terraces, ponds, waterways and erosion control structures, ditches and canals for irrigation drainage, brush removal, well drilling, and purchase of irrigation equipment.[47]

The SCS has a limited role in the administration of the soil bank program, which has had very limited applicability to the conditions of Arizona farming. SCS personnel assist in the selection of land to go in either the acreage reserve or conservation reserve phases of the program.[48]

A great deal of emphasis is being placed on improved irrigation practices, such as leveling of land, more efficient methods of application of water, lining of canals and laterals, and others. For several years, during the later 1950's, farmers were lining their ditches at the rate of 500 miles per year, and leveling their land at a rate of between 40,000 and 60,000 acres per year. They engaged in practices which improved water application at a rate of 70,000 acres per year. With the extension of responsibility for soil and water conservation practices on range lands, the districts, in cooperation with the SCS, contemplate better conservation of soil and water through improved grazing management, reseeding, construction of ditches, dams, and reservoirs.[49] The adoption of conservation crop rotation is considered very desirable but few farmers are willing to adopt such practices under the influence of high prices for staple crops such as cotton. The problems to be overcome in the long-range program include inefficient use of existing water supplies, low infiltration rates of soils, diminishing water supplies in pump-irrigated areas, salinity and alkalinity of soils owing to use of water having high salt content, waterlogged soils, erosion and deposition, and land and crop damage resulting from floods.

The SCS has a complex set of interrelationships with other federal and state agencies owing to the checkerboard ownership of land in Arizona. Federal, state, and private land is interspersed, requiring agreements among landholding agencies on land-management practices. The inclusion of range lands increased the need for cooperation. "Administrators of these public land management agencies, SCS supervisors, and ranchers are developing conservation plans for entire ranching units including public and private lands . . . . This coordination on the ground, where the problems are, results in a more practical and workable conservation program on the public and private property."[50] The SCS is precluded from doing ranch planning on public land, but under the leadership of the soil conservation district supervisors a coordinated ranch plan can be worked out by the private landowners and public agencies. On June 21, 1960, the Arizona SCS office signed a pledge of cooperation with other agencies of the Department of Agriculture, the Department of the Interior, and private conservation groups which recognized the need for teamwork to accomplish their common purposes.

Although the SCS has been stripped of most of its research responsibilities as the result of a reorganization of the Department of Agriculture in 1954, the SCS continues to cooperate with other federal agencies in obtaining hydrologic data. This data will be especially valuable for P.L. 566 projects in Arizona.[51]

The SCS has undertaken one project indicating the feasibility of conservation measures on the watersheds. At White Tanks near Litchfield Park in central Arizona a project has been completed to prevent the serious floods that have afflicted the area in the past. Supported by the local soil conservation district and irrigation district, and the state soil conservation commissioner, the project included the construction of retention dams, dikes, and levees designed to prevent flooding as the result of storms of a 200-year frequency.

In addition to losing its research functions in the 1954 reorganization, the SCS was also internally revamped. Regional offices were abolished and state offices were set up in their stead. The purpose, according to the Secretary of Agriculture, was "to give better service to the farmer through decentralization of many functions" and "to give the States more responsibilities in some areas."[52] The elimination of regional offices has tended, in the view of some, to lessen the effectiveness of the technical advice which can be given by staff people to the line officers in the field since the former now work out of Washington rather than out of regional offices. The field staff are now required to spread their skills among many areas of soil conservation, and there is some feeling that the farmer loses in the kind of advice he gets.

The work of the SCS is actively supported by the Arizona Association of Soil Conservation Districts, the private organization of the public servants who head the soil conservation districts. This organization has provided support for the SCS work and has resisted attempts to diminish its area of activity.

Soil and water conservation has made tremendous advances in Arizona, particularly in the irrigated areas, through the Soil Conservation Service and the educational assistance given by the state agencies. It is too soon to determine what progress will be made on the range lands of the state, but it is clear that some of the biggest problems are there. The SCS and cooperating agencies are trying to slow down the rate of erosion and eventually achieve stabilization. As the head of the SCS stated in 1956, however, "The stabilized conditions cannot be considered as satisfactory but at least you can have some assurance that conditions will not get a great deal worse."[53]

CHAPTER 12

Management for Recreation and Conservation

The provision for recreation and leisure-time activity has become one of the most important responsibilities facing our governmental units in the twentieth century. With the population growing rapidly, with the time actively spent in gainful employment lessening gradually, and with purchasing power increasing, there is a high value placed on areas and facilities which are suitable for recreational activity. While much of the desire for entertainment may be satisfied before the television set, a considerable portion of this need can be provided for only by the reservation of some of our natural resources for recreational pursuits. The needs are truly staggering, and with the population of the Southwest growing more rapidly than other areas of the nation, planning to meet these needs is all the more necessary.[1]

The recognition of the expanded needs of burgeoning populations creates the demand for intelligent planning and management of our land, water, plant, and wildlife resources for maximum benefit, which in itself involves settlement of conflicting claims regarding superiority of use. And in the settlement of such conflicts "recreationists have the right to demand that realistic values be placed on water for recreation and that all plans for its management consider those values."[2] Too frequently, it is charged, water-use planners assign recreation an inferior position or recognize its value only as an afterthought.

MANAGEMENT FOR RECREATION

The potentialities of Arizona as a playground for the Southwestern region and for the United States are increasingly recognized. Some predict that recreation will ultimately be the foremost industry in the state.[3] The National Park Service, in a study of the Colorado River Basin after World War II declared, "It is only natural in a region so endowed that recreation should become one of the major industries," and that "to foster this industry it must be recognized that recreational use of land may, in certain places, be the highest or best use of the land for the general welfare of the people in the basin and, in vast sections of the basin, should be on an equal basis with other uses, such as grazing or production of timber."[4] In some areas, notably northern Arizona, the tourist industry has already become the leader and it is expected that "Northern Arizona's future growth and prosperity are largely dependent upon development of resources which attract out-of-state visitors."[5] The Stanford Research Institute states:

> Northern Arizona has perhaps the major scenic asset in America, the Grand Canyon, plus a unique combination of forest, mountain, and desert scenery, with interesting Indian civilizations and prehistoric ruins. While other individual regions equal or exceed Northern Arizona with respect to one or a few of its attractions (except for the Canyon), probably no other region in the West has as great a diversity and extent of natural vacation resources.[6]

Recreational use of land, trees, and water does not in itself result in consumptive use of water. The maintenance of these recreational values is, however, dependent on management of the water resources. In some instances management for recreational values is in direct conflict with management which would maximize other values. It is clear that the development of maximum hydroelectric energy or storage of maximum amounts of water for irrigation or flood control could interfere with recreational and scenic values in some areas. Today the Grand Canyon is involved in this type of controversy.

A proposal to produce power at Kanab Creek would divert water impounded at Marble Canyon around the Grand Canyon to maximize the hydroelectric possibilities of the river at the place where Kanab Creek enters the Colorado River. Only one-seventeenth of the average flow would pass through the Grand Canyon. The National Park Service is adamantly opposed to such a scheme, stating that it would "take away from the Grand Canyon the very agent that created it; the remaining trickle would be sham and mockery in comparison to the once great force that carved the canyon. . . . Without the normal flow of the river the significance and completeness of the park will be destroyed. . . ."[7] In the estimation of the NPS, management for scenic beauty and geologic significance precludes management for products of economic value such as electrical energy.

The Park Service has been reluctant to give its approval for the construction of Bridge Canyon Dam, an integral part of the Central Arizona Project, because under some forms of the proposed project, the reservoir behind the dam would back into Grand Canyon National

Monument and would inundate certain geologic formations, lessen the grandeur of some deep chasms, and modify some other natural features now seen by a very small percentage of the visitors to the much larger Grand Canyon National Park each year. While not absolutely opposed to such a structure, and recognizing the very real economic issues involved, the NPS is concerned lest the choice is made without determining "how much the people of the United States care about preserving the natural conditions and scenery in the portion of the Grand Canyon selected for such preservation. . . ."[8] The various plans for dam construction at Bridge Canyon are designed to exclude the Grand Canyon National Monument but allow for later reconsideration if it is thought desirable to raise the height of the dam.

In the construction of Glen Canyon Dam such a choice has already been made. This dam will impound waters — forming Lake Powell — in a relatively inaccessible stretch of the Colorado River where only the most hardy have been willing to venture. While the lake will inundate some areas of great geologic interest, it will at the same time make many scenic wonders accessible to the public. Lying athwart the stream near one of the major north-south highways, Glen Canyon Dam promises a great expansion of the recreational facilities for the region. The NPS is engaged in planning the recreational facilities for the dam and lake. As is also customary, NPS archaeologists surveyed the reservoir site and gathered representative archaeological collections prior to the filling of the reservoir. An earlier brief survey indicating the archaeological values of the area was made by the Museum of Northern Arizona.[9]

The development of the Lake Mead Recreational Area is testimony in itself to the potentialities of these reservoirs for recreation. The stretch of the river near Black Canyon was formerly unsuitable for fish because of the heavy silt load. With the sediment deposited at the bottom of the reservoir there has been created a lake renowned for its fishing, swimming, and boating. Over 2,500,000 visitors come to Lake Mead each year to use the recreational facilities made possible by this stored water. Bridge Canyon Dam would also serve these purposes. It was estimated in 1946, on the basis of travel counts along nearby highways and numbers of visitors to adjacent recreational areas, that the construction of a reservoir would attract 365,000 visitors annually.[10] Undoubtedly the figure would be much higher today. The Arizona Game and Fish Department reported completion in 1960 of a study in which it estimates that visitors to Bridge and Marble Canyon areas would spend as much as $24 million annually after building of the dams was completed.[11]

The competition regarding the management of water is most obvious in the case of major streams such as the Colorado, but it is no less real on lesser watersheds where management practices can diminish or increase the supply of usable water. Here the sportsman, the photographer, the camper, and the outdoor enthusiast and lover of nature finds his element. For the latter "the splash and spreading ripple of a beaver on a twilit pond or the fairy-like chimes of a thrush in the cathedral stillness of a forest, are a source of inspiration complete in itself. Their enjoyment is not to be measured in dollars, any more than sunshine or freedom can be so measured."[12]

To many irrigationists in the valleys, however, such "sentimentality" is unimportant compared to the economic values to be achieved by increased streamflow resulting from forest and range vegetative manipulation. Anything which tends to increase consumptive use on the watershed, in their estimation, is to be opposed. For this reason, they frequently oppose the enlargement of recreational areas and even the adoption of some conservation practices on the assumption that such practices do increase consumptive use. In part, this was the reason for the failure of the state legislature to create a park system in Arizona until 1957.

Historically, the federal government has played the only significant role in providing protection for natural recreational areas, scenic wonders, and places of historical and archaeological interest. The role of the state has been negligible. Whether the state's role will increase as the result of the creation of the State Parks Board is still to be seen. The National Park Service administers 20 areas in the state, including such well-known sites as Grand Canyon National Park, Lake Mead Recreational Area, and Chiricahua and Petrified Forest National Monuments.[13] In 1960 there were 5,441,800 visitors to these parks and monuments, an increase of over 100 percent since 1950.[14] Nearly two-thirds of the visitors went to Lake Mead and the Grand Canyon. The NPS predicts that by 1970 there will be nine million visitors annually to Arizona's national park and monuments.[15]

Recognizing the greatly increased pressure on existing facilities the NPS has embarked on its "Mission 66" program by which it hopes to expand and modernize its facilities to meet the demand.[16] Most of the increased appropriations for the program are being devoted to construction of buildings, trails, and other physical facilities, but a small yet significant amount has been allotted for such conservation activities as fire protection and control, soil and moisture conservation, wildlife conservation, purchase of water rights, and development of water supplies. On some Park Service lands there are serious erosion problems and depletion of the land through overgrazing.[17] Already the NPS feels that the program is inadequate for the task before it, and is restudying its program.[18]

The Park Service has recognized its acute water problem, particularly in the West, and considers availability of water "a potent factor in determining types of use and extent of development in many parks."[19] This situation necessitates "added emphasis on the program of finding sources of unappropriated water and in protecting existing rights."[20] In many instances it will be necessary to purchase water rights in order to satisfy the need.

The NPS has been in the process of making more detailed plans for the parks and monuments in Arizona. It has already determined that augmentation of existing water supplies and construction of improved utility systems rank high on the list of priorities.[21]

The national forests of Arizona play an extremely important role in providing recreational facilities for Arizonans, occupying as they do the most scenic and moderate climatic areas of the state. The national forests, and the forests under the jurisdiction of the Bureau of Indian Affairs provide relief from the overpowering summer heat of the desert, much of the best hunting and fishing, and facilities for camping and picnicking. In 1960 the forests welcomed 4,999,800 visitors, of whom 1,589,900 visited picnic areas, 666,000 stayed at forest campgrounds, 311,800 stayed at resorts and hotels, with well over two million visiting other forest areas. The 1956 level of visitation constituted an increase of more than 100 percent over the number of visitors in 1953, while in the single year from 1959 to 1960 the figure rose 7 percent. The use of facilities alone creates increased demand for water in the national forests, and the trend is steadily upward. Each year the Forest Service spends well over $1,000,000 for the construction of recreation-use facilities in addition to lesser amounts for game, fish, and other wildlife.[22]

The Indians are beginning to develop recreational areas on their reservations in hopes of increasing their income. In impounding water for the creation of 4500-acre-foot Hawley Lake on the Fort Apache Reservation, the Apaches have already created opposition from the Salt River Valley Water Users' Association which claims prior water rights, since the stream forming the lake is a tributary of the Salt River.[23] The Apaches have built boat landings, leased homesites, laid out an 18-hole golf course, and provided winter sports facilities. Here again the desire to provide recreational facilities on the watersheds has run afoul of the interests of those who want a maximum amount of water for irrigation and municipal use.

Rucker Lake is typical of the man-made mountain lakes built primarily for recreation. Some of these small lakes draw 5000 visitors on a summer weekend. The state's growing population, 70 percent of which lives in the desert areas, demands expanded recreational facilities. Such expansion requires additional water and often conflicts directly with other uses of water which, under state law, have priority.

—Josef Muench

Only after overcoming the opposition of cattlemen and some irrigationists was the State Parks Board created in 1957.[24] The stockmen had argued that the federal government provided all that was necessary for recreational activity, making the creation of a state park system unnecessary.[25] They had been able to prevent the establishment of such a system for 27 years, and finally consented to the creation of the board only after the proposed legislation was modified to give them increased influence on the board, and to limit any park established by the board to 160 acres. With limited financial resources — only $30,000 was provided for the first year — the board has been able only to devise criteria for the establishment of state parks, monuments, and scenic and recreation areas, and to draw up short- and long-range plans. By early 1961 the board had established monuments at Tubac, Tombstone Court House, and Yuma Territorial Prison, and had plans for W.G.P. Hunt's tomb.[26] These totaled 14 acres in area.

Contrary to the viewpoint expressed by the stockmen of Arizona, the National Park Service has discovered numerous areas suitable for protection or development because of their scenic, historical, or recreational values. As early as 1941 the NPS encouraged the state to embark on a program to provide for the increased population, and it pointed out specific areas which could be developed.[27] While not all of those suggested are suitable for national development, some do have national significance and could be readily developed by the state or local governments because of their local significance, or their value as local recreational areas. Some suggested sites were Fort Bowie, the Kinishba Ruins, Meteor Crater, the Travertine Bridge, Oak Creek Canyon, the Apache Trail, San Xavier Mission, and Ajo copper mine.[28] These sites include Indian ruins, spectacular natural formations, places of importance in Arizona history, and areas of natural beauty suitable for camping.

FISH AND WILDLIFE MANAGEMENT

The last state in the nation to create a state parks agency, Arizona is showing signs of trying to avoid being the last state in providing recreational facilities for its citizens. As with the parks, however, management for hunting and fishing is made difficult by the limited water supply, and by the conflicting interests desirous of using that water, for the availability of water is no less a determining and limiting factor for fish and wildlife than it is for man. With a shortage of water a normal situation, it is a constant struggle in Arizona to maintain certain species of game, and a serious problem to provide a supply of fish and game for those who find their recreation in hunting and fishing. Every attempt to manage the environment, whether for water, forage, or trees has a decided impact on wildlife, and full recognition must be made of this fact in planning for range and watershed management.

The importance of intelligent management is indicated by the steadily increasing pressure on wildlife and fish by sportsmen. In 1944, 59,099 resident permits of all kinds and 2,509 nonresident permits were issued in Arizona. By 1954 these figures had risen to 125,178 resident, and 24,482 nonresident permits.[29] Between the years 1958 and 1959 alone hunting license sales increased 11 percent and fishing licenses 10 percent.[30] Officials of the Arizona Game and Fish Department admit their inability to maintain wildlife numbers in the proportion that existed in relation to population before the great influx of people into the state. The problem is to make the best use of what is available, and to develop whatever potential resources exist. In spite of the problems, the Game and Fish Department officials reported that, through improved management, hunting and fiishing were better in Arizona in 1960 than at the turn of the century.[31]

With water supplies a perennial problem, it is perhaps natural that allocation of water for wildlife and fish would rank low on the list of priorities. In fact, in the determination of the relative importance of the uses to which water may be put, the legislature has not even recognized fish and wildlife. In cases of conflict over a limited supply of water the relative values to the public are domestic and municipal, irrigation and stock watering, and power and mining, in that order; fish and wildlife are not even mentioned.[32] The State Water Code specifically states that "when it becomes necessary for the state land department to determine the relative values to the public of proposed uses of water, wildlife uses, including fish, shall be deemed inferior" to those uses enumerated above.[33] The sportsmen chafe under what they consider to be discriminatory treatment, and are constantly on guard against "water grabs" that might deny water for recreation and wildlife.[34] In view of the economic and social importance of recreation and outdoor life, they seek a status of equality for these uses with the other uses of water. The Arizona Game Protective Association in October, 1956, passed a resolution requesting that the Water Code be amended to recognize recreational use of water as valuable to the public. As one commentator put it, "Even though it is mentioned in fourth rank, that would be better than no mention at all."[35]

The legislature, however, has not looked favorably upon such amendments.

As in the case of other natural resources, the management of fish and game is a joint state and federal responsibility. The significant difference lies in the fact that the state takes an active part in conservation and developmental work relating to wildlife and fish, while in most other resource matters the state remains relatively passive. In common with most other states, the Arizona Game and Fish Department spends far more than any other resource agency. A major portion of departmental activity is supported by revenues from the sale of licenses, with some money coming in from fines and federal grants-in-aid.[36] The legislature makes no appropriation from the general fund for the department, but makes an appropriation from the game and fish fund each year for the operation of the department.[37]

The Arizona Game and Fish Department was created in its present form by legislative action in 1929, at which time the game and fish code was thoroughly rewritten. Previous to that time a game ranger and his deputies had constituted the entire wildlife agency of the state. The Game and Fish Commission was established, consisting of five members, appointed by the governor with the advice and consent of the Senate, who are responsible for the administration of the game and fish laws of the state. Membership on the commission is bipartisan and is required to be representative of different sections of the state. The members are further required to be "well informed on the subject of wild life and requirements for its conservation."[38]

The responsibilites of the commission include the control and management of the propagation and distribution of wild birds, wild animals, including amphibians, and fish. It manages hatcheries, is responsible for the enforcement of laws for the protection of wildlife, prescribes the manner, methods, and devices to be used in killing and capturing game animals, and may establish game refuges.[39] To effectuate its work, this part-time commission appoints a director who has "general knowledge of game and fisheries management, and the requirements for conservation of animals, birds and fish," and who directs the activities of departmental employees.[40]

The department is actively engaged in programs designed to increase the supply of water available for fishing and support of wildlife. The surface-water supply is appropriated almost in its entirety, leaving the streams dry for the better part of the year. Propagation of fish is thus dependent upon fish hatcheries, and fishing as a sport in streams is dependent upon stocking. Even fishing in warm water lakes is greatly dependent upon artificial management by the department. Water in these impoundments behind dams is managed primarily in the interest of irrigation agriculture, and with only secondary regard for the necessities of fish propagation. The only truly dependable water supply for fishing in Arizona is the Colorado River.[41]

To avoid excessive dependence on these "unstable" waters, and also to increase the fishing waters available, the department is actively engaged in seeking suitable sites for the construction of dams for the impoundment of fishing waters. In this activity the department is assisted, as it is in other matters, by federal funds under the Pittman-Robertson Act. During the period 1956 through 1960, the department constructed six new trout fishing lakes in northern Arizona, providing a total of 405 surface acres of fishing water.[42]

The Lands Division of the department has been engaged in extensive engineering, geological, and hydrological investigations of possible lake sites throughout the state. By 1960 the department had conducted preliminary investigations and had begun negotiations on 18 potential impoundments.[43] Frequently these investigations demonstrate only that the site in question is not feasible. The commission stated in 1956 that

> at least nine such sites out of ten do not have an impervious base for a dam, sufficient runoff to provide a lake with an adequate year long depth or the other requirements which will make an earth fill or concrete dam feasible at a cost commensurate with the surface acres of the resulting lake.[44]

When a feasible site is located, it is then necessary to negotiate with the owners of the water rights. Such negotiations do not always result in a permit to construct a dam. Suitable sites have been located in a number of places throughout the state, but are awaiting authorization or the clearance and negotiation of water rights.

The department is facing opposition from water users with prior rights to water in many areas in their development of these recreational sites. Protests have been lodged over applications filed with the state land commissioner for appropriations of water by the Game and Fish Department, and these protests may well result in curtailment of the program.

Since almost all wildlife is dependent on free water for survival, the department is engaged in the construction of catchments and waterholes for the use of game and other wildlife. During the fiscal year 1956 the department constructed 19 rainwater catchments of 2,500-gallon capacity each, 2 potholes, 2 retention dams, and developed 9 springs.[45] The goal, according to one official, is to arrange it so that deer would not have to go more than three miles for water. This, however, is a very long-range goal. The department hopes to provide by these measures a better distribution of the use of the range since overuse results in the area of the catchments. There is apparently a lack of firm knowledge concerning the precise relationship between game numbers and amount and availability of water, but the department has this matter on its list of research projects for the future.

The department works closely with the U.S. Fish and Wildlife Service in its developmental work along the mainstream of the Colorado, as well as projects involving wholly intrastate streams. The department is especially concerned with the Colorado since it is the most reliable sports area in the state, and also since it involves complex relationships among many agencies, both state and federal. The fisheries are dependent upon the water management policies of the agencies operating the dams along the river, particularly the Bureau of Reclamation, and the department contends that these other agencies are not as concerned with sportsmen's interests as they might be. It also contends that there must be unified management of the area for wildlife, and yet at the present time the area is managed by three different states, three federal agencies, and Mexico.[46] Further concern with water management results from the department's statutory responsibility for protecting Arizona's streams against pollution.[47]

The commission form of administration was established for the purpose of extricating the Game and Fish Department from "politics," a disease to which it had fallen prey during the 1920's. The department has become perhaps the leading example of professionalism and use of a merit system in a state not traditionally characterized by either. Supporting the commission and the department in their efforts has been the Arizona Game Protective Association, the organization of sportsmen throughout the state.

The AGPA is the largest and by far the most important sportsmen's group in the state, with its membership located in all the important and many of the small communities of the state. Organized in 1923, its membership at one time reached 6,000, although in recent years it has dropped to around half that number. The AGPA supports "the protection, propagation and proper management of all wildlife," cooperation with the Forest Service principle of multiple-use management of watersheds, conservation education, and a "nonpolitical impartial Game and Fish Commission and technically trained staff...."[48] Since the game and fish activities of the department occur primarily on national forests, the AGPA stands particularly for the multiple-use principle as a protection against management of the forests for single interests, especially for the cattlemen.[49]

The intimacy between the Arizona Game and Fish Commission and the Arizona Game Protective Association is indicated by the fact that members of the commission are usually chosen from the ranks of the executive committee of the AGPA. Many of the officials of the department in the past have also been members of the AGPA. The association retains as one of its officers a "contact man" who maintains liaison between the commission and the association.

The AGPA conceives of its role as that of a "buffer" between sportsmen and the department, transferring the complaints of sportsmen to the department, and attempting to explain the policies of the department to its membership. With the commission members coming from the AGPA, a high degree of harmony can be expected and usually exists.

The AGPA has been intensely concerned with the management policies as they affect the Game and Fish Department. It has assisted the department in its struggles over water rights by hiring lawyers to assist the department in appearances before the state land commissioner, and also in the preparation of legislation. Although the department can utilize the services of the attorney-general's office, it was felt that changes of personnel in the latter agency caused the department to rely on inexperienced aid.[50]

Both the AGPA and the Game and Fish Department, the latter not so openly, have resisted the Barr Report, especially those portions which call for watershed modification practices designed to increase the supply of stream water. On this matter they find themselves in opposition to the major water users, particularly the irrigation water users in the central valley. In this opposition they are defending what they conceive to be the natural habitat for wildlife against the attacks of those who would destroy this habitat for the recovery of water.

Although responsibility for the management of fish and game is, for the most part, under the jurisdiction of the state government, the federal government does share a part of the burden. Most of the areas best suited for hunting and fishing are located in national forests or on Indian reservations. The land management agencies, therefore, must consider the effect of game management policies on their lands and resources. On Indian reservations the fish and game policies are determined by the Indian tribes in cooperation with the Bureau of Indian Affairs so a high degree of coordination is necessary to maintain intelligent programs in accordance with state plans. The programs and policies of the Game and Fish Department on the national forests are made after consultation and agreement with forestry officials.

Through the U.S. Fish and Wildlife Service of the Department of the Interior the federal government takes a more active role in the management of fish and game. The Fish and Wildlife Service is active in Arizona in managing certain game refuges, in providing funds to the Arizona Game and Fish Department for experimentation and research, and serving other federal agencies through its facilities.

In managing refuges in Arizona, the Fish and Wildlife Service finds the availability of water central to its problems. The service manages two important big-game ranges, the Kofa Game Refuge and the Cabeza Prieta Game Range, in cooperation with the Bureau of Land Management. These ranges, located in the extremely arid southwestern section of Arizona, were established in 1939 primarily for the protection of bighorn sheep whose numbers had declined greatly. The areas are also the habitat of significant numbers of deer, javelina, quail, and doves. The principal developmental activity there has been concerned with expanding the water supply by enlarging natural or creating artificial waterholes. It is hoped thereby to create a dependable water supply within reasonable distances.[51] Efforts to this time have indicated a growth in bighorn numbers. Long-range plans call for further water developments. During the mid-1950's some 750 hunters and 3,000 other persons visited the Kofa Game Range each year.[52] The Cabeza Prieta Range has been closed to visitors.

The Fish and Wildlife Service administers three wildlife refuges in Arizona, at Parker and Imperial dams on the Colorado River, and Roosevelt Dam on the Salt River above Phoenix. All three are designed primarily as resting or wintering areas for migratory birds, but also provide recreational opportunities for some 265,000 fishermen, 5,500 hunters, and 80,000 other visitors each year.[53] The development of these refuges was the result of construction of dams for the purpose of storing or diverting water for flood control, irrigation, or power. The refuges consequently suffer at times owing to the management of the reservoirs for these purposes. The Salt River refuge for waterfowl is particularly hard hit at times as the result of the reduction in water area at Roosevelt Lake. Those along the Colorado are threatened by the increased appropriation of water from the river for irrigation. Future programs for improvement of these areas include the improvement of waterfowl habitat through the clearing and eradication of pest plant growth, and replacement of unproductive marsh areas and obnoxious plant growth with more desirable food plants and habitat.

The Fish and Wildlife Service provides money through the grant-in-aid program to the Game and Fish Department for research and development work. Under the Dingell-Johnson and the Pittman-Robertson Acts, the service makes money available to Arizona for developmental work on fishery resources and restoration of wildlife habitats. A considerable portion of this money has been spent on projects for the development of water facilities and for improving the habitat for fishing.[54]

The service serves the other federal agencies, notably the Indian tribes through the Bureau of Indian Affairs, and the state government in propagation of fish. As the Indians improve their streams for fishing purposes, the service will increase its assistance to them in this regard. The service also investigates and evaluates sites for sport fishing, especially along the Colorado River where much improvement is expected in future years. The views of the service are sought on multiple-purpose land and water management programs by other federal agencies. With more intensive management of the water resources, it may be expected that the service will play an increasingly important role. For example, on Public Law 566 projects, the service is required to report its views on the proposed projects as it has done in the past on proposals by the Bureau of Reclamation, the Army Corps of Engineers, and other federal agencies.[55]

The Arizona Cooperative Wildlife Research Unit, established in 1951 at the University of Arizona, is an agency concerned with the water problems of the state, owing to the critical importance of water in sustaining life on the desert. The unit is supported by the University of Arizona, the Arizona Game and Fish Commission, the U.S. Fish and Wildlife Service, and the Wildlife Management Institute.

The activities of the unit are divided among research, student training, and extension work.[56] Under the direction of the unit's director and his assistant, graduate students engage in research designed to provide information "on matters of economic importance to the fish and game management and on basic biological problems . . . ." The program is "directed toward problems not being met elsewhere," the results of which, because of the uniqueness of the arid conditions, will "have potential significance for other arid countries."[57] Because of the extreme variability of the precipitation from year to year, research projects must be carried on over a number of years, for,

to those who understand the harshness of desert drought, the fickleness of Arizona's rainfall and yet the lushness of the desert in times of abundant moisture, it is clear that research over a period of years will alter the results and impressions gained in one year's work. Thus these studies must be carried through dry years as well as "wet" years and continued and expanded until all situations are met.[58]

Nearly every research project is developed within the context of the variable water supply: studies of browse, the desert mule deer, Abert squirrels, doves, cottontails, quail. At least one study has been made by unit researchers on the relationship of waterholes to wildlife. While some animals can survive without free water, others, particularly the larger game animals, cannot. Much valuable information has been obtained about the behavior of these animals, their need for water, the spacing of waterholes, and the animals that benefit therefrom.[59]

The unit is supported by the Arizona Game and Fish Department in providing for operating expenses and fellowship students. The unit has also worked closely with range management agencies, in particular with the Forest Service, in undertaking its research. Although limited in its resources, the unit promises to make a positive contribution to the knowledge of the water needs of wildlife.

Both the U.S. Fish and Wildlife Service and the Arizona Game and Fish Department are much concerned about protecting the place of fish and wildlife in our resource policies. Feeling on the defensive regarding the comparative economic value of wildlife recreation, these agencies have in recent years promoted research and publicity to convince the public that fishing and hunting rank high in economic return. Statistics developed by the Fish and Wildlife Service indicate that people of the Rocky Mountain region devote much more time and money to these two sports than does the average person on a nationwide basis,[60] while the Arizona Game and Fish Department estimated in 1959 that each surface acre of lake in the state was worth $1,000 annually to the economy.[61]

A recent survey conducted by the University of Arizona for the Game and Fish Department indicates that fishing and hunting do indeed make a substantial contribution to the Arizona economy. On the basis of the survey, it is estimated that during 1960 some 183,000 persons spent a total of $40,151,000 in Arizona on these two sports, not including the cost of licenses. Of this total, state residents spent 95 percent, while visitors to Arizona accounted for only 5 percent.[62] The study points out that the actual income to the state is much higher than $40 million, however, since this amount constitutes only initial expenditures, while "the firms who receive these original expenditures redistribute them among other firms for replacement of inventories sold and for services of employees, for interest, rents, taxes and profits. This process of redistribution of income is repeated many times . . . until the total income generated by initial hunting and fishing expenditures is many times the original amount."[63]

In the view of sportsmen and their public representatives, conservation problems should be laid at the doorstep of the stockmen for overgrazing the watersheds and causing erosion. Proposals for vegetation modification and paving of watersheds, as suggested in the Barr Report, for example, have also aroused the ire of the sportsmen. By altering the natural habitat, they argue, the entire wildlife population of the area is threatened. The stockmen and farmers naturally deny these allegations. This dispute constitutes one of the major conservation obstacles to resource management in the Southwest, and its resolution does not appear to be in sight. The sportsmen in Arizona appear to be on the defensive against attempts to downgrade their interests even further, but they do have powerful allies in the federal management agencies who are reluctant to accede to demands for alleged single-interest programs.

As the state's population continues to grow, and the amount of leisure time available to people increases, the demands for adequate facilities for recreation, hunting, and fishing will mount, at the same time that the burgeoning urban areas, the industries that support them, and the farmers and cattlemen will also be making increased demands on the limited water supply. The problem is one that can be resolved satisfactorily only through research and a careful assessment of the alternative values to the public.

CHAPTER 13

Water Research Activities

Like the administration of water resources, the field of water research is also divided among a great many agencies and groups. Such division results from the natural interest which all resource agencies and economic groups dependent on water have in the development and utilization of water. The constitutional division of responsibility between the states and the federal government has served to disperse the research activity even further.

Research that will be applicable to the arid regions of Arizona involves the efforts of many states, the federal government, and even international agencies. Much that is done in other Western states will have a direct impact on further research in Arizona, and perhaps on management practices. For example, the study of watershed management for production of more usable water was partly the result of similar research in Colorado and elsewhere. Recent meetings and publications by the United Nations Educational, Scientific and Cultural Organization, through its Advisory Committee on Arid Zone Research, have indicated a worldwide movement to correlate and share the results of research efforts made in all the arid regions.[1]

The federal government has played a crucial role in the field of water research as it has in management and development. In part, this is because the federal government owns or manages a large portion of the land area of Arizona, making research on that land its primary responsibility. Thus one finds the Forest and Range Experiment Stations, the Agricultural Research Service, the United States Geological Survey, and units within the Bureau of Land Management, the Bureau of Indian Affairs, and the Fish and Wildlife Service playing central roles in the investigation of problems relating to use and management of water. In many areas, such as research on fish and wildlife management and irrigation practices, the state has cooperated fully. In other areas, because of a lack of resources or foresight, the state has either failed to take the initiative or has acquiesced in the assumption of responsibility by the federal agencies.

It cannot be said that the combined efforts of all units of government have been adequate in view of the soil and water problems of Arizona. As Edward Higbee stated, ". . . only about 5 percent of our experiment station research is devoted to problems on the frontiers of the unknown and unpredictable . . . our public research is . . . safe inside the bounds of achievement within some approaching fiscal year. Even the scientist is usually expected to meet a deadline."[2] Others have pointed out the serious deficiencies in our knowledge of basic hydrologic processes and even in the collection of the basic data upon which studies must be based. Linsley contends that since 100 million new acres of agricultural land will be needed by 1975, it is necessary to embark on an accelerated program to accomplish the following:

1. Evaluate accurately the quantities of water involved in each phase of the hydrologic cycle within the area.
2. Determine the most effective development and use of the available water, including reclamation of water lost through natural processes.
3. Consider the feasibility of obtaining water from new resources.
4. Develop the accessory information necessary to the best application of the principles determined under 1, 2, and 3.[3]

Some of the areas needing immediate attention, he asserts, are: increasingly accurate means of measuring precipitation and gross water supply and their variation; improvement in the techniques by which the presently available supply may be measured; study of the processes and the amounts of evaporation and transpiration and consumptive use; augmentation of water supply through various techniques such as nucleation of clouds, utilization of sea water, artificial recharge, and conservation practices.[4]

Similar increased effort was recommended by the President's Water Resources Policy Commission in 1950, and the President's Advisory Committee on Water Resources Policy in 1955. The earlier commission was concerned primarily with the collection and integration of data on a river-basin level. It charged that the "Federal basic information program is essentially fragmented, with many gaps and overlaps . . . ."[5] Essentially it recommended: collection and analysis of basic data should be recognized as an essential element in a resource program with increased allocations of money; collection of data on a river-basin basis with regular reports to Congress; expansion of hydrologic information through increased numbers of gauging stations, surveys, and processing of data; enlarged mapping programs; extended land surveys and classifications; collection of data on fish and wildlife; gathering of data on social and economic conditions relative to water policy; and many others.[6] The Advisory Committee noted similar deficiencies in scope and quality of record keeping and evaluation of data and recommended a doubling of expenditure in the research and data collection work.[7]

The 1950 commission presented a comprehensive analysis of the deficiencies of data collection for the entire United States and then broke this data down into regions and river basins. The following represents the adequacy of data collection facilities in Arizona river basins expressed as a *percentage* of what is considered necessary:

Precipitation: 41-60.

Evaporation: 41-60 in Gila Basin; 0-20 in Lower Colorado; 21-40 in Little Colorado; 81-100 above Hoover Dam.

Snow Course Coverage: 41-60 in Gila Basin; 89-100 in Lower Colorado; 21-40 in Little Colorado; 61-80 above Hoover Dam.

Surface Water Stream Gauging: 41-60 in Gila Basin, Lower Colorado, and above Hoover Dam; 61-80 in Little Colorado.

Ground-water Data: 0-20 for all areas except the central valley, Gila River above Yuma, and Safford Valley where it is 41-60.

Chemical Water Quality: 21-40 for all areas except Little Colorado, where it is 0-20.

Sediment Load: 0-20 in all areas except above Hoover Dam where it is 21-40.

Reservoir Sedimentation: 41-60 in Gila Basin; 21-40 in Lower Colorado and Little Colorado; 0-20 above Hoover Dam.

Topographic Mapping: most of the state unmapped; new maps in some areas, especially in central and southern Arizona; older surveys in some areas.

Geologic Mapping: less than 10 percent of area of state, in isolated tracts, adequately mapped.

Soil Surveys: suitable surveys for present agricultural needs in central valley, Yuma area, and Safford Valley; no surveys for almost all the remainder of state.

Soil Conservation Surveys: limited to very small isolated areas in central valley.[8]

Unquestionably there have been improvements in data collection in some areas such as soil and soil conservation surveys, ground-water data collection, and especially in geologic mapping. This last has been done by the Arizona Bureau of Mines which has recently completed the geologic mapping of the entire state.[9] Despite these advances, the improvements have been nowhere sufficient to make up for the deficiencies cited by the commission.

There have been some indications that the state's interest in research is on the increase. The establishment of the Water Resources Division of the State Land Department may be a harbinger of greater activity. The cooperative effort of the Agricultural Experiment Station, the Salt River Valley Water Users' Association, and the State

Land Department in the work leading to the Barr Report is itself evidence of increased interest.

In the chapters which follow, the work of several important different state and federal agencies and groups which are involved in water research is discussed in greater detail. The fact which needs to be continually emphasized, however, is that too little has been done. One Arizona scientist suggested that this should be the main point of any discussion of scientific effort in the field of water resources. It is his opinion, and one shared by the writer, that this deficiency has transcended the scientific field and become a problem of politics. Only with greater public awareness of the deficiency and of the contributions which science can make, and only with public willingness to provide sufficient means to support full-scale research, will many of the water problems facing Arizona be alleviated.

CHAPTER 14

Research at The University of Arizona

A significant portion of water-resources research being undertaken by non-federal agencies in the state is conducted by various divisions of the University of Arizona. This work is supported not only by state appropriations, but also by grants from the federal government, from private foundations and agencies, and from other state and federal agencies. Increasingly, the emphasis is on the interdisciplinary approach, with personnel from one field or division working in close cooperation with that of another. For this reason, it is appropriate to discuss the exceedingly varied water research going on at the university in a single chapter.

ARIZONA AGRICULTURAL EXPERIMENT STATION

The oldest and one of the most important research institutions in the state is the Arizona Agricultural Experiment Station. Established under a grant from the U.S. Department of Agriculture in 1889, it has been a fertile source of information for the farmer and rancher on every phase of the agricultural and stockraising industries. While not all the work of the experiment station is directly related to soil and water conservation, all of it is done within the framework of a region perpetually faced with a water shortage, and thus the work must be directed in the light of that condition.

During its second decade the station began the investigations of arid region agriculture which it has continued down to this day.[1] The lines of approach naturally concentrated on the problems of irrigation agriculture and the study of means of adapting the techniques and crops found elsewhere to this artificial watering system. Work also began early on the improvement of the cattle and sheep industries in the Southwest.

Like other experiment stations, the personnel of the Arizona station divide their time between research and teaching in the College of Agriculture of the university. The work of the station is as varied as the field of agriculture itself, resulting in the division of research among departments of agricultural biochemistry, agricultural chemistry and soils, agricultural economics, agricultural education, agricultural engineering, agronomy, animal pathology, animal science, botany, dairy science, entomology, home economics, horticulture, plant breeding, plant pathology, poultry science, and watershed and range management.

A sizable share of the support for research in agriculture at the station comes from the federal government through its several grant-in-aid programs. In fiscal 1961-62 the federal government provided $425,874 out of a total of $3,101,819 in receipts. The state, however, is the largest supporter of agricultural research, its appropriations reaching $1,902,224. An additional $773,721 came from private and federal special endowments and gifts.[2]

There has been a decided increase in the regional research program at the station, and more is expected in the future. As an example, a study of the historical changes on Southwestern ranges was undertaken in cooperation with the New Mexico station.

The director of the Agricultural Experiment Station is the coordinator of research and is responsible for the formulation of an annual program of work, including a budget, for the approval of the Office of Experiment Stations in the Department of Agriculture insofar as the program involves the expenditure of federal money. Periodic inspections are made by this office and progress reports are required of the station personnel. In most instances the initiative for station projects comes from local station personnel, although at times the Department of Agriculture will propose projects of interest to the department and having relevance to conditions in many states.

The Arizona Agricultural Experiment Station maintains a very close working relationship with research people in the Agricultural Research Service, of which the federal Office of Experiment Stations is an integral part. This closeness is exemplified by the fact that many of the ARS researchers are physically housed in the College of Agriculture on the campus of the university. There is close consultation on the development of research plans and in the execution of these plans. Graduate students in the college often take part in these projects and some ARS personnel teach in the college also. Many of the projects involve cooperation between station personnel and the ARS employees. The university has 11 experimental farms throughout the state on which experiment station personnel conduct much of their research. These facilities are also utilized by the ARS employees.

Since its very inception the station has provided a steady stream of technical information for research people as well as for the people actively engaged in agriculture. It now publishes "Bulletins" primarily for the use of the farmer, and "Technical Bulletins" which are more useful in the furtherance of future research. The College of Agriculture also publishes *Progressive Agriculture in Arizona* to disseminate the results of experiment station research to the practicing farmer. The station each year publishes *Arizona Agriculture*, a report assessing the general agricultural situation and providing statistical data on the productiveness of agriculture, and on income, costs, and related problems for the previous year.

In irrigation agriculture the condition of the soil has a vital bearing on the productivity of the land. Where water is at a premium it is necessary to determine the correct amount of water to be applied, at what times, and on what kinds of soil. Research in this field has received a great deal of attention by station personnel and with important results.[3] Extensive studies have been made of many kinds of soil and of their reactions to various kinds of fertilization. Particular attention has been paid to the effect on water movements in soil under application of these fertilizing elements.[4] Studies of this kind continue at the present time on both irrigated lands and range lands in order to provide the farmers and ranchers with information on which to base their land management and tillage practices.

The importance of such studies is indicated by estimates that significant savings of water could be made through improved applications of water on ground that had been properly prepared, with the proper degree of fertilization, and at the proper time. Studies have been made of irrigation practices on various crops such as cotton, citrus trees, irrigated pastures, lettuce, and alfalfa, indicating the ways in which the farmer can increase his yields, protect his crops against deleterious effects of various conditions and pests, and thereby provide a higher return on his investment.[5] For example, at one time it was thought that irrigating cotton immediately after planting caused excessive leafy growth; however, experiments conducted at the station have demonstrated that irrigation shortly after planting stimulates plant growth and increases yield.[6]

The station has always played a leading role in the investigation of water supplies in the state. Led by G.E. P. Smith, who contributed greatly to the knowledge of the underground water supplies, the station's early studies stand out as competent analyses of the limits to which agricultural production could be extended.

The Department of Agricultural Engineering has devoted much of its time to the investigation of the water supplies of the surface streams and the ground-water basins. This has involved determining the quantities of water in the streams and basins under varying conditions and over long periods of time, and also developing improved methods for extracting and utilizing the water. For example, studies have been made of the underground water resources and the geologic settings of selected valleys,[7] as well as of precipitation and stream flow in a number of river basins in the state.[8]

With the cooperation of the city and county governments, the station completed a survey of the underground water supplies of the Upper Santa Cruz Valley in which the city of Tucson is located.[9] This study indicated the serious nature of the water problem and pointed to some possible solutions, taking into account the heavy increase of population expected in the area. Limited studies of this kind have also been conducted in the Little Chino Valley near Prescott.[10]

The extremely small percentage of water recovered in Arizona from the total annual precipitation has led station personnel to investigate the possibility of increasing the recovery by means of recharge of underground supplies.[11] Preliminary to actual recharge experiments, a study is being made of the hydrology and water utilization of small drainage areas on the valley slopes of southern Arizona. Total runoff is being studied by means of rain gauges and reservoir measurements. Much of the runoff is lost through evaporation and transpiration and it is hoped that such techniques as waterspreading and recharge through wells will allow a much higher percentage of the storm water to go into the ground-water basins. If successful, such practices could go far toward reversing the long-term trend toward exhaustion of ground water for agricultural use.[12]

Fully as important as finding ways to increase the water supply are studies designed to improve the efficiency of water usage. The farmers of Arizona have made great strides in the management of water by such methods as lining canals and ditches, improving the efficiency of their pumps, and experimenting with sprinkler irrigation. Much of the research work done along these lines has taken place under the auspices of the experiment station.[13]

As noted elsewhere, Arizona has problems that relate not only to quantity of water but also to quality of water. Analyses have been made periodically of the quality of the waters throughout the state, indicating the difficulties encountered in producing crops or providing suitable domestic water in those areas with unfavorable quantities of salts, sediment, or fluorine. Studies of this kind began with the establishment of the station, involving the setting up of standards of quality of water for various purposes, determining the effects of different kinds of water on plants, soils, human beings (as in the mottling of teeth), power plants, and laundry use, and surveying the water of the state in terms of these standards.[14]

The salt problem continues to cause much concern in certain agricultural communities in the state. There is a steady increase in the salt content of the soils of the Safford and Lower Gila valleys, owing to the use and re-use of waters having a high salt content. The increase of pump irrigation has continued to add salt to the soils, with the result that only the most salt-tolerant crops can be grown in some areas. The Salt River Project maintains the general good quality of its waters only by a judicious mixing of its gravity flow and underground waters. The addition of Colorado River water is looked upon as a means of reducing the salt content of Arizona waters, when and if Colorado River water is available. Studies have been undertaken at the Safford experimental farm to determine the effects of the use of water of high salt content, the means of reducing the amount of salt, and the crops that can be grown under high-salt-content conditions.[14]

In 1954 the University of Arizona established the Institute of Water Utilization "to plan and coordinate an over-all program of research in the conservation and utilization of water supplies under semi-arid and arid climatic conditions." Originally proposed on the basis of a program costing $400,000 per year for five years, the institute was slow in taking form and was finally activated in 1957 with an annual budget of approximately $30,000. While its mandate is broad, the institute has tended to concentrate its activities on projects involving artificial recharge and evaporation suppression — projects which will continue to bulk large in the institute's future plans. It is also very much interested in the efficient use of water in irrigation projects and stock tanks. In its work the institute has cooperated with federal and state agencies and university departments in their programs of management, research, and education.[15]

Two important departments in the experiment station are agronomy and watershed management, which includes range management. With the Arizona cattle and sheep industries of great economic importance, the role of research in maintaining and improving the condition of the range is critical. This is particularly true inasmuch as the ranges in this arid country have been so badly depleted through drought and poor management practices. Studies by the station during recent years on two widely separated areas of the desert grassland indicated that there "has been a 20 to 28 percent decrease in desirable forage production and as much as 107 percent increase in undesirable shrubs *during the last* 40 *years alone!*"[16]

Range management investigations have included studies of water requirements for various grasses, and the desirability of different grasses for forage;[17] surveys of various species of grass found naturally in Arizona; experiments with various types of forage grasses, under varying conditions, to increase production;[18] historical studies of the ranges and the causes of their depletion;[19] experiments with many different techniques and materials to control the invasion of undesirable shrubs and plants;[20] surveys of range conditions, and experiments with management practices to maximize production on a long-term basis.[21]

These studies, often undertaken in cooperation with various government agencies, have provided much needed information on methods of caring for the ranges. They have prevented much wasteful experimentation by individual ranchers in the use of inadequate or incorrect techniques on ranges that would not have responded to such treatment. They have provided information on means of conserving forage production during periods when the ranges are depleted because of inadequate rainfall. The job of protecting and salvaging the ranges is far from over, and the experiment station continues to provide money and personnel for this work.

Perhaps the most important field of current research is on the control of noxious shrubs. Work has been conducted on methods of eradicating such invading plants as cactus, chaparral, and others. Studies are also in progress on factors affecting and affected by noxious-shrub control activities on the range, on the effects of shrub control on forage production, and the determination of sites favorable for shrub invasion.[22] Further work has been done on changes in the desert grassland and analyses of the causes. Another study investigated the possibility of providing more water for stock on the ranges by means of paved watersheds.[23] Promising research is being conducted on plants not now being grown in Arizona but which have potential value for forage production and soil building on Southwestern range lands.

Current research at the station falls generally into four categories: noxious-plant control, range reseeding, range management practices, and classification of ranges. Considerable work still needs to be done on range fertilization, development of improved management methods, crossbreeding of new forage plants adaptable to Southwestern conditions, and developing improved plant strains.

Agriculture in Arizona is a high-cost operation, and thus it is natural that the experiment station has placed a great deal of emphasis on research that will have a practical bearing on the economic position of the Arizona farmer in the present-day economy. For this reason most of the research has been the applied kind — improving accepted techniques, improving products of proven value, and finding new uses for these products. This emphasis is in conformity with the general approach to research laid down by the Hatch Act of 1887 which authorized expenditure of funds for the acquisition and diffusion among the people of the United States of "practical information on subjects connected with agriculture, and to promote scientific investigations and experiments respecting the principles and applications of agricultural science."[24] Title I of the Bankhead-Jones Act expanded this authority to include "research into laws and principles underlying basic problems of agriculture in its broadest sense,"[25] but emphasis has remained on "practical" work.

Whether research at the experiment station will continue to emphasize the "practical" aspects is not clear. Several years ago the Dean of the College of Agriculture at the University of Arizona called for a redirection of research in the future, away from work designed to increase production and toward research designed to protect the agricultural resource itself. "We suggest more attention to soil and water conservation, adjustments in management of farm enterprises, and developing new uses for agricultural products."[26]

It should be noted that in recent years some of the personnel of the station have taken leading positions in conservation programs proposed for Arizona watersheds. Dr. George Barr, formerly of the Department of Agricultural Economics, is the leading author of the now-famous Barr Report proposing departures in the management of the forest and chaparral regions in the interest of increased water production. Robert Humphrey has been a leading exponent of prescribed burning in the forests and the use of wildfire on the ranges as a means of increasing forage yield and reducing water consumption by non-beneficial plants.

LABORATORY OF TREE-RING RESEARCH

The Laboratory of Tree-Ring Research has a long-range role in water research. This role is the result of investigations made by A.E. Douglass, an astronomer at Lowell Observatory in Flagstaff, Arizona, who in 1901 became interested in studying the solar and climatic changes as they were recorded by the varying character of tree rings.

Through his observation of pine trees in the Southwest, Dr. Douglass determined that variations in precipitation are reflected in the width of the annual growth rings of trees growing in "sensitive" locations — locations where trees are totally dependent upon precipitation for their moisture. During a wet year these trees form a wide ring, and during a dry year a narrow ring. Dr. Douglass further discovered that widespread variations in the amount of precipitation from year to year cause the same responses in trees over large areas, and he concluded that tree-ring patterns would furnish long-range weather records. Thus the science of dendrochronology, or tree-ring dating, was born.

In 1906 Dr. Douglass joined the faculty of the University of Arizona. Under his aegis research in dendrochronology flourished, and in 1939 the Laboratory of Tree-Ring Research was established. The work of the laboratory has been supported for the most part by the university, but considerable assistance has been given by private agencies and foundations, and more recently by federal agencies such as the National Science Foundation and the Office of Naval Research.

The basic principle of dendrochronology is the principle of cross-identification or cross-dating.[27] By matching the ring patterns of specimens of overlapping periods, a longer pattern than is represented by any one specimen can be built. This chronology can be carried as far back as old wood can be found to extend it. In the process the ring patterns of large numbers of specimens must be compared in order to eliminate growth characteristics which tend to skew the record: false and missing rings, for example. Only by comparisons of this kind can assurance be obtained regarding the authenticity of the records provided by the trees.[28]

An equally important principle is that of sensitivity, which "permits an estimation of the quality of climatic record in a ring series by the magnitude of change in width of successive annual rings."[29] Thus, the more violent the fluctuation of width of rings in a series of cross-dated trees, the most useful is the record contained in the series.[30]

The laboratory has made many important advances. It has found trees in a wide variety of places having the growth characteristics necessary to furnish the needed climatic information. In the High Sierras of California it has found bristlecone pines more than two thousand years old which can thus provide information on climatic history in that area for that period of time; it has given archaeologists information concerning the age of prehistoric sites by applying its methods of cross-dating to archaeological specimens; it has made analyses of more than one million tree-ring radials, and has established a master tree-ring chronology for the Southwest having accuracy as far back as two thousand years; and it has found trees having a higher degree of sensitivity to climatic change.[31]

Early researchers in this field assumed that there was a relationship between tree growth and climate; therefore the early period of study was devoted to the development of growth histories from the trees that were thought to be the best indicators of climatic change. During the decades that followed, two lines relevant to water research became discernable. The first of these was a continuation of the study of climatic history as recorded by the tree rings. This approach emphasizes the acquisition of improved data from which to describe more accurately the climatic pattern of the past. Great effort was made to find trees that were more sensitive and of greater age so that the histories might be pushed back further and therefore be more useful in describing climatic changes and perhaps improve the ability to discern regularities in the fluctuations in the climate.

An important corollary development involved the archaeological dating of ruins of prehistoric cultures and their artifacts. By comparing the tree-ring patterns found in building timbers and in charcoal from fire pits with patterns recorded from the millions of specimens examined, an accurate determination of the dates of a culture and its works can be made.[32]

The second approach was largely an offshoot of the first and concerned analyses of the available data to determine the "physical explanation of such cyclic variation, and the use of this variation to predict changes in the future. Since variations in the amount of annual growth of trees reflect variations in meteorological elements (principally rainfall in the dry Southwest) we have the problem of determining the cycles in climatic phenomena."[33] The value of such analyses is obvious. The existing meteorological records provide no more than a century's history, and in the Southwest generally much less than that. If more precise regularities could be discerned upon which predictions of rainfall and temperature variations could be made, planning of other activities dependent on rainfall and water supply could be rationalized and losses owing to a lack of knowledge regarding water supply could be minimized.

Although the field of dendrochronology holds great promise for the future, there are many problems involved in attempting to derive a general relationship between rainfall and the width of tree rings, owing to the entrance of such disturbing factors as "variation in the distribution of storms within the year; changes from year to year in the frequency of specially intense or of long-lasting storms; differences in rainfall at the tree from that at the rain-gauge, pronounced spottiness of summer rains, occasional major maxima in the relatively inefficient summer rains," and others.[34] The interpretation of river flow as related to rainfall is complicated by such phenomena as "flash floods in intermittent streams which may constitute an appreciable part of the annual flow but have little or no counterpart in tree-growth."[35] In spite of these limitations on the accuracy of the technique, it is hoped to discover more precisely the long-term fluctuations in rainfall and runoff as recorded by the tell-tale trees.

The efforts of the Laboratory of Tree-Ring Research have concentrated on the western part of the United States and the arid region of the Southwest in particular. Definite advances have been recorded in analyzing the climatic fluctuations of the region which are considered to have great potential value in water planning. As yet, however, the laboratory personnel have not been able to work out a method for predicting fluctuations in precipitation based on tree-ring data. Writing in 1951 Schulman said:

> Since the growth records suggest no strong secular trend during the past millenium, the conclusion is inescapable that a series of generally wet years, as in the past, will again occur in such areas. Though the statistical chance of early occurrence is high, it will be obvious that in the absence of a physical explanation for such fluctuations no precise forecasting is possible. It is perhaps widely realized too, even at best a wet interval can only result in temporary alleviations of the increasingly critical problem of water shortage in these regions of rapidly growing population.[36]

After discovering and examining trees of higher sensitivity and longer growth records, Schulman was able to say of the present drought situation in the Southwest that it represents a "major disturbance in the general circulation" over western North America. Whether this represented a fundamental change in the type of climatic fluctuation was not clear and awaited the development of "more extensive, significant dendroclimatic histories."[37] It was his opinion that the current drought is likely the "most severe drought since the late 1200's," which is the period of the most severe drought on record.[38]

The additions to hydrologic knowledge resulting from tree-ring research have broadened immeasurably the field of research and have clarified the historical context within which water investigation and planning can go on. Though still imprecise, and though the future may never bring mathematical certainty in the field of prediction, tree-ring research may bring sufficiently accurate information to guide the construction of major works and economic developments dependent on water supplies.[39] Douglass wrote that he believed that the

Schulman report on *Dendroclimatic Changes in Semiarid America* would "carry over to the managers of hydroelectric and reclamation projects about the world a good idea of the type of information that may be secured from properly selected and analyzed trees."[40]

"Full understanding of the information contained in . . . ring indices must await the construction of similar series for the other dry lands of the earth,"[41] but the Laboratory of Tree-Ring Research continues to add vital information about the historical production of water in the desperately water-short Southwest.

INSTITUTE OF ATMOSPHERIC PHYSICS

During the years 1948 through 1950 cloud-seeding experiments, supported by several groups in the Phoenix area, were conducted on the Salt River watershed with inconclusive results.[42] These efforts stimulated the interest of state officials in the possibility of increasing the water supply through weather modification techniques. The support of Lewis Douglas, the former ambassador to Great Britain and a potent figure in Arizona politics, and President Richard A. Harvill of the University of Arizona was enlisted, and through their efforts the Institute of Atmospheric Physics was established within the structure of the university in 1954.[43]

The establishment of the institute is one of the very hopeful signs of progress in Arizona. The significance of the adoption of this program by the university lies less in the fact that Arizonans have recognized the dire water situation in which they find themselves, than in the realization that if any increased water supply is to be obtained, it will come only after extended and expensive research.

Prior to the 1950's there had been little scientific study of the processes of precipitation. The essential problem facing researchers in this field is the determination of whether any appreciable amounts of rainfall can be induced for human use. The institute stated in 1955:

It has become increasingly apparent from the work of commercial seeders and research workers during the past several years that a very great amount of fundamental research has yet to be done before man can intelligently control the precipitation of clouds, if that is indeed possible at all. The natural variability of rainfall is so great that occasional seeding activities followed by heavy rains are not at all clear-cut evidence of the reality of the seeders' claims to be able to increase rainfall.[44]

Through the experiments currently being conducted the scientists at the institute hope to determine scientifically and conclusively whether cloud-seeding is a possible means of increasing the water supply.

The importance of research in meteorology and climatology is indicated by the fact that this is one major phase of the hydrologic cycle yet to be studied seriously and comprehensively. The research projects on which investigation is needed are legion, but personnel and interest — as indicated by financial support — have been lacking until recently. Although the most direct purpose of research in the field of atmospheric physics is its possible application in controlling the formation and flow of moisture in the atmosphere for the benefit of man, such control may prove unfeasible, as researchers freely admit. But it is felt that the knowledge gained from studies of the atmosphere, moisture flow, and conditions of precipitation will be invaluable in planning the economy of the region.

The institute is one of a small but increasing number of institutions engaged in the study of atmospheric physics. One of these, the University of Chicago's Cloud Physics Project, was instrumental in getting the program into operation at the University of Arizona. Within two years the institute had gained sufficient stature in the field to host in 1956 an international Conference on the Scientific Basis of Weather Modification, a meeting financially underwritten by the Rockefeller Foundation and the President's Advisory Committee on Weather Control. The technicians of the institute have completed numerous projects on which they have reported in scholarly journals and in numerous meetings devoted to their research.[45]

The state of Arizona provides the major proportion of the financial support for the program and it is expected that it will carry an even larger share in the future. In addition, the institute has received grants from the Sloan Foundation, the National Science Foundation, the Office of Naval Research, the U.S. Weather Bureau, and the U.S. Air Force.[46] The permanence of the institute in the fabric of state-supported research is indicated by the incorporation of the institute personnel in the Department of Meteorology at the University of Arizona, with work being offered leading to the master's and doctor's degrees.

The fundamental approach taken by the institute toward experimental work involves a concentration on problems having a state or regional significance. Research conducted elsewhere under similar conditions will be relevant to Arizona research, as will the work done in Arizona be relevant to other areas. But the choice of research projects at the institute is guided by a determination to investigate problems that are thought to be specifically related to climatological and geographical factors peculiar to the Southwest and Arizona. The institute has tried to pick problems for research the results of which would be of value to the Southwest whether weather modification itself is successful or not. An example of this is the study now in progress of the moisture flow over the Southwest, a subject of the utmost importance to an understanding of the maximum possible water supply in the area.

One of the most important projects was carried out in cooperation with the U.S. Weather Bureau. The institute converted to punchcard form all the data available from Arizona weather stations from 1948 back to the late nineteenth century when the first weather stations were established. This data was then made available to the Weather Bureau which, in turn made a duplicate set of punchcards for each year since 1948 when the bureau first began using punchcards as a means of record keeping.[47] The data comes from approximately three hundred climatological cooperative observing stations which report temperature and precipitation data on a daily basis. Some of these stations have been in operation for well over thirty years, and thus provide much-needed long-term information. The same data, in more summary form, was later published as *Arizona Climate*.

As important as this information is to the technician in the field of atmospheric physics, it will also prove of indispensable value to others who are concerned with climate and water supply for scientific or economic reasons. A wealth of varied information can be gleaned from the punchcards, such as winter-summer precipitation ratios, relation between cirrus cloud occurrences and rain, meteorological factors contributing to interannual runoff variability in Arizona, and basic hydrological statistics.[48] Pilot studies have been made of some of the data and these have indicated the design which future larger-scale studies should take, as well as many unanswered problems.

In 1957 the institute began a five-year program of cloud nucleation to determine whether "artificial ice-crystal nuclei would modify the actual cloud and rain processes."[49] A target area was established in the Catalina Mountains near Tucson where rain-gauging stations have been located in order to determine statistically the changes in precipitation resulting from cloud nucleation. After four years of experimentation, Battan and Kassander reported that it had not been possible to demonstrate that seeding with silver iodide could lead to an increase in rainfall.[50] The high variability of rainfall is partly responsible for the inconclusive result, but investigations of the natural precipitation processes strongly suggest that seeding with ice-crystal nuclei is not likely to lead to any increase in rainfall in southern Arizona at least. Experimental work continues in order to learn more about the natural processes of precipitation formation, with particular regard to the possibility of modifying them.

As part of its overall research interest in the water economy of arid regions and as a result of certain fundamental measurements on evaporation rates from dry soils, the institute is testing the feasibility of an improved technique for the desalting of sea water.

Using solar energy to heat water and taking advantage of increased evaporation rates in relatively high velocity turbulent air flow, the institute's system, developed by Carl N. Hodges, differs substantially from previous approaches to this problem. Experiments on a few-hundred-gallon per day model at the institute's Solar Energy Laboratory show promise of producing fresh water at costs that would be attractive for domestic and industrial purposes.

The bulk of the research undertaken by the institute can be divided into three main categories: cloud physics and physical meteorology, synoptic and arid-lands climatology, and radiation studies. Much attention is devoted to physical studies of cloud formation. Aerial photogrammetry techniques have been adapted to photograph clouds for accurate measurements of ranges and heights.[51] A cloud census is being taken by radar and photographic techniques to determine the "frequency of occurrence, aerial distribution, and general characteristics of summer and winter clouds."[52] The institute states that it is "necessary to learn how the clouds initially form from condensing water vapor, how and why they grow, how the prevailing weather situation influences this growth, and finally what conditions are critically important in determining whether such clouds will become sources of precipitation. . . ."[53] For this reason much effort is expended in analysis of the dynamics of clouds, processes and rates of growth, and their propensities for precipitation.[54]

There is a great deal to be learned about the general characteristics of the atmosphere in which precipitation forms, and the broad picture of precipitation in the Southwest. The institute has engaged in pilot balloon studies to determine diurnal variation in wind structures, has used tracers to determine the structure of winds near the mountain barriers and to ascertain whether particles for nucleating clouds reach sufficient heights, and has studied mountain-valley wind circulation. Other work includes a study of climatological methods, particularly as they relate to statistical analysis, the role of precipitable water in Arizona summer rainfall to determine the extent to which variability of Arizona's summer rain is dependent on variations in "moisture aloft," the variation of precipitation in terms of frequency-distribution characteristics, and the hydrologic factors producing variability of runoff in the Gila River system.[55]

In the final category, the institute is investigating solar radiation.

—*Manley Commercial Photography*

Desalted seawater is one of the most promising new sources of fresh water for arid and semiarid lands throughout the world. Shown here is a 200-gallon-a-day prototype model of a seawater desalting plant developed by the Solar Energy Laboratory at the University of Arizona. An experimental pilot plant of 10,000 gallons daily capacity is now producing low-cost fresh water at Puerto Peñasco, Sonora, Mexico.

It is experimenting in the measurement of incoming radiation, engaging in practical experiments with solar radiation for summer cooling and winter heating, and attempting to "establish the relationship between the heat budget of the surface of the earth and the net solar radiation with respect to evapotranspiration from irrigated crops" in the hope of demonstrating means of making substantial economies in water utilization.

THE ARID LANDS PROGRAM

Another promising sign of progress in the field of water research was the establishment of an arid lands program at the University of Arizona. Supported by an initial grant from the Rockefeller Foundation for a three-year pilot study of arid-land conditions on an interdisciplinary basis, scientists from the fields of biology, archaeology, geohydrology, geochronology, and climatology joined forces. Projects undertaken by these scientists have included extensive surveys of the literature and field investigations in Southwestern prehistoric ruins, vegetation-soil surveys and analysis of the Sonoran Desert, comparisons of precipitation and precipitation efficiency in various arid regions of the world, an historical study of arroyo cutting in southeastern Arizona in the 1880's, pollen studies and study of fossils in several areas, investigations of the geology of arid regions, and geologic mapping.[56]

Although the Rockefeller Foundation is no longer giving broad support to the program it is supporting individual projects. The university continues to support the program through the many departments and research agencies on the campus which are undertaking studies of arid-land problems. These studies are coordinated through an arid-lands committee whose purpose is to foster an interdisciplinary and coordinated program of action and to encourage people to work together. The arid lands program is evidence of a clear recognition among the physical scientists of the complexity of the data with which they are dealing, and an attempt to maximize the return on their efforts through a many-faceted attack on arid-land conditions.

PROGRAM IN SCIENTIFIC HYDROLOGY

Newest of the water-research programs at the University of Arizona is the recently inaugurated program in scientific hydrology. Interdisciplinary in nature, the program offers an undergraduate degree through the Department of Geology in the College of Mines, and, through an interdisciplinary committee on the graduate level, the master's and doctoral degrees. Already, through a grant from the National Science Foundation, researchers have developed hydrologic models that demonstrate the flow of water through the rocks. Now underway is a program for developing analog computers to be used as the basis for undertaking to determine the probable ground-water conditions in Egypt. The only full-scale program in scientific hydrology in the nation, it will greatly enhance possibilities in water-resources research in Arizona.

CHAPTER 15

The U. S. Geological Survey

One of the most important agencies in the field of water research throughout the United States and certainly in Arizona is the U.S. Geological Survey. Established in 1879, it has provided the states, other agencies of the federal government, municipalities, and, indirectly, private groups and individuals with information on the available water supply and the quality of the water for more than three-quarters of a century.

Water resources investigations conducted by the Survey include "the systematic collection, analysis, and publication of hydrologic and related data; appraisal of water resources of specific areas; determination of water requirements for industrial, domestic, and agricultural uses; and research and development to improve the scientific basis of investigations and techniques."[1] These investigations are financed by the appropriation of federal money for projects where there is a paramount federal interest such as in areas or along streams where federal activity is contemplated. Other projects are financed from funds transferred from other federal agencies, such as the Bureau of Reclamation or the Bureau of Indian Affairs, to make investigations of interest and value to them. Finally, the Survey makes studies under cooperative arrangements with the states and municipalities on matters of peculiar interest to them and for which they are generally required to provide half the financial support.

The work of the Survey began well before the turn of the century in Arizona with stream-gauging work along the Gila, Salt, and Verde rivers. Although the work was discontinuous during the period from 1890 to 1900, the early beginning along the major streams has contributed greatly to the value of the records in providing basic information on the character of Arizona's streams.[2]

Since 1906 the USGS has divided its water resources investigations into branches, at first into two and later into three — Surface

Water, Ground Water, and Quality of Water. Each of these branches operates in Arizona with the first two having state offices located at Tucson, and other stations located throughout the state.

The Surface Water Branch "measures the flow of streams and collects, compiles, and analyzes other basic data relating to surface-water resources, including utilization and flood and drought studies."[3] The USGS maintains gauging stations on all the main streams in the state, including tributaries and washes. The gauges on the mainstream of the Colorado River are maintained by federal appropriations, but the largest part of the remaining gauges are provided for by cooperative funds provided by the state of Arizona and its subdivisions. In 1956 Arizona made available $36,000 to support the work of the Surface Water division, and has made money available ever since 1912 although there was a gap of two years during the middle thirties when the state failed to support the activity.

The importance of the work of the Surface Water Branch in the development of Arizona is indicated by the work of such early field workers as Arthur Davis and Joseph Lippincott. These men made some of the earliest studies of the water conditions on the Salt and Gila rivers and suggested the sites and conditions for constructing dams on those rivers.[4] Davis' work led directly to the construction of Roosevelt Dam; in fact, his work was commissioned by the group that later became the Salt River Valley Water Users' Association.[5] Some of the later studies were important in providing additional information leading to the construction of Coolidge Dam.[6]

The Surface Water Branch has provided information on floods which have proven essential in the construction of works to protect land and property from their destructiveness.[7] Recently, the branch engaged in work of this kind for the Corps of Engineers for the Painted Rock Reservoir on the Gila River.

One of the first general surveys of the Colorado River basin and general assessments of its potentialities for irrigation agriculture was made in 1916. Preliminary surveys were made of prospective sites for dams and power works and problems faced in capturing control of this mighty stream.[8] While inaccurate on many matters, this compilation of historical and general information brought much needed data together.[9] USGS records provided the information necessary for the compilation of such reports as the historical records of stream flows in the Colorado River Basin as found in Water Supply Paper 1313, published in 1954,[10] and the Bureau of Reclamation's *Report on Water Supply of the Lower Colorado River Basin,* published in 1952.[11]

The stream-gauging facilities of the branch are of great importance to the state and its people inasmuch as there is need for exact
determination of water quantities in order to apportion the water
among users. In some instances, as on the Gila and Salt, the water is
apportioned by court decree and it is therefore necessary to have these
records to carry out the mandates of the decrees. Gauging is also
important in the determination of the flow of the Colorado and its tributaries in carrying out the obligations of the United States under its
treaty with Mexico. These stream-flow records naturally play a very
important part also in the litigation over the waters of the Colorado.

Stream gauging also has an important role to play in the construction of dams and other works for the utilization of water. Accurate stream-flow data is basic for the planning of dams, flood channels,
irrigation works, spillway and detention structures, and other works
designed to bring the water to the places of ultimate use. It is necessary to know the quantity of water to be expected under normal flow,
peak flow, and the storage facilities necessary to provide protection
from flooding. The USGS is presently engaged in measuring the flow
characteristics of the San Pedro River near Buttes, preliminary to
the long-hoped-for authorization of a dam at that location for irrigation and flood control purposes.

While the greater part of the work of the Surface Water Branch
is taken up with gauging on the more important streams, considerable
effort is expended in cooperation with other federal agencies on experimental work having a long-range importance to Arizona. One of its
cooperative projects is on the Beaver Creek watersheds in northern
Arizona where the Forest Service is attempting to determine the
effects of several forest and woodland management practices on
stream flow. The USGS is installing the weirs from which the desired
information is to be taken. It also is providing the stream-gauging
work on some detention structures built by the Bureau of Land Management to ascertain the flow of water and sediment through these
structures which are designed to control erosion. On the Cottonwood
Wash project near Kingman the Surface Water Branch is trying to
determine the effects of removing broadleaf riparian vegetation.[12]

Stream gauging is important for the operations of dams and irrigation districts. The gauges transmit information on stream conditions and provide information on which to base the management of
reservoirs and canals and ditches. For this work the Surface Water
Branch has entered into cooperative agreements with a number of
cities and irrigation districts, including the city of Tucson and the
Salt River Valley Water Users' Association.

The records maintained by the Surface Water Branch are of great importance for the conduct of further research in hydrology. The collection of this basic data is essential in establishing correlations with rainfall, underground water supplies, use of water by phreatophytes, the effects of alterations in plant cover, and other relationships. Such information has been useful, for example, in making studies of the feasibility of recharging the ground-water supply by infiltration methods.

As early as 1897, F. H. Newell wrote the following regarding the collection of basic data along the mainstream of the Colorado River:

> When one has noted the interminable lawsuits over water rights and appreciated the fact that most of these are due to lack of precise knowledge as to past conditions, it appears almost incredible that greater care is not taken to ascertain exactly how much water is flowing at different points in the natural streams.[13]

In view of the judgment of the President's Water Resources Policy Commission in 1950 that such records were only fractionally adequate today, Newell's statment has a contemporary significance.

In recent years interest in the USGS in Arizona has centered on the Ground Water Branch whose responsibilities are to investigate "the waters that lie below the surface — their sources, occurrence, quantity, and head; and their conservation and utilization by means of wells and springs."[14] This shift of interest is the result of the fact that ground water has surpassed surface water in the quantity used in Arizona. Ground water presently provides 60 percent of the water used for all purposes in the state. While the mere investigation of ground-water basins and the recording of water supplies will not increase the quantity of water available from this source, it is hoped that ground-water investigations will provide more information concerning the underground supplies and lead to improved utilization of these supplies as well as better estimates of the quantity of subsurface water that can be exploited.

Ground-water studies have been conducted in Arizona since Territorial days. In 1904 and 1905 studies were made of the underground water supplies and the conditions under which they might be used in the Gila and Salt River valleys. This was long before pumping of ground water became a major factor in agricultural enterprise in the state. These early studies did not deal in any detailed way with the geology and structure of the underlying rock formations, but did provide information on the history of water utilization from this source, the flow of water, its quality, the cost of pumping, and other factors pertinent to its use. Later studies, such as those conducted by Meinzer and others, emphasized to a much greater extent the geologic factors and therefore provided much better information upon which to estimate the nature of the water supply.[15]

These earlier studies were conducted entirely through the support of the federal government. Some were the result of the need to provide information for Indian tribes, such as the Papagos, or to improve the conditions for travel on the desert when supplies of water were uncertain.[16]

It was in 1939 that the state of Arizona embarked upon a cooperative program to provide information on the ground-water resources of the state. During the previous decade greatly increased pumpage had occurred and most observers foresaw the need for a ground-water code to control what everyone admitted was a limited supply. The legislature provided less than $10,000 per year for the first few years, an amount clearly insufficient for any extensive or detailed studies such as would be necessary for adequate information on the basins and their supplies.[17] In 1945, with the ground-water situation rapidly becoming worse, the legislature appropriated sufficient money for extensive, though not comprehensive and detailed, surveys of the major ground-water basins. "These investigations were designed primarily to indicate the current ground-water conditions in each basin and the nature of the problems that would be encountered in framing a workable and equitable ground-water code."[18] It was anticipated "that more detailed investigations will be needed to evaluate the ground-water resources of each basin and to define precisely the conditions of ground-water occurrence."[19]

This phase of the Ground Water Branch program in Arizona, supported by both state and federal moneys, has for its purpose the collection of basic hydrologic data, including in its objectives the following:

> (1) To aid in the prediction of trends in ground-water levels as related to present and future ground-water supplies; (2) to delineate the present areas of greatest development as well as to establish the virgin ground-water conditions of areas of potential future development; (3) to aid in determining the geologic and hydrologic characteristics of areas as related to the ground-water regime; (4) to determine the changes in quality of water as to use and the "salt balance" problem; and (5) to provide continuous records of fluctuations of water levels in representative wells.[20]

With the greatly increased appropriations, the Ground Water Branch undertook a series of surveys of the major ground-water basins, indicating the geologic structure of the aquifers, the history of pumping, and the annual safe yield of the basins.[21] In almost every case where there had been extensive pumping operations, these reports indicated that there was a large overdraft and that the safe yield was being exceeded by large amounts. It was on the basis of this information, and in response to the pressure from the Bureau of Reclamation, that the Ground Water Code of 1948 was finally enacted. This survey program continues today and consumes better than half of the funds available to the Ground Water Branch each year.[22]

The inadequacy of the code in protecting the limited water supply has already been discussed. Owing to the continued overdraft of the ground-water supply, the state appropriated $300,000 for the Underground Water Commission to make a study of proposed changes in the code with a view to strengthening it. The USGS provided the commission with technical information in a report compiled in 1952 entitled, *Ground Water in the Gila River Basin and Adjacent Areas, Arizona — A Summary.* This comprehensive report traced the history of irrigation development in the state, outlined the problems faced in each of the major tributaries to the Gila as well as the Gila itself, and suggested the possible means of improving the situation. Since the publication of this report the Ground Water Branch has supplied the state each year, under its cooperative agreement, with a report on pumpage and ground-water levels throughout the state.[23] These reports amply testify to the steady decline in the water tables in Arizona.

In addition to the work done cooperatively with the state, the Ground Water Branch also enters into cooperative agreements with municipalities for investigations of their water supplies. Studies of this kind have been done for Phoenix, Globe,[24] Flagstaff, Safford, and other Arizona cities.

The Ground Water Branch, in cooperation with the federal land management agencies, makes investigations of ground-water conditions in areas subject to their jurisdiction. In recent years it has conducted ground-water surveys on the Indian reservations, notably the Navajo, Hopi, and Papago reservations.[25] In the recent past studies have been made for the Army Corps of Engineers[26] and the Bureau of Reclamation.[27]

Besides the general surveys of ground-water conditions in the broad basins of the state, the branch conducts studies of an intensive nature in some areas, often as the result of newly discovered problems. Studies of this kind have been made in recent years of the Douglas Basin and the Ranegras Plain area near Yuma.[28] Studies have also been completed of the Harquahala Plains, the section of the Bill Wil-

Two kinds of water research: Measuring moisture content of the soil (above), and an experimental pit (below) for recharging wells with flood waters which, if not captured, are lost to evapotranspiration.

—Ray Manley

iams Valley near a newly developed manganese deposit,[29] and others, the reports on which are on open file in the Tucson office.[30] Many of these areas, such as the Harquahala Plains and the Bill Williams River Valley, and more recently the McMullen Valley, have only lately been subjected to economic development, and it is hoped that information can be obtained about their water resources before the limits of feasible development have already been exceeded.

Another important area of inquiry concerns specific hydrologic and geologic problems. Studies of this kind have been made periodically. An earlier example of this facet of the branch's work was the study of the possibility of artificially recharging the underground aquifers by means of flood-control dams along the most important rivers.[31] In view of the tremendous water losses involved, the studies of the use of water by phreatophytes in the Lower Safford Valley are of particular significance, in spite of the fact that the investigation failed to point out methods which could be used economically to eradicate the undesirable vegetation.[32] Studies are currently in progress in the Salt River Valley in an attempt to determine the productivity of deep aquifers and the quality of such water. Analyses are being made of the hydrographic and geologic data collected since 1903 in the Casa Grande area.[33]

The Quality of Water Branch, whose office is located in Albuquerque, New Mexico, is responsible for the collection and analysis of data "concerning the quantity and character (organic and inorganic substances in solution and in suspension) and the physical properties of surface and ground waters."[34] This branch takes on increasing importance as the salt problem becomes more serious. Most of the studies of the ground-water situation in the state include analyses of the quality of the water. These studies have indicated very clearly the continued damage to the land, owing to the use and re-use of water pumped from the subsurface facilities. This problem of salty water in the Salt River Valley has already been noted, although it is being ameliorated by means of mixing gravity-flow and underground water. In some areas, such as the Safford Valley and the Lower Gila, the problem is aggravated each year by continued use of saline waters.[35]

Eleven quality-of-water stations have been established in Arizona along both the mainstream of the Colorado and along its tributaries, where samples are taken each day throughout the year.[36] In the view of the USGS, "It is becoming increasingly apparent that more judicious use must be made of available water for maintaining suitable quality and removing accumulations of salt."[37] In order to provide maximum beneficial use of the limited water supply, however, it is essential to have available continuous records of the chemical quality of surface water at key stations on the main streams that are used for irrigation. These continuous long-term records will assist in the determination of quality of water prior to irrigation development, the extent of impairment of water quality due to drainage return, requirements for maintaining proper salt balance and the equitable division of water between projects, States, and adjoining nations.[38]

In a different area, the Quality of Water Branch recently studied the rates of sedimentation of small reservoirs in order to find a method for determining the origin of sediment and rate of movement from various geologic formations.

Although the USGS is performing a vital function in the state, there are some who feel that not all is being done that should be done to improve the water situation through research. For example, at least 50 percent of the work load of the Ground Water Branch is devoted to the collection of data. This adds, to be sure, to the available quantity of information on ground-water resources, but does little to increase the basic knowledge concerning the resource, the conditions surrounding its occurrence, and the most suitable means of exploitation and replenishment. It is in the area of basic research in this field that much more must be done. Harshbarger, in describing the use of ground water in the state, asserted in 1957:

> Attempts to arrive at a quantitative value for the effective storage or the specific yield of a basin are rather difficult because of the number of variables that control these factors. It is impossible to arrive at any reasonable estimate of storage without detailed knowledge of geologic conditions, such as the interrelationships of the various strata and their structural setting. . . . With sufficient knowledge of geologic conditions and factors, it would be posible to determine the amount of water that could be withdrawn within certain depths.[39]

Only with such information, he contended, by describing the geologic conditions controlling the movement of water and the differences of permeability of the basins, could the proper decisions be made concerning the mining of water.

Harshbarger, in an interview with the author, suggested two important areas of research in which little or nothing is being done. First and of primary importance is the study of the subsurface structural formations and the geologic controls which determine the presence and the movement of ground water. Under present conditions, it is possible only to estimate the quantities of water beneath the surface, the movement of the water, and the response of the aquifer to recharge. With better information it would be possible to develop geologic models from which time responses could be determined.[40]

The second area of study having much promise concerns recharging ground-water reservoirs by means of wells.[41] Again, there is need of information concerning the substructure of the earth in order to know where such practices might be feasible. This method of recharge, asserts Harshbarger, holds the possibility of contributing 100 percent of the surface water put into the wells to the underground basins where the supply would be totally unaffected by evaporation. Some work on this was done in Arizona with optimistic reports. Referring to experiments on Queen Creek in central Arizona a USGS report stated:

> In the two years of record about half of the flow of Queen Creek at the mouth of its canyon was recharged to ground water. It might be possible, however, by the construction of a flood-control dam to put practically all the water into ground-water storage. . . . It should be possible to extend this limit [of ground-water supplies] in some localities by artificial recharge of water, thus utilizing water that otherwise would be wasted by evaporation on the desert.[42]

In spite of these favorable results, there is little prospect that the Ground Water Branch will be able to turn its attention to such matters without a fundamental reversal of attitude on the part of the state. Harshbarger has proposed a study of the possibilities of recharge in the Tucson area, where water shortages are predicted, but thus far has been unable to obtain adequate support.[43]

It is clear from the statistics on ground-water levels in Arizona that the subsurface water is being mined. There is little evidence that this situation could be altered without radical departures from present practices and disastrous effects on investments. The crucial failure, according to some technicians, lies in the fact that the water is not being mined intelligently. With better information it would be possible to exploit a maximum volume of water by determining where wells should be dug, how many should be dug, what distance from each other, and how much water should be pumped. It is the unnecessary waste, say these technicians, that is lamentable. And yet, despite what these experts view to be incontrovertible evidence, there are indications that the state is reluctant to continue the cooperative program with the Geological Survey even at the present relatively low level.

CHAPTER 16

The U. S. Department of Agriculture

A great deal of research that is pertinent to Arizona's soil and water problems is carried on by various agencies of the United States Department of Agriculture. In the majority of cases, there is close cooperation between the personnel of these agencies, as well as with research workers in other departments of the federal government and state and local agencies.

THE FOREST SERVICE

The overwhelming importance of the forested regions in the West to the production of water for domestic and irrigation uses makes the research into the water-giving characteristics of the forests of the highest priority. In forestry research, the federal government, through the Forest Service, plays the leading role in Arizona inasmuch as the state administers virtually no forested lands. The Water Resources Division of the Arizona State Land Department has encouraged experimentation along the lines suggested by the Barr Report, but is hardly equipped to enter the experimental field itself. State agencies like the Arizona Game and Fish Department, and the Salt River Valley Water Users' Association cooperate on many projects.

Operating through regional experimental stations, the Forest Service conducts research in the following general areas which are directly related to the management of forest and range for the production of water: forest economics, including studies of the economics of timber, range, and watershed management; watershed management research, directed toward improving soil and cover conditions, practices to alleviate flood and sediment problems, and improvements in streamflow and ground-water recharge; and range management research involving development of methods for maximization of forage production for livestock and game.[1] The research activities of the Forest Service in Arizona are administered through the Rocky Mountain Forest and Range Experiment Station with headquarters at Fort Collins, Colorado. There are three research centers in Arizona, each of which is located at the site of a major educational institution of the state, at Tucson, Tempe, and Flagstaff.

Like other research agencies concerned with natural resources, the Forest Service sees large areas in which expansion should be undertaken, or inquiry initiated. In the field of hydrology, there is seen a need to expand the research to additional areas and to improve the methods of studying hydrologic relationships and effects of certain practices on water supply; there is a need for improved methods of routing infiltrated water through the soil and rock mantle into the streams and ground-water basins. Intensified work should be undertaken on methods of checking erosion and improving land already eroded; increased study should be made of various processes which may contribute to erosion such as harvesting of trees and grazing of big game; more work should be done in the study of soils.[2]

The Rocky Mountain Forest and Range Experiment Station in Arizona has embarked upon a greatly expanded program of watershed research for the purpose of discovering the interrelationships among soil, plants, and water.[3] The Fort Valley station near Flagstaff and the Sierra Ancha station at Tempe are both engaged in studies which will contribute to understanding of these relationships. The importance of this work is indicated by the fact that the Salt River Valley Water Users' Association is contributing to the efforts at Sierra Ancha in the hope that the results of the experiments will indicate the means by which water production can be increased. They seek to answer the following complex questions:

> What are the relationships between climate (precipitation, temperature, evaporation, etc.), vegetation, and the water and sediment yield of a watershed? How do combinations of different plants, soils, and slopes affect water yield and erosion?
> What changes in water yield and erosion and soil take place on a watershed with a good cover of vegetation when that cover is affected in varying degrees by fire, drought, or grazing by livestock and game?
> What are the minimum vegetation requirements for adequate soil protection? Can these be attained under proper grazing, good timber management, good logging and road-building practices?
> What practical methods are there for restoring the vegetation cover on deteriorated watersheds under harsh desert conditions?
> How can watersheds be managed for maximum amounts of water and minimum amounts of silt without endangering the value of the watershed as a natural resource for forage, timber, and other uses?[4]

Experimental work on several types of water-yielding areas has been going on for a number of years: high water-producing and low sediment-producing areas in the pine-fir forest; intermediate water- and sediment-producing areas of the pinyon-juniper type where temperatures are moderate and rainfall less than in the forest areas; and low water- but high sediment-producing areas of semidesert and desert type where precipitation is low and evaporation is high.[5] Although much remains to be done, significant results have been achieved through these controlled experiments. For example, it has been found that winter storms contribute the greatest amount of usable water while most of summer precipitation is lost through evapotranspiration; water losses from bare soil equal those from soil having plant cover, but bare soil invites the invasion of unpalatable shrubs which use much more water than better forage grasses; perennial grasses use least water during water-yielding periods, while deteriorated-watershed shrubs use the most water; maximum yields of usable water come from watersheds having good plant cover, maximum yields of silt from those with poor plant cover; chief sources of usable water are subsurface runoff in winter and surface runoff in summer; there is a wide variation in the rate at which soils erode, thus requiring a variety of conservation practices; granitic soils at the lower elevations erode more easily, and once the plant cover is removed it is difficult to re-establish.[6]

These results have suggested new lines of approach, particularly concerning the manipulation of the plant cover to reduce erosion and to increase the water yield. One such project, begun in 1953, involved the cutting of timber to ascertain changes in water yield. Although nearly 50 percent of the merchantable timber was removed, it was impossible to determine significant changes in water yield owing to wide variations in precipitation. "More years of record will establish whether the type of cutting does influence water yield."[7]

One of the most ambitious experimental projects is currently in progress on the Wet and Dry Beaver creeks in northern Arizona under a cooperative arrangement among the Forest Service, the Rocky Mountain Forest and Range Experiment Station, and the Surface Water Branch of the U.S. Geological Survey.[8] This project, covering more than 250,000 acres, is perhaps the first to undertake vegetative manipulation on both an extensive and intensive scale, and involves one of the most important tributaries to the Salt River system. After an initial period of calibration, when measurements were made of precipitation, runoff, and sedimentation, the Forest Service began applying a wide variety of treatments including eradication of juniper which abounds in certain parts of the area, extensive thinning of the pine forest to remove the jack-pine in particular, logging of mature timber, and elimination of the understory. Cleared areas have been reseeded successfully. The Forest Service is studying the relationship of watershed treatment to grazing. The USGS is installing the necessary equipment to measure the changes resulting from the treatment and for comparison with control areas. Although this is considered a demonstration project, supported by increased appropriations for the Forest Service, it is hoped that the results will provide scientific evidence regarding broad-scale treatments in the forests which are demanded by many ranchers and irrigationists.[9]

Experimental work of a similar nature is being carried out on higher elevations of Arizona's watersheds also. One of these is at Workman Creek in the Sierra Ancha Experimental Forest in a mixed conifer forest. Streamflow has been measured since 1938 and two forks of the stream were calibrated between 1938 and 1953. Since 1953 Douglas fir has been replaced with grass on one fork, but as yet the results of this cut are inconclusive with regard to water production, although the results appear promising.[10] Other projects will involve block and strip cutting along lines suggested by experiments conducted in the Fraser Experimental Forest in Colorado.[11] While the mixed conifer area is relatively small in extent, it is in the highest precipitation area in Arizona. Researchers hope to discover whether removal of forest cover will influence the deposition of snow with resulting increased runoff.[12]

In the ponderosa pine forests the Forest Service is calibrating experimental watersheds on the Apache and Coconino national forests at elevations from 7,500 to 8,000 feet. Assuming that interception by leaves may affect the disposition of precipitation, it is felt that thinning and pruning practices may be developed which will increase streamflow.[13]

In addition to the Beaver Creek project the Forest Service is conducting several studies of watersheds in the pinyon-juniper and chaparral areas. Using varied techniques, such as control or removal of brush by chaining or cabling, chemical treatment, or wildfire, the service hopes to determine the effects of such practices on water yield and sedimentation.

The Bureau of Indian Affairs is cooperating with the enlarged watershed-study program in Arizona on two of its projects at Corduroy and Cibecue watersheds. Its techniques have included pinyon-juniper removal, prescribed burning in dense ponderosa stands, and reseeding of cleared areas. Although better forage and savings in cattle round-ups resulted, little has been discovered about water production.[14]

The losses of water attributed to phreatophytes are tremendous.[15] It is estimated that at least 1,280,000 acre-feet of water are lost from 405,000 acres in Arizona. The Sierra Ancha station has studied the problem with a project designed to determine the results of the removal of these shrubs upon stream flow. Supported by the Arizona Game and Fish Department, the Salt River Valley Water Users' Association, and the Forest Service, this study involves two basic approaches: the ecological and the physiological. The ecological phase involves studies of the development of these water-loving shrubs, their spreading capacities under various conditions, and means of eradication. The physiological phase includes efforts to measure actual water loss due to these plants and to relate these losses to such environmental factors as temperature, humidity, and types of soil.[16] After an initial calibration period, various kinds of treatment will be undertaken, including burning, chemical applications, and mechanical treatment to determine the effects of such treatment upon water production and soil movement. The Game and Fish Department will attempt to analyze the effects of these practices on wildlife population.

Another important area of research relates to erosion control. Efforts are being made to find the most efficient ways to control gullying and other forms of erosion, through reseeding, rehabilitation of gullies, and replacement of chaparral with grasses.[17] Results so far indicate that mere protection of eroded land will not in itself restore the land to its former condition. With reseeding, however, severely eroded lands have been stabilized.

The Forest Service has been under criticism in Arizona because of the reductions in cattle numbers under grazing permits. The cattlemen charge that the foresters are managing the forest for timber and are ignoring grazing values by allowing jack-pine to grow up, crowding out the forage. They demand a clearing away of the thick mantle of slash and litter which tends to reduce forage also. These cattlemen become the natural allies of the irrigationists who also want thinning of the forests in hopes of increasing streamflow.

The method for eradication of these unwanted trees and forest litter, assert the cattlemen and some forestry practitioners and scientists, is prescribed burning. They argue that burning, under carefully controlled conditions, would not eliminate the competition for forage and water, but would be a major step forward in the protection of the forests from wildfires. Some of the basic evidence for this position has come from experience of foresters working in the forests under the jurisdiction of the Bureau of Indian Affairs.[18] The technicians at the Fort Apache Indian Reservation reported in 1955 that "prescribed burning may be credited with an 82-percent reduction in number of fires, a 94-percent reduction in area burned, and a 65-percent reduction in size over the 3 years [1951-1953]."[19] Those who oppose burning claim that accelerated erosion will result from burning. Those conducting the experiments, however, claim the evidence is to the contrary. Weaver reported in 1952 that "prescribed burning has led to *no accelerated* erosion." More recent tests in California have tended to confirm this experience.[20] The soil appeared to retain its high infiltration capacity and damage appeared to be limited to those areas where the fire burned with great heat.[21] Damage to healthy trees, seedlings, saplings, and poles was felt to be negligible.[22]

For the most part, Forest Service officials in Arizona and the rest of the Southwest take a much dimmer view of the sufficiency of evidence concerning the advisability of using fire in the forest. They assert, and even the advocates of burning agree, that the conditions under which burning can be undertaken in the Southwest are extremely limited. Prescribed burning requires a "cool" fire and the suitable periods of the year and the days within those periods are relatively few. The 1950 conditions, when the best burns occurred, were considered ideal. It was concluded that only 5 out of 46 years would permit cheap, extensive, late November burning, and in 3 of those years, reduction in forest fuel would be less than in 1950.[23] While admitting that under ideal conditions prescribed burning may reduce forest fuels, upon analyzing the Indian-forest burns, the Forest Service researchers found the damage to desirable growth excessive and recovery of the forage too limited. They concluded that "controlled fire may be useful to reduce fuels in ponderosa pine in the Southwest, but the conditions under which it can be done without excessive damage are probably narrow and are not yet defined."[24] They are not convinced about the possibility of restoring grass to the burned area, either by natural reproduction or by reseeding. They assert that if burning results only in baring the soil, the intense storms are likely to erode the land and destroy all of its beneficial properties. At the present time little experimental work is being conducted along this line except on the Indian reservations in spite of powerful pressures from water users, cattlemen, and many technicians.

The Forest Service has jurisdiction over some of the finest open grazing land in the state. In recent years, however, these ranges have been invaded on a large scale by noxious plants which threaten to crowd out the forage and which also create conditions that accelerate erosion. All concerned recognize the critical nature of the problem but there is much dispute over methods of control and res-

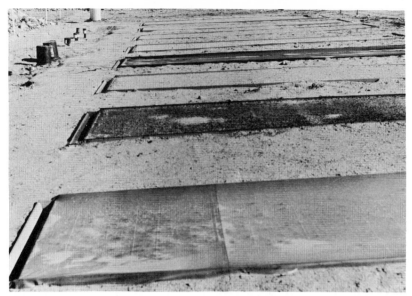

*Small artificial catchment basins on the watersheds can increase the
amount of runoff which can be recovered and put to beneficial use.
At Granite Reef, small plots test the efficiency and durability of
various materials (above). Below, aluminum foil-covered and asphalt-
treated experimental catchments on the Walapai Indian Reservation.*

—*U.S. Water Conservation Laboratory,
Agricultural Research Service*

toration. As a result, the Rocky Mountain Forest and Range Experiment Station, particularly the Santa Rita station near Tucson, is conducting research on the problem in cooperation with technicians at the Arizona Agricultural Experiment Station.

The Fort Valley Experiment Station has a project in progress to study the ecology of the pinyon-juniper trees, the grossest offenders in some areas. The station seeks to determine the reasons for the invasion of these trees into previously grassy areas. The Sierra Ancha station has conducted some experimental work on the use of fire to eradicate some woody species. Workers found some success but discovered that some plants, notably scrub oak, sprout even stronger after repeated burnings.[25] The station studied a wildfire burn on Mingus Mountain (an experiment with prescribed burning that broke out of control) and found that reseeding was almost a complete failure. Ashes and grass seed were lost to the rains and tremendous quantities of the soil were moved. The shrubs, however, quickly sprouted again.[26]

Others who have studied Arizona ranges at length, notably Robert R. Humphrey of the Arizona Agricultural Experiment Station, consider fire to be the answer to shrub invasion, particularly in the desert areas where mesquite is the central problem. He maintains that the control of fires has been instrumental in allowing the shrubs to invade.[27] He considers a burn conducted in the Pinal Mountains to be proof of the efficacy of the technique,[28] while the head of the Watershed Management Division of the Forest Service pointed to the burn as evidence that "burning is not a permanent solution to the problem of encroaching brushlands."[29] The Rocky Mountain Forest and Range Experiment Station concluded in 1955 that as a result of burning

soil is bared to rainfall, and damaging erosion is a consequence. Most of Arizona's brush species such as scrub oak come back quickly after fire. Grasses recover slowly and lose out to the vigorous brush. The end result in this sequence is a period of erosion and then the brush is back again. Fire may be extremely useful in improving chaparral areas, but it will take more knowledge than is now available to use it properly.[30]

The station has also conducted experimental work on the use of chemicals for eradication of shrubs. Such chemicals as 2,4-D and 2,4,5-T, as well as diesel oil have been applied by various methods to these plants with varying degrees of success.[31] Bulldozing and chopping by hand have also been tried, as have chaining and cabling. Attempts at reseeding have been made on both mesquite and pinyon-juniper areas with some success.[32]

In the existing drought situation, the range technicians have been much concerned with establishing range capacities. Measure-

ments of rainfall, herbage production, and livestock numbers on the
Santa Rita Experimental Range have indicated fluctuation in forage
according to the annual and summer rainfall, with the numbers of
livestock having to be reduced by 50 percent in the fall of 1956 because of inadequate forage.[33] Evidence indicates that "stocking should
be 40 per cent below the long time average about 35 per cent of the
time when droughts reach moderate and severe intensity. Drought
reductions necessitate rapid adjustments in animal numbers to maintain proper utilization."[34]

The Rocky Mountain station is attacking the erosion problem
by undertaking studies of the desert and forest soils. It is trying to
determine the productivity of soils in the hope that such information
will provide guide lines in the management of plant cover.[35] It is also
attempting to analyze the effects of the application of fertilizer to the
range: the effects on the soils, the methods, times, frequency, and comparative responses of different plants and soils.[36]

At the request of the Director of the Rocky Mountain Forest
and Range Experiment Station, there has been established a Forest
Research Advisory Committee consisting of members of the Arizona
Water Resources Committee (which is backing the increase in watershed studies as a result of the Barr Report), and representatives of
the State Land Department, the Interstate Stream Commission, the
Arizona Cattle and Wool Growers associations, the Arizona Game
Protective Association, and most of the public resource agencies in the
state. The purpose of this committee is to provide researchers with
"advice . . . on the problems of the state, and the direction which the
research division's studies should take."[37] In view of the controversy
that has developed around the proposals for vegetative manipulation
on the forests and ranges, this arrangement appears to be an attempt
to provide a measure of consultation and cooperation among the private interest groups and the governmental agencies responsible for
finding the answers to resource questions and managing these resources. The formation of the committee has been supported by even
the most outspoken opponents of manipulation of the watersheds.
Ben Avery, columnist for the *Arizona Republic* and a leading oppon-
ent of the Barr Report, wrote:

> Creation of the Forest Research Advisory Council should prove a great
> step forward in bringing all Arizonans together around the table to solve
> the problems of obtaining maximum human use of our natural resources
> without destroying them . . . . In setting up the council all groups were
> given representation. That is a very pleasant contrast to setting up the
> so-called Watershed Conservation Committee to produce the Barr Report

wherein only men known to have been sold on the idea that a tin-roof watershed is the salvation of mankind were permitted to participate.[38]

One thing is certain regarding forest and range research and its relation to water production and conservation. Many more years of research and careful analysis of research results are needed before definitive answers can be given to problems of increased water production, soil erosion, and forage restoration. And those who seek immediate action on these problems are growing impatient.

AGRICULTURAL RESEARCH SERVICE

The most important arm of the Department of Agriculture in the field of soil and water conservation is the Agricultural Research Service. Created in 1954, it combined many of the diverse research operations previously under the jurisdiction of other units in the department for the purpose of "making a coordinated research effort in solving some of the major problems of production and utilization that farmers face."[39] The research work of the Soil Conservation Service, for example, was transferred to the ARS although the Forest Service research was retained by that agency. SCS personnel, including those in Arizona, were not in sympathy with the transfer, attributing it in great part to the antipathy of some of the new top-level administrators toward the SCS and in favor of the Agricultural Extension Service.

The branch of the ARS concerned with soil and water research is the Soil and Water Conservation Research Division. This branch "conducts a broad program of research in soil science and in water management and conservation."[40]

The work of the ARS cannot be considered except in the context of the research conducted throughout the entire Southwest. The ARS maintains branches in Utah and New Mexico where problems and conditions are found having great similarity to those in Arizona. In each state are centers for crop research, range reseeding, soil and water conservation studies, production economics, and other activities related to maximization of production and conservation.[41]

In Arizona the work of the ARS is divided among three centers which devote their efforts toward solving soil and water conservation problems. The service maintains a center for hydrologic investigations and soil and water management at Tucson in connection with the Agricultural Experiment Station. There are also soil and water conservation agencies at Yuma and Phoenix. In various other cities, in addition to Phoenix and Tucson, there are researchers in horticulture, field crops, weed control, and regulatory activities, all of whom labor under conditions dictated by the perpetual shortage of water.[42]

There is a considerable amount of cooperation and consultation between the ARS personnel and related technicians in other federal and state agencies. Some of the researchers in the Soil and Water Conservation Research Division are also members of the Arizona Agricultural Experiment Station;[43] they teach some classes in the University of Arizona, utilize graduate students in their research, and correlate their research with the experiment station's work. The ARS workers utilize the experimental farms at the university, at Mesa, and elsewhere in the state. The ARS field men also consult frequently with the field men of the SCS, reviewing their problems locally and contributing their information to the solution of these problems.[44]

Much of the research effort by the personnel of the Soil and Water Conservation Research Division is taken up with analysis of irrigation practices and techniques. The importance of such research cannot be overemphasized. It is estimated that less than 10 percent of the total precipitation actually becomes runoff in Arizona. Tremendous losses are recorded between the point of diversion and the farmer's headgate. It was estimated by one field worker that for every 1.8 acre-feet diverted at Granite Reef Dam on the Salt River, only one acre-foot arrived at the headgate. The present head of the SCS in Arizona, Robert Boyle, estimated in 1952 that three out of every four acre-feet of irrigation water in the Western states was wasted. Only one out of four acre-feet reached the plant root and was used by the plant, for an efficiency of 25 percent.[45] With such a low level of efficiency he suggested that "if, by applying water conservation practices, the over-all efficiency could be increased to 50 per cent, additional water would be provided the irrigation farms of the West more than equal that expected from all the storage reservoirs now being planned."[46]

In the past few years greatly increased understanding has been achieved concerning the interrelationships between soil and water, and extensions of the previously completed work continue along this line. Studies in progress include: analyzing the factors affecting evapotranspiration, and depletion of soil moisture by various crops; relationships between soil moisture levels and plant spacing and their effect on yield and quality of cotton fiber; soil structure and its influence on water intake and movement; tillage practices and water applications; and methods for measuring these relationships.[47]

Further work is being done on the relationship of mineral soil amendments and crops and soils in irrigated land. In view of the alkalinity problem in Arizona, some attention is being devoted to the reclamation of alkali soil and its management, and studies are in progress of the adaptation of various crops to different kinds of soil.[48]

A great deal is being done in Arizona in analyzing irrigation practices and techniques. Study is being made of the problems involved in establishing design criteria for the layout of irrigation systems. The Soil and Water Conservation Research Division is conducting studies of irrigation practices designed to provide maximum efficiency of water use, uniform distribution of water throughout the root zone, minimum erosion, and other related desirable conditions. Closely related investigations are being made by other branches, notably the Field Crops Branch, on the crops most suited to conditions of irrigation agriculture.

The interrelationship between plant management and erosion control has long been recognized; in Arizona, where the balance between protection of the soil and plant cover and accelerated erosion is always critical, attention to this subject is mandatory. Areas under investigation include proper use of forage, including the determination of optimum livestock or game grazing that will encourage maximum sustained forage and animal production consistent with watershed and other uses; development of reliable and practical criteria and standards for judging conditions and potentials of ranges; range recovery following brush eradication, including effectiveness of fire, required frequency of burning, and resulting vegetational and ecological changes; grazing management of ranges; range fertilization including times, methods, and frequency of applications and their results;[49] breeding and adaptation of native and introduced grasses in the Southwest; reseeding of non-arable lands, including methods and conditions of reseeding, value of various kinds of grasses on each site, and longevity; and brush and weed control by means of various chemicals and herbicides under a variety of conditions.[50]

Closely related to the technical problems involved in plant and soil management are economic considerations. While some practices may be technically feasible they may be economically impossible. The ARS has a project designed to determine the costs and returns to be expected from various conservation practices in certain woodland watershed areas. Although the areas are relatively less important in terms of their water production, they are extremely important in terms of their propensity toward erosion. The purpose of the study is to determine the possibility, in terms of the costs involved, of establishing a grass cover in areas invaded by pinyon-juniper and other brush.

Studies of the economics of noxious-shrub control are also in progress in the Agricultural Research Service. Costs of certain kinds of control have already been indicated by previous research.[51] In the view of one rancher who is faced with shrub invasion, ". . . there is no method of control in use anywhere that is effective enough, inexpensive enough or fast enough to adequately meet the issue."[52] A great deal more research is necessary in this field, not only to improve forage but to protect against the erosion which occurs as the grass is displaced by worthless shrubs.

Two research projects are underway in Arizona in the field of watershed hydrology. On a Soil Conservation Service plot near Safford and on an experimental watershed near Tombstone continuous records are being kept of peak flows and water yields. Future studies are contemplated for the development of procedures by which evaluation can be made of the effects of watershed treatment on streamflow over both large and small areas. Closely related to these studies is an undertaking by the Field Crops Branch in Flagstaff and Phoenix to measure the evapotranspiration from such heavy users of water as salt cedar and juniper.[53]

At the Tombstone project and a related one in New Mexico, the Soil and Water Conservation Research Branch is trying to find the answer to the question that has excited irrigationists, recreationists, and conservationists for many years: What effect do conservation practices have on streamflow? Exact answers are sought to the following:

Under the present conditions of use, what proportion of the rain that falls on the area leaves as streamflow? What happens to the rest? That is, how much evaporates, how much replenishes the groundwater supply, etc.? What effect will range conservation measures have on these values? How will they differ with changes of vegetation from poor range plants to good ones?[54]

During the first five-year period emphasis was placed on calibration: observing the watershed without any modifications in order to measure rainfall, evaporation and transpiration, soil moisture, ground-water levels, runoff, channel losses, and sediment. Soil and vegetation surveys were conducted and the results of these measurements and surveys are being studied. During the next period, while the measurements and surveys are continuing, conservation measures are being introduced, such as reseeding, brush control, stocking adjustments, and erosion-control structures. From these studies it is hoped that information "on amounts of water and sediment contributed by low-rainfall areas to irrigation projects and the effects of range conservation on the net usable water" can be obtained.[55] Only with such results will it be possible to convince irrigationists that good conservation practices on the ranges and watersheds, while improving the condition of the range and perhaps reducing sedimentation, will not decrease the quantity of water available for growing crops.

There remain fields of inquiry as yet relatively untouched by technicians and for which there is no money available. One example is sedimentation. The ARS at present has no studies involving sediment production, sedimentation of reservoirs, or sediment-delivery rates in Arizona in spite of the fact that sedimentation is one of the state's most serious problems. "The high sediment content of many Arizona streams has been widely recognized . . . the San Pedro water is sometimes as much as 25 per cent silt, so thick it will scarcely run. It has carried 150,000 cubic yards daily into canals and laterals."[56]

In 1959 the ARS established at Tempe, Arizona, the U.S. Water Conservation Laboratory (at first called the Southwest Water Conservation Laboratory) as part of a nationwide effort to accelerate research on methods for conserving the existing water supplies and increasing water yields. Research facilities, specifically designed for new research techniques, will enable workers to obtain information that it has not been possible to get before. The most unique installation is a set of electronic weighing lysimeters, or plant growth tanks. Located outdoors, these lysimeters can detect and record a weight change of as little as two ounces in seven thousand pounds of soil.

To avoid duplication of effort, all research programs are planned in consultation with the experiment stations in Arizona, Nevada, New Mexico, and Utah, and with the Soil Conservation Service. Current research includes investigations of such varied aspects of the hydrologic cycle as the development and testing of chemicals to reduce evaporation, studies of the use of hexadecanol for lessening transpiration by plants, and the development and testing of soil sealants to increase precipitation runoff.

Of equal importance is research aimed at increasing irrigation efficiency. Irrigation water applied to fields has rarely been measured accurately; studies now in progress will help determine the minimum amount of water necessary on a given crop with a particular soil and under varying conditions of temperature and humidity to achieve maximum production. A more complex problem is the study of flow phenomena to develop mathematical equations "to describe the flow of water in thin sheets or small streams over rough, porous surfaces," since the only equations currently available "describe the flow of water in pipes or large channels and which do not apply to the kind of flow which occurs in an irrigated field."[57] Through research of this kind engineers will be able to design irrigation systems of far greater efficiency.

CHAPTER 17

Conclusions

One of the leading students of the water problems of Arizona once quoted an old saying to the effect that "life ain't in holding a good hand, but in playing a pore hand well."[1] The question for brief consideration here is a tentative judgment on how well Arizona is playing its "pore hand" of water.

Since the beginning of intensive settlement of the Territory by Anglo-Saxon peoples, water has been the crucial factor in sustaining the economy. In addition to its necessity for sustaining human life, water provided the means by which fertile but otherwise worthless land became highly productive. It was essential in the mining industry. It sustained the grazing economy. In spite of the wastage of water and abuse of the soil and forage resources which protected the water supply, the appropriation, development, storage, and application of these waters stimulated the economic growth of the region. Agriculture, including stockraising, became one of the leading industries of the state, and with mining, provided the principal outlet for economic endeavor.

The people of Arizona have periodically faced critical water shortages which were alleviated by helpful public policies and by technological developments. When the surface-water supplies in central Arizona were fully appropriated and no longer sufficiently reliable to supply all of the lands available for development, the reclamation policies of the federal government provided the financial means by which major structures could be installed for the storage of water. When even stored water proved inadequate, technological developments in pumps made them economically feasible for use in extracting water from the vast underground reservoirs.

Arizona has reached another critical period regarding its water supply. Not only are surface and stored waters fully exploited, but the pumpage of underground water has gradually reached the point

253

where it is exceedingly costly to pump the water to the surface because of the great depths from which it must be lifted. Water is literally being "mined." It is partly for this reason that there has been a reduction in agricultural acreage in the central valley in recent years. Unless other sources of water can be tapped, it appears certain that further reductions will be made. The rapidly growing populations in the major metropolitan centers are competing for this water. In the short run this relieves the pressure on the water base since many more urban people can survive on a given acre of ground using the water for domestic purposes than is the case when the water is used for irrigation. Nevertheless in the long run the competition of urban populations for water threatens seriously the agricultural base unless new supplies are found.

Public policy and technology are again looked upon as the means of salvation. Few are willing to allow the economics of the market place to allocate water to the highest bidder. For this reason the state, and particularly the residents of the central valley, have sought acceptance of the Central Arizona Project by the Congress. Now assured by the Supreme Court that Arizona will receive title to the 1,200,000 acre-feet of Colorado River water it claims, state officials are ready to press their case before Congress, buttressed by a favorable report by the Bureau of Reclamation. Less spectacular but nonetheless efficacious are the water-saving techniques utilized by Arizona farmers under various assistance programs such as the Agricultural Conservation Program and the Small Watershed Conservation Act Program. Moreover, several agencies apply a number of approaches toward water conservation through removal of undesirable growth and more effective channeling of water.

There of course remains the long-range hope that some technological breakthrough will alleviate the pressing water shortage without the need for extraordinary expenditures or for the deprival of anyone's water rights. At the present time interest in vegetation manipulation is at its highest but there remain many unresolved questions. These include the amounts of water which watersheds can be induced to produce when their cover has been changed, whether the problem of erosion can be effectively controlled, whether the costs of such manipulation justify the investment and if so, who will pay for the treatment. Much remains to be done before any firm conclusions can be drawn.

Some have placed great hopes in the development of practicable methods of cloud-seeding. But additional experimentation has only added to the skepticism of the atmospheric physicists that such techniques will ever prove fruitful. Desaltation of sea water may in the near future prove economically justifiable on a large scale, but it is doubtful that this method will help Arizona unless the costs become phenomenally low, since water so produced would still require transportation and elevation to the water-short areas of central Arizona.

It is entirely possible and indeed probable that Arizona will experience another period of wet years following the extended drought which now prevails. The Tree-Ring Laboratory asserts that there is little evidence of a progressive trend toward desiccation of the region; thus it may be expected that wet years will follow dry years according to the pattern laid down over the centuries. It would appear unwise, therefore, in spite of the temptations that will be presented, to base expansion of the economy on temporary increases in the water supply, since there is almost certain expectation that the future will bring renewed periods of drought. The wiser course would seem to require recognition of the minimum supplies available in the long run, and development of the economy on that basis.

With a projected population of more than two million in 1970, it is clear that the state will find it necessary to make major decisions involving the utilization of its water supply. Under present conditions it has been necessary to divert water from agriculture to municipal use in Phoenix, and similar diversions are now being required near Tucson. Other cities, such as Flagstaff, have experienced shortages of water. The rapid increase in job seekers cannot be absorbed by the agriculture and mining industries, for these industries are becoming more mechanized all the time. The economic future for the majority of Arizona's increasing population would appear to lie in other pursuits such as manufacturing, services, tourism, and perhaps in government employment. These activities will also require increased amounts of water which can only come from those supplies now devoted to agriculture unless some other source of water is obtained.

The population growth and the influx of industry will increase the requirements for electrical energy, some of which will come from hydroelectric developments along the Colorado River. Without the addition of this power in the very near future, Arizona will face critical power shortages comparable to those she has faced in the past.

These rapid changes in the nature of the Arizona economy invite forethought and planning of the use of water and the other resources that are involved in the production of water for beneficial use. The establishment of a resource planning office, or reorganization of state administration, while each is desirable in itself, will not supply this forethought and planning. Effective control may be extremely difficult because of the multitude of private decisions in the market place which determine how water is used. Nevertheless, both the state and federal governments, through their various agencies, carry a major share of the responsibility for ensuring intelligent use of the limited supplies.

In spite of the old adage, "When the well is dry, they know the worth of water," it does not appear that Arizonans have fully recognized their dependence on this vital commodity. Many assert the need for planning and increased adoption of the practices needed to conserve the limited water supply, and yet the state has done too little in taking the initiative in planning and integrated forecasting. In fact, some argue against planning *because* of the present shortages, asserting that it is unacceptable since it would mean the deprivation of some existing interests. Some agencies, such as the Interstate Stream Commission, the Arizona Power Authority, and the Water Resources Division of the State Land Department do engage in limited planning, but the planning by these agencies is often predicated upon as-yet-unreliable premises such as Congressional approval of the Central Arizona Project, and the technical feasibility of the manipulation of vegetation.

The state and federal government have both supported scientific research in water development and utilization and their efforts continue. It is doubtful, however, that enough of the scientific thinking makes a sufficient impact on those whose responsibility it is to make policy. The scientific information which might save the state untold economic hardship is too often filed away and ignored by those policy makers.

Instead, economic development dependent on the utilization of water resources and the ecological factors surrounding these resources goes on in an essentially haphazard fashion. Effective measures have been and are being taken by farmers and irrigation districts to utilize more efficiently their limited supplies. But attempts at regulation of the use of water have failed in recent decades as state agencies, executive, legislative, and judicial branches bow before the power of vociferous economic groups who see their positions threatened by needed adjustments. Moreover, the decisions that should have been made decades before, as in the case of ratification of the Colorado River Compact and the development of Hoover Dam power, are delayed until the situation becomes desperate. The decisions that should have been made in the 1940's in regard to ground water still had not been made in the early 1960's.

The law of the market place unquestionably has its virtues in regard to many economic questions. But when the question involves the fundamental basis of the economy, as the question of water does in Arizona, it would appear necessary to take the long view, to place the short-run economic interests in the context of the future economy of the state, and to provide the leadership in gaining acceptance of the policies which will bring most efficient utilization of the existing water supply and whatever adjustments might be required to meet changing economic conditions.

Efforts are being made to overcome the lack of interest in over-all planning of water resources development, because of the recognition that the shortages are becoming more severe and difficulties and complexities of solving the shortages more apparent. And at least with regard to watershed management there is an attempt to integrate carefully policy planning with technical data.

It may not be that the highly developed civilization of the desert will collapse for lack of water.[2] But it is possible that the state may ignore the advice given by Paul Burgess of the Agricultural Experiment Station in 1940: "Better a moderate area of permanent homes than a promotional development that must be shortlived."[3] The state must take steps to maximize the use of the water it has, and to plan for its needs in the context of the changing economy. Public agencies now confer on common problems, but there must be leadership which focuses attention and develops programs which take into account all of the highly complex factors of interrelated resources. Until this minimum for policy-making is attained, it appears that this most precious resource, the *sine qua non* for life on the desert, will not be put to maximum use in the support of the varying needs of man in this arid country.

—U.S. Bureau of Land Management

Only a vigorous conservation and reclamation program can restore the Southwestern rangelands denuded by overgrazing and drought (above). The few streams which have not been developed for urban, industrial, or agricultural use, such as Aravaipa Creek (below), help to feed the underground water tables, retarding the depletion of this supply. They also help to preserve natural vegetation and Arizona's wildlife.

—Chuck Abbott

—Chuck Abbott

Summer rainfall in the desert areas (above) contributes very little to the usable water supply due to excessive evaporation and aridity. Winter precipitation, especially snow in higher elevations (below), is the principal source of the state's surface-water supply. Much of the spring runoff is stored in reservoirs on the major river systems. The rest not lost to evaporation helps replenish underground supplies.

Notes to Chapters

The abbreviations listed below have been used throughout the Notes and the Bibliography which follow:

AAES—Arizona Agricultural Experiment Station
ACPS—Agricultural Conservation Program Service
AGPA—Arizona Game Protective Association
AISC—Arizona Interstate Stream Commission
APA—Arizona Power Authority
ARS—Agricultural Research Service
ASLD—Arizona State Land Department
BIA—Bureau of Indian Affairs
BLM—Bureau of Land Management
CAPA—Central Arizona Project Association
FPC—Federal Power Commission
FS—Forest Service
F&WS—Fish and Wildlife Service
NPS—National Park Service
SCS—Soil Conservation Service
UA—University of Arizona
USDA—U.S. Department of Agriculture
USDI—U.S. Department of the Interior
USGS—U. S. Geological Survey

Chapter 1

[1]Richard J. Hinton, *The Hand-Book of Arizona: Its Resources, History, Towns, Mines, Ruins, and Scenery* (Tucson: Arizona Silhouettes, 1954), p. 320.
[2]Joseph Miller, *The Arizona Story* (New York: Hastings House, 1952), pp. 304-306.
[3]Anthropologists are not yet convinced regarding all the reasons for the decline of the Hohokam culture, but all see the water problem as central. See Odd S. Halseth, *Reclamation Era*, Vol. 33, No. 12 (Dec. 1957), p. 251; O. A. Turney, "Prehistoric Irrigation," *Arizona Historical Review*, Vol. 2, Numbers 1, 2, 3, 4 (April, July, October, 1929, and January, 1930).
[4]W.G.V. Balchin and Norman Pye, "Recent Economic Trends in Arizona," *Geographical Journal*, Vol. 120, Part 2 (June 1954), p. 161.
[5]William D. Sellers (ed.), *Arizona Climate* (Tucson: University of Arizona Press, 1960), climatological tables for Phoenix and Yuma (unpaged).
[6]*Ibid.*, p. 12.
[7]*Ibid.*, climatological table for Crown King, and p. 20.
[8]*Ibid.*, p. 10.
[9]*Compilation of Records of Surface Waters of the United States through September 1950*, USGS Water-Supply Paper 1313 (1954), Part 9, *Colorado River Basin*, pp. 526, 604, 675, 695.
[10]*Estimated Use of Water in the United States, 1955*, USGS Circular 398 (1957).
[11]*The Underground Water Resources of Arizona* (Arizona Underground Water Commission, 1953), p. 66; *Pumpage and Ground-Water Levels in Arizona in 1955*, by P.W. Johnson, N.D. White, and J.M. Cahill of the USGS (ASLD, Phoenix, Oct. 1956), pp. 8-9. See also *Ground Water in the Gila River Basin and Adjacent Areas, Arizona — A Summary*, USGS, in cooperation with the Arizona Underground Water Commission (1952).
[12]*Pumpage and Ground-Water Levels in Arizona in 1955*, p. 38.

[13]Jack L. Cross, Elizabeth H. Shaw, and Kathleen Scheifele (eds)., *Arizona: Its People and Resources* (Tucson: University of Arizona Press, 1960), p. 111.

[14]See C. Warren Thornwaite, C.F. Steward Sharpe, and Earl F. Dosch, *Climate and Accelerated Erosion in the Arid and Semi-Arid Southwest with Special Reference to the Polacca Wash Drainage Basin, Arizona*, USDA Tech. Bull. No. 808 (May 1942), p. 127; Luna Leopold, "Vegetation of Southwestern Watersheds in the Nineteenth Century," *Geographical Review*, Vol. 41, No. 2 (April 1951).

[15]"Journal of Captain A. R. Johnson" in W.H. Emory, *Notes of A Military Reconnoissance, from Fort Leavenworth in Missouri to San Diego, in California, ...* 30th Cong., 1st Sess., House Exec. Doc. No. 41 (1848), p. 593.

[16]Report of Lieut. Col. P. St. George Cooke in Emory, p. 555.

[17]Thornwaite, Sharpe, and Dosch, p. 103.

[18]Harold S. Colton, "Some Notes on the Original Condition of the Little Colorado River: A Side Light on the Problems of Erosion," *Plateau*, Vol. 10, No. 6 (1937), pp. 17-20. The testimony is mixed to be sure, some finding the country to be desolate and unrewarding to the extreme. See Lorenzo Sitgreaves, *Report of an Expedition down the Zuni and Colorado Rivers*, 32d Cong., 2d Sess., Sen. Exec. Doc. No. 59 (1853), and Joseph C. Ives, *Report Upon the Colorado River of the West*, 36th Cong., 1st Sess., House Exec. Doc. No. 90, 1861.

[19]Patrick Hamilton, *Arizona, For Homes, For Health, For Investments* (Phoenix, 1886), p. 13.

[20]Hinton, p. 273; H. C. Hodge, *Arizona As It Is* (New York, 1877), p. 42.

[21]Patrick Hamilton, *The Resources of Arizona* (San Francisco: Bancroft, 1884), p. 361.

[22]*Ibid.*, pp. 401-402.

[23]*Ibid.*, p. 261.

[24]Sylvester Mowry, *Arizona and Sonora* (New York: Harper, 1864), p. 193.

[25]James R. Hastings, "Vegetation Change and Arroyo Cutting in Southeastern Arizona During the Past Century," *Journal of the Arizona Academy of Science*, Vol. 1, No. 2 (Oct. 1959), pp. 60-67.

Chapter 2

[1]Paul Sears, "Science and Natural Resources," *American Forests*, Vol. 63, No. 3 (March 1957), pp. 56-57.

[2]Leopold, *Conservation and Protection*, USGS Circular 414-A (1960), p. 2.

[3]*Ibid.*, p. 3.

[4]U.S., Senate Select Committee on National Water Resources, *Report*, Report No. 29, 87th Cong., 1st Sess., 1961, p. 4.

[5]*Resources for Freedom*, President's Materials Policy Commission, Vol. 5 (June 1952), p. 84.

[6]*Ibid.*

[7]H. V. Smith, *The Climate of Arizona*, AAES Bull. No. 279 (1956), p. 28.

[8]K.A. MacKichen and J.C. Kammerer, *Estimated Use of Water in the United States, 1960*, USGS Circular 456 (1961).

[9]Richard A. Harvill, "The Economy of Arizona," *Arizona Business and Economic Review*, Vol. 6, No. 1 (Jan. 1957), p. 6.

[10]H.C. Schwalen and R.J. Shaw, *Groundwater Supplies of the Santa Cruz Valley of Southern Arizona Between Rillito Station and the International Boundary*, AAES Bull. 288 (October 1957) .

[11]*Statistical Abstract of the United States: 1956*, Bureau of the Census, p. 13.

[12]*United States Census of Population, 1960: Number of Inhabitants* (Bureau of the Census), p. 4—10.

[13]*Current Population Reports: Population Estimates*, Bureau of the Census, Series P-25, No. 160 (Aug. 9, 1957), p. 11. In addition, Bureau of the Census predictions for Arizona place the population at over four million by the year 2000. U.S., Senate Select Committee on National Water Resources, *Population Projections and Economic Assumptions*, 86th Cong., 2d Sess., Committee Print No. 5, 1960, pp. 6-7.

[14]*Statistical Abstract of the United States: 1956*, p. 15.

[15]Harvill, p. 6.

262   THE POLITICS OF WATER IN ARIZONA

[16]*Census of Manufactures, 1947,* and *ibid., 1958,* Bureau of the Census; see also Hiram S. Davis, "New Manufacturing Plants Strengthen the Phoenix Economy," *Arizona Business and Economic Review,* Vol. 6, No. 2 (Feb. 1957), p. 1.

[17]*Arizona Statistical Review,* 1961, p. 3.

[18]*A Survey of the Recreational Resources of the Colorado River Basin,* NPS (Washington, D.C., 1950), p. 21.

[19]L.W. Casaday, *Arizona: An Economic Report, 1953* (University of Arizona, Bureau of Business Research, Special Studies No. 7, 1953), p. 82. The Valley National Bank reported that tourism contributed $290,000,000 to the Arizona economy in 1960. *Arizona Statistical Review,* 1961, p. 3. See also Robert Waugh, "Tourism — A Billion Dollar Industry," *Arizona Business and Economic Review,* Vol. 8, No. 4 (April 1959).

[20]*Some Economic Implications of the Tourist Industry for Northern Arizona,* Stanford Research Institute (Phoenix, Feb. 19, 1954), p. 3.

[21]The Park Service considers the availability of water "a potent factor in determining types of use and extent of development in many parks." *Mission 66 for the National Park System* (1957), p. 113. See mimeographed reports on the following areas: *Mission 66 for the Pipe Springs National Monument* (undated), p. 6; *Mission 66 for Organ Pipe Cactus National Monument* (undated), p. 7; *Mission 66 for Chiricahua National Monument* (undated), p. 5; *Mission 66 for Saguaro National Monument* (undated), p. 5.

[22]See chapter on land management.

[23]*A Survey of the Recreational Resources of the Colorado River Basin,* pp. 136-37.

[24]*A Study of the Park and Recreation Problem of the United States,* NPS (Washington D.C., 1941); and *A Survey of the Recreational Resources of the Colorado River Basin.*

[25]Letter from William L. Bowen, Regional Chief, Division of Recreational Resource Planning, National Park Service, Region 3, Santa Fe, New Mexico (Aug. 25, 1957); Albert H. Schroeder, *A Brief Survey of the Lower Colorado River from Davis Dam to the International Border,* NPS, Region Three, Reproduced by the Bureau of Reclamation, Region Three (Boulder City, Nevada, 1952).

[26]*Arizona Farmer-Ranchman,* editorial, Vol. 36, No. 6 (March 16, 1957), p. 4.

[27]The legal and constitutional problems created by the federal system cannot simply be wished away, as apparently suggested by Lois G. Forer, "Water Supply: Suggested Federal Regulation," *Harvard Law Review,* Vol. 75, No. 2 (Dec. 1961), p. 332.

[28]Ernest A. Englebert, "Political Aspects of Future Water Resources Development in the West," *Research Needs and Problems* (Western Agricultural Economics Research Council, Committee on the Economics of Water Resources Development of the West, Report No. 1, March 2-3, 1953), p. 89.

[29]Concern first developed as a result of the case of *Federal Power Commission* v. *Oregon* (340 U.S. 345) over the construction of a dam on an Indian reservation without a state permit.

[30]*Arizona Daily Star,* Aug. 10, 1957, p. 1.

[31]*Los Angeles Examiner,* Dec. 6, 1957, p. 22.

[32]*Ibid.*

[33]H.R. 8325, 84th Cong., 1956.

[34]U.S. Senate, *Water Rights Settlement Act,* Hearings before the Subcommittee on Irrigation and Reclamation of the Committee on Interior and Insular Affairs on S. 863, 84th Cong., 2d Sess., 1956, p. 321.

[35]*Resources for Freedom,* Vol. 5, p. 94.

[36]*Water Law with Special Reference to Ground Water,* USGS Circular 117 (June, 1951), p. 1. Adapted from a report prepared for the President's Water Resources Policy Commission.

[37]*Bristor* v. *Cheatham,* 73 Arizona 228, and 75 Arizona 227. See also Ch. 4.

[38]See Justice Udall's dissent in *Ernst* v. *Superior Court of Apache County,* 82 Arizona 17, 307 P.2d 911 in which he makes this point with regard to the Little Colorado River.

[39]See Frank J. Trelease, "Preferences to the Use of Water," *Rocky Mountain Law Review,* Vol. 27 (Feb. 1955), pp. 138-60.

[40]Englebert in *Research Needs and Problems,* p. 90.

[41]For a general discussion of this problem, see Gilbert E. Hyatt, "Western Water Rights: May They be Taken Without Compensation?" *Montana Law Review*, Vol. 13 (Spring 1952), p. 102.

[42]G. Gordon Robertson, "Percolating Waters — Ownership Rule Restated in Arizona," *Rocky Mountain Law Review*, Vol. 26 (1954), p. 107.

[43]*Ibid.*

[44]William Bredo, "A Framework for Water Pricing Policy," *Research Needs and Problems*, p. 69.

[45]Charles L. Hamman, "Water Policy and Western Industrial Development," *ibid.*, p. 79.

[46]John F. Timmons, "Theoretical Considerations of Water Allocation Among Competing Uses and Users," *Journal of Farm Economics*, Vol. 38, No. 5 (Dec. 1956), p. 1257.

[47]See J. Karl Lee, "Economic Analysis of Water Resources Policy," in *Research Needs and Problems*.

[48]Luna B. Leopold, *The Conservation Attitude*, USGS Circular 414-C (1960), p. 17.

[49]*Ibid.*, p. 18.

[50]Jack Hirshleifer, James C. De Haven, and Jerome Milliman, *Water Supply*: *Economics, Technology and Policy* (Chicago: University of Chicago Press, 1960).

[51]John V. Krutilla and Otto Eckstein, *Multiple-Purpose River Development* (Baltimore: Johns Hopkins Press, 1958).

[52]For a recent symposium see Franklin S. Pollak, ed., *Resources Development*: *Frontiers for Research*, Western Resources Conference, 1959 (Boulder: University of Colorado Press, 1960).

[53]H.H. Wooten, *Supplement to Major Uses of Land in the United States*, USDA Tech. Bull. 1082 (Oct. 1953), pp. 80-81.

[54]*Future Needs for Reclamation in the Western States*, USDI Bureau of Reclamation (Committee Print No. 14 for the Senate Select Committee on National Water Resources, Washington, D.C., 1960), 86th Cong., 2d Sess., p. 12.

[55]Wooten, pp. 87-88. The Senate Select Committee asserts that one acre of irrigated cropland will produce the equivalent of three acres of non-irrigated cropland. *Future Needs for Reclamation*, p. 2.

[56]Morris E. Garnsey, "Water: West," *The Annals of the American Academy of Political and Social Science*, Vol. 281 (May 1952), p. 163.

[57]Paul B. Sears, "Comparative Costs of Restoration and Reclamation of Land," *ibid.*, p. 126; see Roy E. Huffman, "Public Water Policy for the West," *Journal of Farm Economics*, Vol. 35, No. 5 (Dec. 1953), p. 724, for a discussion of several factors that must be considered.

[58]Michael Strauss, *Why Not Survive?* (New York: Simon and Schuster, 1955), p. 72.

[59]Arthur Carhart, *Water — or Your Life* (New York: Lippincott, 1951), p. 214.

[60]*Ibid.*, pp. 518-19.

[61]*Recovering Rainfall: More Water for Irrigation*, Arizona Watershed Program (Tucson, 1957), pp. 16ff. See also Edward N. Munns, "Yield and Value of Water from Western National Forests," *Journal of Forestry*, Vol. 50, No. 5 (June 1952), p. 464.

[62]See Charles F. Cooper, *Yield and Value of Wild-Land Resources of the Salt River Watershed, Arizona*. University of Arizona M.A. Thesis (1956). Written by a member of the original Barr committee staff.

[63]Raymond Seltzer, "Cooperative Financing Possibilities in Arizona," in *Water: Preliminary Economic Considerations of the Arizona Watershed Program*, Proceedings of the Second Annual Meeting, Arizona Watershed Program, Sept. 22, 1958.

[64]Harvey O. Banks, "Bases of an Adequate State Water Program," *State Government*, Vol. 33, No. 2 (Spring 1960), p. 134.

[65]Merrill Bernard, "The Appraisal of Water Resources in the United States: Analysis and Utilization of Data, Water Supply and Flood Forecasting," in *Proceedings of the United Nations Scientific Conference on the Conservation and Utilization of Resources*, Vol. IV, *Water Resources* (Lake Success, N.Y., 1951), p. 57.

[66]*Arizona Revised Statutes* (1955), 45-102.

[67]Griffenhagen and Associates, *Report on General State Organization* (3 Vols.; Chicago, 1949). Prepared for the Special Legislative Committee on State Operations, 19th Legislature, 1st Special Sess. Cited hereafter as Griffenhagen Report.

264   THE POLITICS OF WATER IN ARIZONA

⁶⁸Arizona Resources Board, *Annual Report* (Dec. 31, 1920).
⁶⁹U.S. National Resources Board, *State Planning: A Review of Activities and Progress* (Washington, D.C., 1935).
⁷⁰*Ibid.*, p. 17.
⁷¹*Ibid.*
⁷²Arizona State Planning Board, *Reports*, in cooperation with the National Resources Committee, Works Progress Administration (Dec. 1936). See Vol. 2 for a discussion of water and soil conservation problems. The conclusions reached at this early date are remarkably similar to those reached by later investigators of the resource situation.
⁷³Arizona. Legislature. *Journal of the House of Representatives*, 14th Legislature, 1939, p. 15. Hereafter cited as *House Journal*.
⁷⁴*House Journal*, 15th Legislature, 1941, p. 19.
⁷⁵Arizona State Resources and Planning Board, Post-War Planning Committee, *Preliminary Post-War Plans for Arizona Agriculture* (Phoenix, Feb. 1944).
⁷⁶Griffenhagen Report, pp. 52ff.
⁷⁷*Ibid.*, p. 134.
⁷⁸*Ibid.*, p. 87.
⁷⁹*Ibid.*, p. 90.
⁸⁰Griffenhagen Report, p. viii.
⁸¹R.E. Riggs, *Administrative Reorganization in Arizona*, University of Arizona M.A. Thesis (1952), p. 88.
⁸²Griffenhagen Report, p. 88.
⁸³*Arizona's Growth and Its Future*, Arizona Legislature, House of Representatives, Committee on Arizona Development (mimeo., Jan. 12, 1950).
⁸⁴*An Analysis of Arizona's Agricultural Productive Capacity*, USDA in cooperation with the U. of A. (1951), p. 5.
⁸⁵One official of the State Land Department stated that all legislative proposals coming from the department now go through the Legislative Council.
⁸⁶Arizona Development Board, *Report on Preliminary Survey of Selected Natural Resources*, conducted by Arizona Research Consultants, Inc. (Phoenix, 1955), and *Preliminary Survey and Recommendations Relating to the Establishment of a State Parks and Recreation Board*, prepared for the Legislative Council (Phoenix, 1955).
⁸⁷For newspaper accounts of the formation of these groups, see *Arizona Daily Star*, Oct. 25, 1953, p. 12; Oct. 21, 1954, p. 12; Nov. 21, 1954, p. 2.
⁸⁸*Arizona Republic*, Sept. 19, 1957.
⁸⁹Arizona. Legislature, Senate. *Journal of the Senate*, 22d Legislature, 2d Reg. Sess., 1956, p. 20. Hereafter cited as *Senate Journal*.
⁹⁰Leopold, *Conservation and Protection*, p. 4.

Chapter 3

¹For a recent discussion of water rights in Arizona see Fred C. Struckmeyer and Jeremy E. Butler, *A Review of Water Rights in Arizona* (Phoenix, 1960).
²For authoritative discussions of legal theories and practices involving water as they relate to the West, see Samuel C. Wiel, *Water Rights in the Western United States* (San Francisco, 1912); and Clesson S. Kinney, *A Treatise on the Law of Irrigation* (San Francisco, 1912).
³Kinney, p. 1012, par. 588.
⁴*Clough* v. *Wing*, 2 Arizona 371 at 381 (1888).
⁵Struckmeyer and Butler, p. 17.
⁶Arizona (Ter.), *Howell Code*, Bill of Rights, Article 22 (1864). Hereafter cited as *Howell Code*.
⁷Arizona. *Acts, Resolutions, and Memorials*, 1919, Ch. 164, par. 1, p. 278. Also called *Session Laws*, and hereafter cited as such.
⁸*Session Laws*, 1921, Ch. 64, par. 1; *Arizona Revised Statutes* (1955), 45-101A.
⁹See *Wattson* v. *U.S.*, 260 F. 506 (1919); and *Brewster* v. *Salt River Valley Water Users' Association*, 27 Arizona 23, 229 Pac. 929 (1924).
¹⁰*Session Laws*, 1912, Ch. 38; *Arizona Revised Statutes* (1955), 45-12001.
¹¹*Howell Code*, Article LV, sections 2,3.

[12]*Ibid.,* Article LV, section 17.
[13]*Ibid.,* Article LXI, section 7.
[14]*Clough* v. *Wing,* 2 Arizona 371 at 380 (1888).
[15]*Laws of the Territory of Arizona,* 1885 [Session Laws], P. 133, no. 68.
[16]Arizona (Ter.), *Revised Statutes of Arizona* (1887), Title LXIII, Ch. 1, par. 3198, sec. 1.
[17]*Boquillas Land and Cattle Company* v. *Curtis et al.,* 11 Arizona 128, 80 Pac. 504 (1907).
[18]*Boquillas Land and Cattle Company* v. *Curtis et al.,* 29 Sup. Ct. 493, 213 U.S. 339 (1909).
[19]Act of July 26, 1866, par. 9, 14 Stat. 251, 253.
[20]Act of March 3, 1877, par. 1, 19 Stat. 377, 43 U.S. Code 321.
[21]Act of June 17, 1902, 32 Stat. 388.
[22]*Stewart* v. *Verde River Irrigation and Power District,* 49 Arizona 531, 38 P.2d 329 (1937).
[23]2 Arizona 371 at 380 (1888).
[24]*Session Laws,* 1893, Ch. 86, section 1.
[25]*Slosser* v. *Salt River Valley Canal Co.,* 7 Arizona 376, 65 Pac. 332 (1901).
[26]*Gould* v. *Maricopa Canal Co.,* 8 Arizona 429, 76 Pac. 598 (1904).
[27]*Slosser* v. *Salt River Valley Canal Co.,* 7 Arizona 376 at 388.
[28]R.H. Forbes, *Irrigation and Agricultural Practice in Arizona,* AAES Bull. No. 63 (1911), p. 57.
[29]*Ibid.,* p. 59.
[30]*Wormser et al.* v. *Salt River Valley Canal Company* (1892) Federal District Court for Arizona, Phoenix.
[31]Forbes, *Irrigation and Agricultural Practice in Arizona,* p. 59.
[32]7 Arizona 376 at 394.
[33]8 Arizona 429.
[34]See Kinney, p. 3156.
[35]Act of June 17, 1902, 32 Stat. 388, 43 U.S.C. 391.
[36]*Hurley* v. *Abbott et al.,* No. 4564 Decree. In Federal District Court of 3rd Judicial District of Territory of Arizona in and for county of Maricopa (March 1, 1910).
[37]*Ibid.,* p. 15.
[38]Arizona State Water Commissioner, *First Biennial Report* (July 15, 1919, to Dec. 31, 1920), p. 16.
[39]*Ibid.,* pp. 30-31.
[40]*Session Laws,* 1919, Ch. 164, par. 1.
[41]*Ibid.*
[42]*Ibid.*
[43]*Arizona Revised Statutes* (1955), 45-141A.
[44]*Session Laws* 1941, Ch. 84, par. 1; *Arizona Revised Statutes* (1955), 45-141C.
[45]*Arizona Revised Statutes* (1955), 45-142.
[46]*Ibid.,* 45-143.
[47]Struckmeyer & Butler, p. 27.
[48]*Salt River Valley Water Users' Association* v. *Norviel,* 29 Arizona 360, 241 Pac. 583 (1925).
[49]*Arizona Revised Statutes* (1955), 45-146.
[50]*Ibid.,* 45-147.
[51]*Ibid.,* 45-150.
[52]*Ibid.,* 45-154.
[53]*Ibid.,* 45-231.
[54]*Ibid.,* 45-239.
[55]*Ibid.,* 45-102.
[56]*Ibid.,* 45-171.
[57]*Ibid.,* 45-103.
[58]*Ibid.,* 45-105, 106.
[59]*Stuart* v. *Norviel,* 26 Arizona 493, 226 Pac. 908 (1924).
[60]*Ibid.*
[61]*Stewart* v. *Verde River Irrigation and Power District,* 49 Arizona 531.
[62]Struckmeyer and Butler, p. 7.

Chapter 4

[1]See MacKichen and Kammerer.

[2]Harry F. Blaney and Martin R. Huberty, "Irrigation in the Far West," *Agricultural Engineering*, Vol. 38, No. 6 (June 1957); Schwalen, "Arizona's Water Problem," *Progressive Agriculture*, Vol. 6, No. 2 (July, Aug., Sept., 1954).

[3]*The Underground Water Resources of Arizona*, p. 66; *Pumpage and Ground-Water Levels in Arizona in 1955*, pp. 8-9. See also *Ground Water in the Gila River Basin and Adjacent Areas, Arizona — A Summary.*

[4]Schwalen, "Arizona's Water Problem," *Progressive Agriculture*, Vol. 6, No. 2, p. 3.

[5]*Arizona Agriculture, 1959*, Ariz. Agric. Exten. Serv. Circ. 270, p. 5.

[6]*Pumpage and Ground-Water Levels in Arizona in 1955*, p. 21.

[7]*Ibid.*, pp. 21 and 38.

[8]The cash cost for water in the production of upland cotton varied from $13.50 for 4 acre-feet in the Salt River Project to $37.00 for the same amount in 350-foot-lift areas. See *Arizona Agriculture, 1959*, p. 8.

[9]*Howell Code*, Statutory Bill of Rights, Article 22, 1864.

[10]*Session Laws*, 1919, Ch. 164, par. 1, p. 278.

[11]Kinney, pp. 2150-53.

[12]*Howard* v. *Perrin*, 8 Arizona 347, 76 Pac. 460 (1904).

[13]*Howard* v. *Perrin*, 200 U.S. 71 (1906).

[14]G. E. P. Smith, *Groundwater Law in Arizona and Neighboring States*, AAES Tech. Bull. No. 65, 1936.

[15]*McKenzie* v. *Moore*, 20 Arizona 1, 176 Pac. 568 (1918).

[16]*Proctor* v. *Pima Farms Co.*, 30 Arizona 96, 245 Pac. 369 (1926).

[17]*Ibid.*, 30 Arizona 96 at 100.

[18]*Ibid.*, at 103.

[19]G.E.P. Smith, *Groundwater Law in Arizona and Neighboring States*, p. 63.

[20]*Maricopa County Municipal Water Conservation District* v. *Southwest Cotton Company et al.*, 39 Arizona 65, 4 Pac. 369 (1926).

[21]*Ibid.*, 39 Arizona 65 at 71.

[22]*Ibid.*, at 79.

[23]*Ibid.*, at 80.

[24]*Ibid.*, at 83.

[25]*Fourzan* v. *Curtis*, 43 Arizona 140, 29 P.2d 722 (1934).

[26]*Campbell* v. *Willard*, 45 Arizona 221, 42 P.2d 403 (1935).

[27]*Parker et al.*, v. *McIntyre et al.*, 47 Arizona 484, 56 P.2d 1337 (1936).

[28]G.E.P. Smith, *The Groundwater Supply of the Eloy District in Pinal County, Arizona*, AAES Tech. Bull. No. 87 (1940).

[29]G.E.P. Smith, *Groundwater Law in Arizona and Neighboring States*, p. 87; see also the Introduction by Paul Burgess to Smith, f.n. 28, *supra.*

[30]Arizona State Water Commissioner, *Eleventh Biennial Report* (1939-1940), p. 75.

[31]*House Journal*, 15th Legislature, 1st Special Sess., 1942, H.B. 13; *Senate Journal*, 17th Legislature, Regular Sess., 1945, S.B. 109; *House Journal*, 17th Legislature, Regular Sess., 1945.

[32]See Governor Osborn's message to the legislature, *House Journal*, 15th Legislature, 1st Special Sess., 1942, p. 23.

[33]ASLD, *Thirty-Second Annual Report* (July 1, 1943 to June 30, 1944), p. 5.

[34]*Preliminary Post-War Plans for Agriculture*, p. 6.

[35]*Senate Journal*, 17th Legislature, 1st Special Sess., 1945, p. 15.

[36]*Session Laws*, 1945, 1st Special Session, Ch. 12.

[37]*A Summary of Ground Water Legislation in Arizona*, ASLD Bull. No. 301 (1954), p. 7.

[38]*Session Laws*, 1945, 1st Special Sess., Ch. 12, sec. 1.

[39]*House Journal*, 17th Legislature, 1st Special Sess., 1945, H.B. 193.

[40]*Ibid.*, 18th Legislature, Regular Sess., 1947, H.B. 8.

[41]Wells A. Hutchins, *Selected Problems in the Law of Water Rights in the West*, USDA, Miscellaneous Publications, No. 418 (1942), p. 159.

[42]*House Journal*, 17th Legislature, 4th Special Sess., 1947, H.B. 1.

[43]*Ibid.*, 17th Legislature, 6th Special Sess., 1947, H.B. 2.

[44]ASLD, *Thirty-fourth Annual Report* (July 1, 1945 to June 30, 1946), p. 5.

[45]*Arizona Daily Star*, Jan. 31, 1947, p. 1.

[46]*Arizona. Daily Star*, Feb. 6, 1948, p. 1.

[47]*Arizona Farmer*, Vol. 26, No. 4 (Feb. 26, 1947), p. 6.

[48]See *Geology and Ground-Water Resources of the Salt River Valley Area, Maricopa and Pinal Counties, Arizona*, USGS (mimeo., Feb. 4, 1947); and Samuel F. Turner, *Further Investigations of the Groundwater Resources of the Santa Cruz Basin, Arizona*, USGS Duplicate Report (March 11, 1947).

[49]*House Journal*, 18th Legislature, 2nd Special Sess., 1947, p. 13.

[50]Other sessions were called in the meantime for other reasons. The third special session on ground water was the sixth special session of the biennium. A seventh was yet to come.

[51]See editorial, *Arizona Daily Star*, March 27, 1948.

[52]*Arizona Republic*, March 26, 1948.

[53]*Session Laws*, 1948, 6th Special Sess., Ch. 5, sec. 3.

[54]*Ibid.*, sec. 2.

[55]*Ibid.*, sec. 6b.

[56]*Ibid.*

[57]*Ibid.*, sec. 8.

[58]See, *A Summary of Ground-Water Legislation in Arizona;* also *Session Laws*, 1948, 6th Special Sess., Ch. 5.

[59]*Session Laws*, 1948, 6th Special Sess., Ch. 5, sec. 2.

[60]*Ibid.*, sec. 16.

[61]See Paul Kelso, "The Arizona Ground Water Act," *Western Political Quarterly*, Vol. 1, No. 2 (June 1948), pp. 181ff. for an excellent summary of these constitutional objections.

[62]*Ibid.*, p. 181.

[63]ASLD, *Thirty-sixth Annual Report* (July 1, 1948 to June 30, 1949), p. 6.

[64]See ASLD, *A Summary of Ground Water Law in Arizona*, Bull. No. 302 (1957), p. 18, for dates when areas were declared critical.

[65]See the annual reports of the State Land Department for fiscal years 1950 through 1954.

[66]Reported in *Arizona Agriculture*, the annual report of the condition of the state's agriculture by the Agricultural Experiment Station, Tucson. See the reports for 1950 and 1954, Bulletins 226 and 253. If the span of years included 1947 and 1948 it would show an increase of nearly double the acreage, from approximately 700,000 acres in 1947.

[67]*Arizona Republic*, March 19, 1953.

[68]*Pumpage and Groundwater Levels in Arizona, 1952;* and *Ground Water in the Gila River Basin and Adjacent Areas, Arizona — A Summary.*

[69]*Arizona Republic*, Oct. 12, 1951, p. 1.

[70]For the best available summary of the committee's recommendations, see *Arizona Republic*, Nov. 2, 1951, p. 1.

[71]*House Journal*, 20th Legislature, 2d Regular Sess., 1952, p. 14.

[72]*Bristor v. Cheatham*, 73 Arizona 228, 240 P.2d 185 (1952).

[73]*Ibid.*, 73 Arizona 228, at 234.

[74]19 Stat. 377, 43 U.S.C. 321.

[75]*California Oregon Power Co. v. Beaver Portland Cement Co.*, 295 U.S. 142 (1936).

[76]73 Arizona 228 at 234.

[77]*Ibid.*, at 235.

[78]*Ibid.*, at 243.

[79]An interesting feature of *Bristor v. Cheatham* was that former Supreme Court Justice Lockwood, who had written the opinion of the Court in the Southwest Cotton Co. case in 1931, recommended reversal of his own opinion. His views followed closely the arguments presented by the majority opinion above. *Arizona Republic*, Jan. 24, 1952, p. 1.

[80]*Arizona Republic*, Jan. 16, 1952, p. 1.

[81]*Ibid.*, Jan. 18, 1952, p. 1; Jan. 20, 1952, p. 1.

[82]*Arizona Daily Star*, Jan. 15, 1952, p. 1.

[88]*Arizona Republic,* Jan. 22, 1952, p. 1.
[84]*Ibid.,* Jan. 31, 1952, p. 1.
[85]*Ibid.,* Feb. 4, 1952.
[86]*Arizona Daily Star,* Jan. 18, 1952, p. 1.
[87]*Ibid.,* Feb. 2, 1952.
[88]*Ibid.,* Feb. 12, 1952, p. 1.
[89]*Ibid.,* Feb. 11, 1952, p. 1; April 23, 1952, p. 14B.
[90]*Arizona Republic,* Jan. 27, 1952, p. 1.
[91]*Senate Journal,* 20th Legislature, 2d Sess., 1952, S.B. 56.
[92]*Arizona Republic,* Feb. 19, 1952, p. 1. and *Arizona Daily Star,* Feb. 19, 1952, p. 1.
[93]*Arizona Republic,* Feb. 28, 1952, p. 6.
[94]*Arizona Farmer,* Vol. 31, No. 6 (March 15, 1952), p. 6.
[95]*Arizona Daily Star,* March 8, 1952, p. 1.
[96]*Arizona Farmer,* Vol. 31, No. 8 (April 12, 1952), p. 2.
[97]*The Underground Water Resources of Arizona,* Appendix C, *passim.*
[98]*Ibid.,* p. 3.
[99]*Arizona Daily Star,* Jan. 14, 1953.
[100]*Arizona Republic,* Jan. 13, 1953, p. 6.
[101]*Ibid.*
[102]*Bristor* v. *Cheatham,* 75 Arizona 227, 255 P.2d 173 (1953).
[103]*Ibid.,* 75 Arizona 227 at 234-35.
[104]See Kinney, p. 2160.
[105]75 Arizona 227, at 243.
[106]*Arizona Republic,* March 15, 1953, p. 1.
[107]*Senate Journal,* 21st Legislature, 1953, Senate Bill 109.
[108]See *Arizona Daily Star,* Dec. 6, 1953. The issues of this case are discussed below in *Southwest Engineering Co.* v. *Ernst.*
[109]*Senate Journal,* 21st Legislature, 2d Regular Sess., 1954, Senate Bill 90.
[110]*Ibid.,* p. 15.
[111]*Arizona Daily Star,* Feb. 16, 1954; *ibid.,* p. B-1; *ibid.,* letter by R.H. Forbes, Feb. 25, 1954, p. B-12.
[112]*Ibid.,* March 25, 1954, p. 1.
[113]*House Journal,* 21st Legislature, 2d Regular Sess., 1954, p. 654.
[114]*Arizona Daily Star,* April 11, 1954, p. 1.
[115]*House Journal,* 21st Legislature, 2d Regular Sess., 1954, p. 760.
[116]*Arizona Daily Star,* April 13, 1954, p. 12-B.
[117]*House Journal,* 22d Legislature, 1st Regular Sess., 1955.
[118]*A Summary of Ground Water Law in Arizona,* p. 18.
[119]ASLD, *Forty-Sixth Annual Report* (July 1, 1957 to June 30, 1958), p. 25.
[120]*Southwest Engineering Company* v. *Ernst,* 79 Arizona 403, 291 P. 2d 764 (1955).
[121]*Ibid.,* 79 Arizona 403 at 410.
[122]*Ibid.,* at 413.
[123]*Ernst* v. *Collins,* 81 Arizona 178, 302 P.2d 941 (1956).
[124]*Vance* v. *Lassen,* 82 Arizona 188, 310 P.2d 510 (1957).
[125]*Arizona* v. *Anway,* 87 Arizona 206, 349 P.2d 774 (1960).
[126]*Session Laws,* 1959, Ch. 109.
[127]See Schwalen and Shaw, p. 2.

Chapter 5

[1]Quoted in Abel Wolman, "Utilization of Surface, Underground and Sea Water," in *Proceedings of the United Nations Scientific Conference on the Conservation and Utilization of Resources,* Vol. IV, *Water Resources,* p. 98.
[2]B.P. Herber, *An Analysis of the Economy of Arizona,* University of Arizona M.A. thesis (1955), p. 27.
[3]Casaday, p. 92.
[4]*Ibid.,* pp. 92-93.
[5]A perusal of several years' issues of such journals as the *Arizona Farmer-Ranchman, Arizona Stockman,* and *Arizona Cattlelog* is evidence of this feeling.

[6]For a discussion of the role of organized labor in the writing of the Arizona Constitution, see Tru A. McGinnis, *The Influence of Organized Labor on the Making of the Arizona Constitution,* University of Arizona M.A. thesis (1930).

[7]V.D. Brannon, *Employers' Liability and Workmen's Compensation in Arizona,* University of Arizona Social Science Bull. No. 7 (1934), p. 7.

[8]V.S. Griffiths, *State Regulation of Railroad and Electric Rates in Arizona to 1925,* University of Arizona M.A. thesis (1931).

[9]*Ibid.,* p. 7.

[10]See Neal D. Houghton, "Arizona's Experience with the Initiative and Referendum," *New Mexico Historical Review,* Vol. 29, No. 3 (July 1954).

[11]Joseph Stocker, "Arizona's Maverick Conservative," *Frontier,* Vol. 1, No. 16 (July 1, 1950), p. 5.

[12]Stocker, "Arizona's New Liberal Movement," *Frontier,* Vol. 1, No. 18 (Aug. 1, 1950), p. 5.

[13]Waldo E. Waltz, "Arizona: A State of New-Old Frontiers," in Thomas C. Donnelly, *Rocky Mountain Politics* (Albuquerque: University of New Mexico Press, 1940), p. 282.

[14]Henry F. Dobyns, "Shift in Arizona," *Frontier,* Vol. 7, No. 7 (May 1956), p. 13.

[15]Dean E. Mann, "The Legislative Committee System in Arizona," *Western Political Quarterly,* Vol. 14, No. 4 (Dec. 1961).

[16]*Arizona Republic,* March 4, 1955.

[17]Evidence was presented that O.C. Williams sold state land in such a fashion as to give a single owner monopolistic control over other leased land, thus precluding sales of the leased land. *Arizona Daily Star,* March 4, 1949, p. 1.

[18]See Riggs, p. 88.

[19]Stocker, "Arizona's New Liberal Movement," *Frontier,* Vol. 1, No. 18, p. 5.

[20]*Ibid.*

[21]*Arizona Daily Star,* Sept. 30, 1945, p. 1; Jan. 21, 1947, p. 1.

[22]*Arizona Daily Star,* June 20, 1947, p. 26.

[23]Stocker, "Arizona's Maverick Conservative," *Frontier,* Vol. 1, No. 16, p. 5.

[24]See Virginia L. Brown, *Some Aspects of the History of the Arizona Education Association,* University of Arizona M.A. thesis (1952).

[25]*Arizona Daily Star,* June 20, 1947, p. 26.

[26]Arizona Tax Research Association, *Newsletter,* Phoenix, Oct. 1956.

[27]Joseph Wood Krutch, *The Voice of the Desert* (New York: Sloane, 1954), p. 194.

[28]Krutch, *The Desert Year* (New York: Sloane, 1952), p. 93.

[29]*Ibid.,* p. 97.

[30]*Ibid.,* p. 126.

[31]Nell Murbarger, "Dam in Glen Canyon," *Desert,* Vol. 20, No. 4 (April 1957), p. 4.

[32]Frank Lloyd Wright, "Plan for Arizona State Capitol," *Oasis,* Scottsdale [?], Ariz., Feb. 17, 1957.

[33]Harold James, "Keep the Desert Living," *Pacific Discovery,* Vol. 9, No. 2 (March-April 1956), p. 3.

[34]*Arizona Park, Parkway and Recreational Area Plan: Progress Report,* Arizona Highway Commission, Arizona Resources Board, Arizona State College, Tempe, in cooperation with the National Park Service (Feb. 1941), p. 159.

[35]Speech reported in part in "Conservation and a Farm Editor," *Arizona Wildlife-Sportsman,* Vol. 27, No. 12 (Dec. 1956), p. 16.

[36]*Ibid.*

[37]*Ibid.*

[38]*Ibid.*

Chapter 6

[1]See Garnsey, *America's New Frontier: the Mountain West* (New York: Knopf, 1950), pp. 246-50.

[2]*The Problems of Imperial Valley and Vicinity,* 67th Cong., 2d Sess., Sen. Doc. 142, p. 220.

[3]*The Central Arizona Project,* Hearings before the Committee on Interior and

270   THE POLITICS OF WATER IN ARIZONA

Insular Affairs, House of Representatives, 82d Cong., 1st Sess., on H.R. 1500 and 1501, Parts I and II (1951); *Ten Rivers in America's Future,* Vol. II of the *Report* of the President's Water Resources Policy Commission (Washington, D.C., 1950), p. 391.

[4]Malcolm Parsons, *The Colorado River in Arizona Politics,* University of Arizona M.A. thesis (1947), pp. 194-95.

[5]*Ibid.,* p. 36.

[6]*Ibid.,* p. 62.

[7]*Ibid.,* p. 35.

[8]*Ibid.,* pp. 60-66.

[9]Quoted in Parsons, p. 62.

[10]*Ibid.,* p. 78.

[11]Houghton, "Problems of the Colorado River as Reflected in Arizona Politics," *Western Political Quarterly,* Vol. 4, No. 4 (Dec. 1951), p. 634.

[12]*Ibid.,* p. 638.

[13]Parsons, p. 86.

[14]*Ibid.,* p. 89.

[15]*Ibid.,* p. 87.

[16]Parsons, p. 123.

[17]Parsons, pp. 98-103; see also Parsons, "Party and Pressure Politics in Arizona's Opposition to Colorado River Development," *Pacific Historical Review,* Vol. 19, No. 1 (Feb. 1950), p. 47.

[18]Parsons, *The Colorado River in Arizona Politics,* pp. 123-26.

[19]*Arizona* v. *California,* 283 U. S. 423 (1931).

[20]*Arizona* v. *California,* 292 U.S. 341 (1934).

[21]*United States* v. *Arizona,* 295 U.S. 174 (1935).

[22]*Arizona* v. *California,* 298 U. S. 338 (1936).

[23]Colorado River Commission of Arizona, *Report* (Feb. 2, 1933-March 3, 1935).

[24]*Arizona Water Resources,* Hearings on Sen. Res. 304, U.S. Senate Committee on Irrigation and Reclamation, 78th Cong., 2d Sess. (1944), p. 42.

[25]Parsons, *The Colorado River in Arizona Politics,* p. 144.

[26]See G.E.P. Smith, "Future Water Supply and Irrigated Agriculture in Arizona," speech delivered Dec. 5, 1941, included in H.H. D'Autremont, *More Data on the Colorado River Question* (Tucson, 1943), p. 38ff.

[27]See Fred Colter, *Highline Book* (Phoenix, 1934).

[28]G.E.P. Smith, in D'Autremont, p. 44.

[29]Quoted in Parsons, *The Colorado River in Arizona Politics,* pp. 156-57.

[30]*Session Laws,* 16th Legislature, 2d Sess., 1944.

[31]*Senate Journal,* 16th Legislature, 1st Special Sess. (1944).

[32]*Central Arizona Project,* 81st Cong., 1st Sess., House Doc. 136 (1949).

[33]See Benton J. Stong, "Washington Report," *Frontier,* Vol. 1, No. 9 (March 15, 1950), p. 11.

[34]See *The Central Arizona Project,* Hearings, 82d Cong., 1st Sess., 1951, Part II, p. 585ff.

[35]*Colorado River Compact* (Dept. of State Doc. No. 6241, 1922), Article III(b).

[36]See D'Autremont, pp.24-37.

[37]For a discussion of the positions of Arizona and California, see Arizona's brief before the Supreme Court of the United States (October Term, 1952) *State of Arizona* v. *State of California, et al.,* Motion for Leave to File Bill of Complaint and Bill of Complaint; and California's Answer of Defendants to Bill of Complaint (filed May 19, 1953), and Summary of Controversy Exhibit A (filed April 5, 1954).

[38]For a discussion of this point and others see Charles A. Carson, "Arizona's Interest in the Colorado River," *Rocky Mountain Law Review,* Vol. 19 (June 1947), p. 352; and Desmond G. Kelly, "California and the Colorado River," *California Law Review,* Vol. 38, No. 4 (Oct. 1950), p. 696.

[39]*Water for Arizona,* CAPA, Phoenix (Spring 1954), p. 1.

[40]*State of Arizona* v. *State of California et al.,* Statement on Behalf of the California Defendants. In the Supreme Court of the United States (October Term, 1955), No. 10 Original, pp. 19-20.

[41]See also *Arizona Daily Star,* August 7, 1957; and *Arizona* v. *California et al.* Amended Bill of Complaint by Arizona, for the Supreme Court of the United States, October Term 1957.

NOTES   271

[42]For a discussion of the California position see Charles E. Corker, "The Issues in Arizona v. California: California's View," in Pollak.
[43]*Arizona Daily Star*, Aug. 10, 1957.
[44]*Arizona* v. *California et al.,* Statement on Behalf of the California Defendants, p. 24.
[45]*Arizona Republic,* Oct. 6, 1957.
[46]AISC, *Twelfth Annual Report* (July 1, 1958 to June 30, 1959), p. 1.
[47]*Ibid.,* p. 118.
[48]Draft Report, Simon H. Rifkind, Special Master, in the Supreme Court of the United States, October Term, 1959 (May 5, 1960), *State of Arizona* v. *State of California, et al.*

Chapter 7

[1]*Arizona Power Survey,* FPC (March 1942), p. 1.
[2]*Salt River Project: Major Facts in Brief,* Salt River Valley Water Users' Association (Phoenix, 1956); Stephen Shadegg, *The Phoenix Story: An Adventure in Reclamation* (Phoenix, 1958), p. 40.
[3]APA, map, "Generating Stations of Arizona," no date, *circa* 1957.
[4]V.S. Griffiths, pp. 28-29.
[5]*Ibid.,* Ch. 2.
[6]Arizona Constitution, Article XV, Sec. 2.
[7]Griffiths, p. 84.
[8]*National Power Survey: Principal Electric Utility Systems in the United States,* FPC, Power Series No. 2 (1935).
[9]Rollah E. Aston, *Boulder Dam and the Public Utilities,* University of Arizona M.A. thesis (1936), pp. 169-173.
[10]Houghton, "Problems in Public Power Administration in the Southwest — Some Arizona Applications," *Western Political Quarterly,* Vol. 4, No. 1 (March 1951), p. 124.
[11]*Ibid.*
[12]*Electric Power and Government Policy,* Twentieth Century Fund, (New York: 1948), p. 212; data taken from *Average Electric Bills,* FPC (1939), p. 7.
[13]*Average Typical Residential Bills,* FPC, Electric Rate Survey, Rate Series No. 3 (Jan. 1, 1935), p. 16.
[14]Act of December 12, 1928, Ch. 42, par. 1, 45 Stat. 1057.
[15]See Houghton, "Problems of the Colorado River as Reflected in Arizona Politics," *Western Political Quarterly,* Vol. 4, No. 4, p. 634; also Houghton, *Western Political Quarterly,* Vol. 4, No. 1, p. 116ff.
[16]USDI, *Annual Report of the Secretary,* 1939, p. 196.
[17]*Arizona Power Survey,* p. 11.
[18]*Ibid.,* p. 6.
[19]*House Journal,* 15th Legislature, Regular Sess., pp. 23-25.
[20]*Senate Journal,* 16th Legislature, Regular Sess., p. 88.
[21]*Ibid.,* p. 90.
[22]*Ibid.*
[23]*Senate Journal,* 16th Legislature, Senate Bill 74, 1943.
[24]*Arizona Daily Star,* March 1, 1943, p. 1.
[25]*Ibid.,* Feb. 17, 1943, p. 1.
[26]*Ibid.,* Feb. 11, 1943, p. 1.
[27]*Senate Journal,* 16th Legislature, Regular Sess., 1943, p. 437; *Arizona Daily Star,* March 2, 1943, p. 1.
[28]*House Journal,* 16th Legislature, Regular Sess., 1943, p. 16.
[29]*House Journal,* 16th Legislature, 2d Special Sess., 1944.
[30]*Arizona Daily Star,* Feb. 29, 1944, p. 1.
[31]*Ibid.,* March 4, 1944, p. 1.
[32]*Arizona Daily Star,* March 10, 1944, p. 8.
[33]*Ibid.,* Feb. 29, 1944, p. 1.
[34]*Ibid.,* Feb. 16, 1944, p. 1.
[35]*Session Laws,* 1944, Ch. 32, par. 3; *Session Laws,* 1947, Ch. 139, par. 2; *Arizona Revised Statutes* (1955), 30-121A.

[36]*Arizona Revised Statutes* (1955), 30-121B.

[37]See APA, *Seventh Annual Report* (1950), p. 29.

[38]This contract and many other documents pertaining to APA operations are found in APA, *Fifth and Sixth Annual Reports* (Dec. 30, 1947 to Dec. 31, 1949).

[39]APA, *Ninth Report* (July 1, 1951 to June 30, 1953), pp. 10, 11.

[40]*Senate Journal*, 17th Legislature, 1st Special Sess., 1945, p. 17.

[41]See *Arizona Daily Star*, Sept. 25, 1945.

[42]*Session Laws*, 1945, 1st Special Sess., Ch. 11.

[43]*Ethington et al. v. Wright et al.*, 66 Arizona 382, 189 P. 2d 209 (1948).

[44]*Arizona Daily Star*, Feb. 15, 1947, p. 1.

[45]*House Journal*, 18th Legislature, Regular Sess., 1947, H.B. 135.

[46]*Arizona Daily Star*, Feb. 4, 1957, p. 1.

[47]*Arizona Daily Star*, Feb. 15, 1947, p. 1.

[48]This opinion is confirmed by Houghton in *Western Political Quarterly*, Vol. 4, No. 1, p. 125.

[49]*Session Laws*, 1947, ch. 139.

[50]APA, *Fifth and Sixth Annual Reports* (Dec. 30, 1947 to Dec. 31, 1949), p. 36.

[51]APA, *Eighth Annual Report* (Jan. 1, 1951 to June 30, 1951), p. 12.

[52]APA, *Fifth and Sixth Annual Reports*, pp. 25-26.

[53]*Ibid.*, p. 124.

[54]*Ibid.*, pp. 36-38.

[55]APA, *Ninth Report* (July 1, 1951 to June 30, 1953), p. 11.

[56]APA, *Sixteenth Annual Report* (1959-1960), pp. 22-23.

[57]APA, *Thirteenth Annual Report* (1956-1957), p. 5.

[58]APA, *Tenth Annual Report* (July 1, 1953 to June 30, 1954), p. 18.

[59]*Ibid.*, p. 11

[60]APA, *Eleventh Report*, p. 18.

[61]*Resume of the Use of Electric Power and Energy in the State of Arizona*, APA, Annual Power Survey, for the year 1954 (mimeo., Aug. 31, 1956).

[62]*Ibid.*, p. 11.

[63]APA, *Fifteenth Annual Report*, p. 3, and *Fourteenth Annual Report*, p. 9; also *Colorado River Development Within the State of Arizona: Colorado River Projects. Preliminary Planning Report.* Harza Engineering Co., Chicago, 1958. Prepared for the APA.

[64]APA, *Thirteenth Annual Report*, p. 19.

[65]*Arizona Republic*, Sept. 6, 1956, and Sept. 23, 1956; *Tucson Daily Citizen*, Sept. 13, 1956.

[66]See "Those Glen Canyon Transmission Lines — Some Facts and Figures of a Bitter Dispute," a Special Report by Rep. Morris K. Udall, multilithed [1961].

[67]*Session Laws*, 1956, 2d Reg. Sess., Ch. 15.

[68]APA, *Twelfth Annual Report*, p. 5.

[69]APA, *Fourteenth Annual Report*, p. 11.

[70]*Ibid.*, p. 19.

[71]APA, *Colorado River Project: Bridge Canyon Development, Marble Canyon Development, Little Colorado River Development*, Amendment to Application for License Project 2248 (Phoenix, November 1959), pp. 1-15.

[72]Anthony Wayne Smith, "Campaign for the Grand Canyon," *National Parks Magazine* (April 1962), pp. 12-15.

[73]APA, *Eleventh Annual Report* (July 1, 1954 to June 30, 1955), p. 7.

[74]*Arizona Republic*, April 10, 1957.

[75]USDI, Bur. of Reclamation, "News Release" (Peterson — Interior 4662), April 4, 1962.

[76]*Arizona Daily Star*, Dec. 4, 1954, p. 1, and editorial Jan. 6, 1955.

[77]*Arizona Daily Star*, April 29, 1955, p. 1.

[78]*Ibid.*, June 4, 1955.

[79]See Houghton, "Problems in Public Power Administration in the Southwest — Some Arizona Applications," *Western Political Quarterly*, Vol. 4, No. 1, pp. 127-129.

[80]APA, *Second Annual Report*, cited in the *Fifth and Sixth Annual Reports* (Dec. 10, 1947 to Dec. 31, 1949), p. 23.

Chapter 8

[1]Arizona Department of Mineral Resources, *Inventory of Arizona Lands as of June 30, 1961* (Phoenix, Aug. 1962).
[2]See Cross, Shaw, and Scheifele, pp. 116-31, for a summary of the irrigation projects in the state.
[3]*Session Laws*, 1957, Ch. 99. See also *Arizona Republic*, Feb. 22, 1957.
[4]Griffenhagen Report, p. 19.
[5]Until 1959 its responsibilities also included oil and gas conservation. In that year these were transferred to an independent oil and gas commission.
[6]*Session Laws*, 1942, Ch. 28, sec. 5.
[7]State Water Commissioner, *Third Biennial Report* (1923-24), p. 5; ASLD, *Thirty-first Annual Report* (July 1, 1942 to June 30, 1943), p. 4.
[8]Griffenhagen Report, p. 134.
[9]*Arizona Revised Statutes* (1955), 37-131, B., C.
[10]*Ibid.*, 45-102, A.
[11]*Ibid.*, 45-102, B.
[12]*Session Laws*, 1921, Ch. 181, sec. 55A.
[13]State Water Commissioner, *Second Biennial Report* (Jan. 1, 1921 to Dec. 31, 1922), p. 52.
[14]*Ibid.*
[15]See State Water Commissioner, *Third Biennial Report*, p. 128.
[16]ASLD, *Forty-fourth Annual Report* (July 1, 1955 to June 30, 1956), p. 38.
[17]*Ibid.*
[18]*Ibid.*, pp. 38-39.
[19]State Water Commissioner, *Tenth Biennial Report* (1937-1938), p. 66.
[20]ASLD, *Forty-seventh Annual Report* (July 1, 1958 to June 30, 1959), p. 41.
[21]ASLD, *Forty-fourth Annual Report*, p. 31.
[22]*Ibid.*, pp. 31-32.
[23]*Ibid.*, p. 32.
[24]*Ibid.*
[25]State Water Commissioner, *Fourth Biennial Report* (1925-1926), p. 84.
[26]State Water Commissioner, *Eighth Biennial Report* (1933-1934), p. 5.
[27]State Water Commissioner, *Third Biennial Report*.
[28]ASLD, *Fortieth Annual Report* (July 1, 1951 to June 30, 1952), p. 14.
[29]ASLD, *Forty-second Annual Report* (July 1, 1953 to June 30, 1954), p. 14.
[30]The land commissioner reported that over 100 had done so through June 1956. ASLD, *Forty-fourth Annual Report*, p. 35.
[31]*Session Laws*, 1951, Ch. 130; other evidence of the depository character of the department is the Soil Conservation Division discussed elsewhere.
[32]ASLD, *Forty-third Annual Report*, p. 26; *Forty-fourth Annual Report*, p. 28; *Forty-seventh Annual Report*, p. 19.
[33]*Senate Journal*, 22d Legislature, 2d Regular Sess., 1956.
[34]*Recovering Rainfall: More Water for Irrigation.* See also Chapter 10 for an extended discussion of proposed changes in watershed management practices.
[35]*Arizona Republic*, Jan. 4, 1957.
[36]See, for example, *Progress in Watershed Management*, Proceedings of Third Annual Meeting [Arizona Watershed Program], Sept. 21, 1959.
[37]ASLD, *Forty-seventh Annual Report*, p. 22.
[38]*Session Laws*, 1948, 18th Legislature, 1st Special Sess., Ch. 14.
[39]*Ibid.*, par. 2; *Arizona Revised Statutes* (1955), 45-506 B.
[40]*Arizona Revised Statutes* (1955), 45-502, 503.
[41]AISC, *First Annual Report* (Feb. 1 to Dec. 31, 1948), p. 4.
[42]"Petition Asks Discharge of Stream Commission," *Arizona Daily Star*, May 23, 1957.
[43]USDI, Bureau of Reclamation, *Proposed Report of the Bureau of Reclamation*, Jan. 26, 1948.
[44]For a display of the arguments and tactics used, see *The Central Arizona Project*, 82d Cong., 1st Sess., 1951.
[45]For a brief summary of the preliminary motions, see AISC, *Ninth Annual Report* (July 1, 1955 to June 30, 1956), pp. 86-90.

[46]AISC, *Eighth Annual Report* (July 1, 1954 to June 30, 1955), p. 29.
[47]Upper Colorado River Basin Compact, executed in Santa Fe, New Mexico, Oct. 11, 1948, and ratified by Arizona, *Session Laws*, 1949, Ch. 4, sec. 1.
[48]AISC, *First Annual Report*, p. 30.
[49]AISC, *Ninth Annual Report*, p. 66.
[50]*Session Laws*, 1956, Ch. 150, sec. 1a.
[51]AISC, *Ninth Annual Report*, p. 66.
[52]See AISC, *Eighth Annual Report* for a complete list of cooperating organizations.
[53]CAPA, *News of Arizona's Water Fight* (June 1, 1955), p. 1.
[54]CAPA, *Report to Directors and Members*, by John M. Jacobs, Pres. (Dec. 2, 1955), p. 1.
[55]CAPA, *News of Arizona's Water Fight*, p. 2.
[56]Vincent Ostrom, "State Administration of Natural Resources in the West," *American Political Science Review* (June 1953), pp. 478-93.

Chapter 9

[1]See the *Articles of Incorporation of the Salt River Valley Water Users' Association* (signed 1903).
[2]*Salt River Project: Major Facts in Brief*. For a recent popular account of the growth of the Phoenix area see Shadegg, *The Phoenix Story: An Adventure in Reclamation*.
[3]For a brief description of the project in its inception and historical development see USDI, Bur. of Reclamation, *General Information Concerning the Salt River Project, Arizona*, (April 15, 1941); U.S. Reclamation Service, *Salt River Irrigation Project* (Oct. 1, 1909).
[4]*Major Irrigation Developments in Region 3, USDI*, Bur. of Reclamation, Region 3 (Boulder City, Nevada: June 1953), p. 6.
[5]Salt River Valley Water Users' Association, *Salt River Project: Annual Report* (1956), pp. 8, 10.
[6]R.J. McMullin, "Rehabilitation and Betterment Pays a Dividend," *Reclamation Era*, Vol. 41, No. 1 (Feb. 1955), pp. 20ff.
[7]*Session Laws*, 1935, 1st Special Sess., Ch. 10, par. 15; *Arizona Revised Statutes* (1955), 45-902.
[8]*Major Irrigation Developments in Region 3*, p. 8; USDI, *Annual Report of the Secretary*, 1960, p. 139.
[9]For a dated but generally accurate description of the Yuma Project, see *Yuma: Federal Reclamation Project*, Bur. of Reclamation (1936).
[10]*How Reclamation Pays*, USDI, Bur. of Reclamation (Washington, 1947), p. 324.
[11]USDI, *Annual Report of the Secretary*, 1956, p. 11; *Annual Report of the Secretary*, 1960, p. 139.
[12]For general information see *Gila Project*, USDI, Bur. of Reclamation, Region 3 (Boulder City, Nev., Feb. 1953).
[13]USDI, *Annual Report*, 1956, p. 11; *Annual Report of the Secretary*, 1960, p. 139.
[14]Frank Barsalou, John R. Gale, and James Gillies, *Yuma: Its Economic Growth and Land Use Potential*, Stanford Research Institute (Project No. I-1823 [Menlo Park, Calif.] 1956), p. 17.
[15]For an excellent summarization of the Hoover Dam Project and its development, see *Boulder Canyon Project Final Reports*, USDI, Bur. of Reclamation (Washington, D.C., 1948). See chapters on power and recreation.
[16]*Davis Dam and Power Plant*, USDI, Bur. of Reclamation (1955), p. 3.
[17]*Reclamation Project Data*, USDI, Bur. of Reclamation (1948), p. 345.
[18]*The Colorado River*, USDI, Bur. of Reclamation (Washington, D.C., 1946).
[19]"Glen Canyon Dam," *Reclamation Era*, Vol. 44, No. 1 (Feb., 1958), p. 13. The data for potential projects is taken from *The Colorado River*, and *Ten Rivers in America's Future*.
[20]For an illuminating example, see the exchange between Governor Pyle and Raymond Moley, *Newsweek* (June 18, 25, Sept. 3, 10, 1951).
[21]See *Central Arizona Project*, 81st Cong., 1st Sess., 1949, House Doc. 136.
[22]*Ibid.*, pp. 153-154.

²³*The Central Arizona Project,* Hearings, 82d Cong., 1st Sess., Part I, p. 68.
²⁴*Colorado River Development Within the State of Arizona: Colorado River Projects. Preliminary Planning Report.* Harza Engineering Co., Chicago, 1958. Prepared for the Arizona Power Authority, Phoenix.
²⁵*Ibid.,* pp. X-1ff.
²⁶APA, *Colorado River Project: Bridge Canyon Development, Marble Canyon Development.* Application for License before the Federal Power Commission. Phoenix, Arizona, July 1958; and APA, *Colorado River Project.* Amendment to Application for License Project 2248.
²⁷*Ibid.,* p. H-4.
²⁸*Arizona Daily Star,* June 25, 1959.
²⁹*Appraisal Report, Central Arizona Project,* USDI, Bureau of Reclamation, Region 3 (Boulder City, Nev., Jan. 1962), p. 9.
³⁰See *Power Market Survey — Colorado River Storage Project,* Federal Power Commission (June 1958).
³¹*Appraisal Report, Central Arizona Project,* pp. 16-17.
³²Walter B. Langbein, *Water Yield and Reservoir Storage in the United States,* USGS Circular 409 (1959), p. 4.
³³*Water Resources Development by the U.S. Army Corps of Engineers in Arizona,* U.S. Army Corps of Engineers, South Pacific Division (Jan. 1, 1961), p. 7; *Report on the Gila River and Tributaries Below Gillespie Dam,* 81st Cong., 1st Sess., House Doc. 331, 1949; U.S. Dept. of the Army, *Annual Report of the Chief of Engineers,* 1956, Vol. 2, p. 1538, and *Annual Report . . . 1959,* Vol. 2, p. 1557.
³⁴U.S. Department of the Army, *Annual Report of the Chief of Engineers,* 1959, Vol. 2, p. 1156.
³⁵For a brief summary of the dispute see, "The Buttes Dam Story," *Arizona Farmer-Ranchman,* June 9, 1956, pp. 6-7ff.
³⁶For the viewpoint of the Safford Valley water users see "Gila Valley Officials Deny Block of Buttes Dam," *Arizona Republic,* May 11, 1956, p. 2.
³⁷U.S., BIA, Phoenix Area Office, *Soil and Moisture Conservation Annual Report, Colorado River (consolidated) Reservation,* 1956.
³⁸Arizona Commission on Indian Affairs, *Annual Report,* p. 13.
³⁹William H. Kelly, Peter Kunstadter, and Robert A. Hackenberg, *Social and Economic Resources Available for Indian Health Purposes in Five Southwestern States,* Book II: *Arizona,* Part 1. University of Arizona, Bureau of Ethnic Research (June 15, 1956), p. 67.

Chapter 10

¹Thornwaite, Sharpe, and Dosch, p. 125.
²See the following: Arizona Department of Mineral Resources, *Inventory of Arizona Lands As of June 30, 1961* (Phoenix, Aug. 1962); Louis S. Meyer, *State of Arizona: A Brief Analysis of Land Status and Utilization,* Arizona State University, Bureau of Government Research, Research Study No. 6 (Tempe, 1962); ASLD, *Forty-ninth Annual Report;* and "Arizona Land Ownership," a report prepared by the Service Division of the ASLD, Sept. 1, 1962 (multilithed).
³Edward Higbee, *The American Oasis* (New York: Knopf, 1957), p. xi.
⁴*Arizona Revised Statutes* (1955), 37-105, 211, 253, 321, 341, 481, and 721.
⁵ASLD, *Fortieth Annual Report,* p. 5.
⁶For the best discussion of this matter, see Joseph Stocker, "Saying No to Three-Cents-an-Acre," *Survey* (Sept. 1949), p. 465; see also his "The Big Grab in Arizona," *Frontier,* Vol. 1, No. 7 (Feb. 15, 1950), p. 3.
⁷*Session Laws,* 1950, 1st Special Sess., ch. 58.
⁸*Arizona Daily Star,* Feb. 24, 1950, p. 1.
⁹*Arizona Daily Star,* Feb. 25, 1950, p. 1.
¹⁰G.E.P. Smith, *Groundwater Supply of the Eloy District in Pinal County, Arizona.*
¹¹Edward N. Munns, "Yield and Value of Water from Western National Forests," *Journal of Forestry,* Vol. 50, No. 5 (June 1952), p. 464.
¹²*Timber Resources Review,* USDA, FS, Ch. 11, Appendices, p. 6.

276   THE POLITICS OF WATER IN ARIZONA

[13]For the best history of Forest Service administration in Arizona, see M.E. Lauver, *A History of the Use and Management of the Forested Lands in Arizona, 1862-1936,* University of Arizona, M.A. thesis (1938); see also J.J. Wagoner, *History of the Cattle Industry in Southern Arizona, 1540-1940,* University of Arizona Social Science Bull. No. 20 (1940).

[14]See House Joint Memorial No. 6, Arizona State Legislature, Feb. 17, 1931; also "Public Domain States Knuckled Under," *Arizona Stockman,* Vol. 14, No. 1 (Feb. 1948), p. 9.

[15]*Arizona Stockman,* Vol. 14, No. 6 (July 1948), p. 10.

[16]See Munns, p. 467, for statistics on economic importance of several forest products.

[17]*National Forest Facts: Southwestern Region, 1957,* USDA, FS (mimeo), p. 3; USDA, FS, *Report of the Chief of the Forest Service, 1960,* p. 39.

[18]*Timber Resources Review,* p. 123.

[19]*National Forest Progress: Southwestern Region, 1960,* USDA, FS, pp. 4-5.

[20]USDA, FS, *Report of the Chief of the Forest Service,* 1960, p. 40.

[21]*National Forest Facts: Southwestern Region, 1954,* p. 7, and *1957* p. 7.

[22]G.R. Salmond and A.R. Croft, "The Management of Public Watersheds," *Water: The Yearbook of Agriculture, 1955,* p. 195.

[23]Forest Service, *Program for the National Forests,* USDA, FS (May 1959), p. 4.

[24]*National Forest Facts: Southwestern Region, 1957,* p. 17.

[25]Jack Williams [then mayor of Phoenix], "Water," *Arizona Cattlelog,* Vol. 13, No. 4 (Dec. 1956), p. 10; Kenneth Wingfield, "The Stockman's Viewpoint on Brushland Control," *ibid.,* Vol. 11, No. 9 (May 1956), p. 12; Cecil Miller, "Multiple Use of Range Lands in Arizona," *ibid.,* Vol. 7, No. 10 (June 1952), p. 17.

[26]H.G. Wilm et al., "The Training of Men in Forest Hydrology and Watershed Management," *Journal of Forestry,* Vol. 55, No. 4 (April 1957), p. 268; Raphael Zon, "Forestry Mistakes and What They Have Taught Us," *ibid.,* Vol. 49, No. 3 (March 1951), p. 182; A.R. Croft and Marvin D. Hoover, "The Relation of Forests to Our Water Supply," *ibid.,* Vol. 49, No. 4 (April 1951), pp. 245-249.

[27]*Recovering Rainfall,* Part 1, p. 6.

[28]*Ibid.,* p. 19.

[29]*Ibid.,* p. 7.

[30]*Ibid.,* Part 2, p. 175.

[31]*Ibid.,* Part 2, pp. 216-17.

[32]*Ibid.,* Part 1, p. 19.

[33]*Ibid.,* Part 1, p. 26.

[34]*Progress in Watershed Management,* p. 4.

[35]USDI, BLM, *Statistical Appendix to the Annual Report of the Director, Bureau of Land Management for the Fiscal Year, 1960,* p. 5.

[36]J. Russell Penny and Marion Clawson, "Economic Possibilities of the Public Domain," *Land Economics,* Vol. 29, No. 3 (Aug. 1953), p. 189.

[37]H.C. Fletcher and L.R. Rich, "Classifying Southwestern Watersheds on the Basis of Water Yields," *Journal of Forestry,* Vol. 53, No. 3 (March 1955), p. 202.

[38]Norman H. French, "A New Look at Erosion Control in Arizona," *Journal of Range Management,* Vol. 10, No. 5 (Sept. 1957), p. 234.

[39]E. Louise Peffer, *The Closing of the Public Domain* (Stanford, Calif.: Stanford University Press, 1951), p. 224.

[40]Quoted in Richard J. Morrissey, "The Early Range Cattle Industry in Arizona," *Agricultural History,* Vol. 24, No. 3 (July 1950), p. 155.

[41]Wagoner, pp. 37-65.

[42]*Ibid.,* p. 39; also see Hastings.

[43]For examples of the attitudes of some stockmen, see Gladys and Charles Niehuis, "Greed and Grass," *Arizona Teacher-Parent* (Dec. 1946), pp. 11-12.

[44]Rich Johnson, "Transfer of Public Domain," *Arizona Farmer,* Vol. 26, No. 23 (Nov. 15, 1947), p. 20.

[45]USDI, BLM, Area 2, Arizona, *Annual Narrative Report — Range Management,* fiscal year 1956, p. 3.

[46]*Ibid.,* fiscal year 1961.

[47]USDI, BLM, *Report of the Director of the Bureau of Land Management, 1961: Statistical Appendix,* p. 130.

[48]USDI, BLM, Area 2, Arizona, *Annual Narrative Report — Range Management,* fiscal year 1956, p. 4.
[49]USDI, BLM, *Report of the Director . . . 1956: Statistical Appendix,* pp. 79-80.
[50]Edward Rowland, Director, Arizona State Office, Bureau of Land Management, "The Public Domain and the Bureau of Land Management: Brief History and Problems." Speech given at Arizona State College, Flagstaff, June, 1956.
[51]USDI, Office of the Solicitor, Opinion M-36263 (Feb. 23, 1955).
[52]Rowland, p. 14.
[53]USDI, BLM, *Soil and Moisture Conservation Operations* (undated), pp 7-8.
[54]Penny and Clawson, *Land Economics,* Vol. 29, No. 3, p. 196.
[55]USDI, BLM, *Report of the Director . . . 1956: Statistical Appendix,* pp. 90-95
[56]USDI, BLM, Area 2, Arizona, *State Summary of 20-Year Program for Sub-basins* (1956).
[57]French, *Journal of Range Management,* Vol. 10, No. 5, p. 225.
[58]Ibid., p. 227.
[59]USDI, BLM, Area 2, Arizona, *Detailed Plan: Railroad Wash Community Watershed,* LM-2-0-6-4; LM-29-0-6-A4 (1956).
[60]Robert A. Darrow, *Arizona Range Resources and Their Utilization, I: Cochise County,* AAES Tech. Bull. 103 (1944), p. 318.
[61]Norman A. French, "Silt is a Major Thief" (typewritten, June 13, 1957).
[62]Kenneth W. Parker and S. Clark Martin, *The Mesquite Problem on Southern Arizona Ranges,* USDA Circular No. 908 (1952), pp. 67-68.
[63]Robert R. Humphrey, *Forage Production on Arizona Ranges, III: Mohave County,* AAES Bull. 244 (1953), p. 11.
[64]Humphrey, "The Desert Grassland, Past and Present," *Journal of Range Management,* Vol. 6, No. 3 (May 1953), p. 159.
[65]Public Law 167, 84th Congress.
[66]USDI, *Annual Report of the Secretary, 1956,* p. 250.
[67]USDI, BLM, Area 2, Arizona, *Annual Narrative Report — Range Management,* fiscal year 1961, p. 7.
[68]See Edward Woosley, "Our Public Domain," *Arizona Cattlelog,* Vol. 11, No. 7 (March 1956), p. 45.
[69]U.S. Dept. of Commerce, Bureau of the Census, *Statistical Abstract of the United States: 1962,* p. 29.
[70]This figure does not include the Navajo Reservation which is managed from Gallup, New Mexico. The information was provided by the Phoenix Area Office and thus includes operations in Utah and Nevada, but work there is slight compared to Arizona.
[71]For the best summary of resources on the Arizona reservations, see Kelly, Kunstadter, and Hackenberg, Book II, Part 1.
[72]USDI, BIA, Phoenix Area Office, Sells Agency (Papago), *Soil Conservation Program: 20-Year Period* (1954), p. 2.
[73]Papago Tribal Council, *The Papago Development Program: 1949,* p. 41.
[74]USDI, BIA, Phoenix Area Office, Sells Agency (Papago), *Soil Conservation Program: 20-Year Period,* p. 8. Evidence that the conditions on the reservation are not new is found in *Irrigation and Flood Protection: Papago Indian Reservation, Arizona,* 62d Cong., 2d Sess., Sen. Doc. 973 (1913).
[75]Arizona Commission on Indian Affairs, *Annual Report,* p. 13.
[76]USDI, BIA, Phoenix Area Office, Branch of Land Operations, *Outline of Work Necessary to Complete Soil and Moisture Conservation Program: Colorado River Indian Agency* (Parker, Dec. 21, 1954).
[77]USDI, BIA, Phoenix Area Office, *Soil Conservation Activities, Branch of Land Operations* (1955).
[78]*Arizona Republic,* June 27, 1957.
[79]Kelly, Kunstadter, and Hackenberg, Book II, Part 1, p. 52.
[80]Hackenberg, p. 75.
[81]Ibid., pp. 83-86; also Edward H. Spicer, *Cycles of Conquest: The Impact of Spain, Mexico, and the United States on the Indians of the Southwest, 1533-1960* (Tucson: University of Arizona Press, 1962), pp. 148-51.
[82]Kelly, Kunstadter, and Hackenberg, Book II, Part 1, p. 67.
[83]*The San Carlos Apache Indian Reservation,* Stanford Research Institute (Stanford, Calif., 1954).

[84]USDI, BIA, Branch of Land Operations, *Outline of Work and Funds Necessary to Complete Soil and Moisture Conservation Development Program*: San Carlos Apache Reservation (1955).
[85]*The San Carlos Apache Indian Reservation*, p. 130.
[86]Arizona Commission on Indian Affairs, *Annual Report*, p. 7.
[87]Clyde Kluckhohn and Dorothea Leighton, *The Navajo* (Cambridge, Mass.: Harvard University Press, 1946), p. 16.
[88]*Ibid.*, pp. 16-17.
[89]*The Navajo: Long-Range Program for Navajo Rehabilitation*, Report of J.A. Krug, Secretary of the Interior, USDI, BIA (1948).
[90]Kluckhohn and Leighton, p. 17.
[91]Public Law 474, 81st Cong., 1950.
[92]Robert W. Young, comp., *The Navajo Yearbook of Planning in Action, 1955*, USDI, BIA, Navajo Agency (Window Rock, Ariz.), p. 40.
[93]*Ibid.*, p. 102; *ibid.*, *1958*, p. 61.
[94]*Ibid.*, *1958*, p. 87.

Chapter 11

[1]*Soil Conservation in Arizona*, USDA, SCS, Region Six (Albuquerque, N.M., 1947), pp. 1-2.
[2]Robert V. Boyle, "Southwest Gets Thirstier as Water Problem Grows," *Soil Conservation*, Vol. 17, No. 9 (April 1952), p. 195.
[3]U.S.C.A., par. 341-348.
[4]*Session Laws*, 1915, Ch. 23; *Arizona Revised Statutes* (1955), 3-121.
[5]P.H. Ross, *Twenty Years of Agricultural Extension Work in Arizona*, Arizona Agricultural Extension Service, Extension Project Circular No. 15 (June 1935).
[6]*Session Laws*, 1921, Ch. 67, par. 3; *Arizona Revised Statutes* (1955), 3-124.
[7]*Arizona Revised Statutes* (1955), 3-125, 126.
[8]*Ibid.*, 3-128.
[9]*Ibid.*, 3-123.
[10]P.H. Ross, p. 102.
[11]Arizona Agricultural Extension Service, *Annual Report of County Extension Agents* (Dec. 1, 1954 to Nov. 30, 1955).
[12]*Ibid.*, p. 15.
[13]*Ibid.*, p. 7.
[14]*Irrigating in Arizona*, Circular 123 (June 1944); Charles Hobart and Karl Harris, *Fitting Cropping Systems to Water Supplies in Central Arizona*, Circular 127 (Nov. 1950); and James E. Middleton, *Water Management*, Circular 205 (Oct. 1952).
[15]Agricultural Extension Service, *Annual Report of County Extension Agents* (1954-55), p. 31.
[16]See Charles Hardin, *Politics of Agriculture* (Glencoe, Ill.: Free Press, 1952).
[17]*Session Laws*, 1941, ch. 43, par. 2; *Arizona Revised Statutes* (1955), 45-2001.
[18]*Session Laws*, 1945, ch. 31, par. 4; *Arizona Revised Statutes* (1955), 45-2013.
[19]For a more detailed listing of the everyday duties of the Soil Conservation Division, see ALSD, *Forty-fourth Annual Report*, pp. 24-25.
[20]*Soil and Water Conservation in Arizona*, USDA, SCS (Phoenix, March 1961).
[21]*Session Laws*, 1954, Ch. 38.
[22]*Report of Land Use and Ownership in Soil Conservation Districts: Arizona Summary — All Districts*, USDA, SCS, Arizona State Office (Phoenix, June 30, 1956).
[23]USDA, SCS, *Administrator's Memorandum*, SCS-84 (May 12, 1955), p. 1.
[24]*Arizona Revised Statutes* (1955), 45-2054A, 1-6.
[25]ASLD, *Forty-seventh Annual Report*, p. 14.
[26]*Session Laws*, 1941, ch. 43, par. 2; *Arizona Revised Statutes* (1955), 45-2001.
[27]*Arizona Revised Statutes* (1955), 45-2033B, 45-2054C; see D.A. Dobkins, "Soil Conservation in Arizona," *Arizona Highways*, Vol. 32, No. 8 (Aug. 1956), p. 6.
[28]*Arizona Revised Statutes* (1955), 45-2033.
[29]*Ibid.*, 45-2035A, B.
[30]Public Law 46, 74th Cong., 1935; Public Law 534, 78th Cong., 1944 (Flood Control Act of 1944); Public Law 566, 83d Cong., 1954 (Watershed Protection and Flood Prevention Act of 1954); and Agricultural Appropriations Act of 1954, providing funds under P.L. 46, 1935.

[31]USDA, SCS, *Administrator's Memorandum*, SCS-71 (Dec. 2, 1954).
[32]*Ibid.*
[33]*Ibid.*
[34]USDA, SCS, *Administrator's Memorandum*, SCS-84, pp. 24-25.
[35]*Arizona Revised Statutes* (1955), 45-2055B.
[36]*Long Range Program of Soil and Water Conservation Operations: Arizona,* USDA, SCS, Arizona State Office (1956), p. 6.
[37]*The Arizona Soil Conservationist*, Vol. 8, No. 2 (April 1, 1956), p. 3.
[38]See R.V. Keppel and Joel E. Fletcher, "Runoff from Rangelands of the Southwest," Agricultural Research Service, typewritten, 1960, p. 23.
[39]*The Arizona Soil Conservationist*, Vol. 8, No. 2, p. 1. He is reported elsewhere to have said, in substance, "Give me the water and we'll take the silt." See also ASLD, *Forty-fourth Annual Report*, p. 26; *House Journal*, 23d Legislature, 2d Sess., 1959, pp. 281-82.
[40]Senate Bill 204, 24th Legislature, 1st Sess., 1960.
[41]*Soil and Water Conservation in Arizona*, p. 7; see also Dobkins, p. 7.
[42]*Long Range Program of Soil and Water Conservation Operations: Arizona*, p. 6.
[43]*Facts About the Agricultural Conservation Program*, USDA, Agricultural Conservation Program Service, PA-272 (1955).
[44]*Agricultural Conservation Program: Arizona Handbook for 1956*, USDA, ACPS (Washington 1955), p. 1.
[45]*Ibid.*, p. 3.
[46]*Statistical Summary of Major Range and Farm Practices Completed Under the Agricultural Conservation Program for the Years 1936-1955*, USDA, Agricultural Stabilization and Conservation Service, Arizona (typewritten, 1956).
[47]USDA, *Report of the Secretary*, 1955, pp. 52-53; see also Robert G. Craig, "Farmers Home Administration Loans," *Arizona Cattlelog*, Vol. 12, No. 2 (Oct. 1956).
[48]USDA, SCS, *Administrator's Memorandum*, SCS-109, pp. 2-6.
[49]See Boyle, "Southwest Gets Thirstier as Water Problem Grows," *Soil Conservation*, Vol. 17, No. 9, p. 6.
[50]Dobkins, "Soil Conservation in Arizona," *Arizona Highways*, Vol. 32, No. 8, p. 8; see also, for example, *Memorandum of Understanding between the Soil Conservation Service, USDA, and the Bureau of Land Management, USDI, Relative to Inter-Agency Cooperation within Soil Conservation Districts in Arizona*, USDA, SCS (mimeo., Aug. 1, 1956).
[51]See, for example, *Memorandum of Understanding between the Soil Conservation Service, the Forest Service, and the Agricultural Research Service, relating to Inter-Agency Coordination of Programs in Watersheds as authorized by Section 6 of Public Law 566, 83d Cong.*, USDA (mimeo, Feb. 2, 1956).
[52]USDA, *Report of the Secretary*, 1954.
[53]*Soil Conservation*, Vol. 8, No. 3 (July, 1956), p. 3.

Chapter 12

[1]Marion Clawson, "The Crisis in Outdoor Recreation," *American Forests* (March and April, 1959).
[2]John H. Sieker, "Planning for Recreational Use of Water: A Plea," *Water: The Yearbook of Agriculture, 1955*, p. 577.
[3]See *Arizona's Growth and Its Future*.
[4]*A Survey of the Recreational Resources of the Colorado Basin*, p. 21.
[5]*Some Economic Implications of the Tourist Industry for Northern Arizona*, p. 3.
[6]*Ibid.*, p. 4.
[7]*A Survey of the Recreational Resources of the Colorado River Basin*, p. 142.
[8]*Ibid.*, p. 136.
[9]Gene Foster, "A Brief Archaeological Survey of Glen Canyon," *Plateau*, Vol. 25, No. 2 (Oct. 1952), pp. 21-26. See also Albert H. Schroeder.
[10]*A Survey of the Recreational Resources of the Colorado River Basin*, p. 144.
[11]APA, *Sixteenth Annual Report*, p. 11.
[12]*A Survey of the Recreational Resources of the Colorado River Basin*, p. 11.

[13]For a description of all these monuments, see *Arizona's National Monuments*, Southwestern National Monuments Association, Popular Series No. 2 (Santa Fe, N.M., 1945).

[14]*Public Use: National Parks and Related Areas*, USDI, NPS (Washington, D.C., Dec. 1960).

[15]*Forecast of Total Visits, 1966-1970*, USDI, NPS (mimeo. 1960?).

[16]*Mission 66 for the National Park System*.

[17]*Preservation of Natural and Wilderness Values in the National Parks*, USDI, NPS (March 1957), pp. 15-16.

[18]USDI, *Annual Report of the Secretary*, 1960, p. 325.

[19]*Mission 66 for the National Park System*, p. 113.

[20]*Ibid.*, p. 114.

[21]See the mimeographed reports, *Mission 66 for the Pipe Springs National Monument; Mission 66 for the Organ Pipe Cactus National Monument; Mission 66 for Chiricahua National Monument; Mission 66 for Saguaro National Monument* (all undated but presumably 1957).

[22]USDA, FS, *Report of the Chief of the Forest Service, 1960* (Washington, D.C., 1961), p. 42; *National Forest Facts, Southwestern Region, 1954*, p. 8, and *1957*, p. 9.

[23]Jack Karie, "Apaches' New Lake First Step in Creating Vacation Paradise," *Arizona Republic*, June 9, 1957.

[24]House Bill 72, 23d Legislature, 1st Session.

[25]*Arizona Republic*, March 15, 1956, p. 2.

[26]See Lester N. Inskeep, "State to Embark on Park Development," *Arizona Daily Star*, April 25, 1957, and *State Parks, Areas, Acreages and Accommodations, 1960*, USDI, NPS (March 1961), p. 11.

[27]See *A Study of the Park and Recreation Problem of the United States*.

[28]*Ibid.*, p. 139; *A Survey of the Recreational Resources of the Colorado River Basin*, p. 219.

[29]*Wildlife News*, Vol. 4, No. 1 (Jan. 1957), pp. 1 ff.

[30]"30 Years of Progress," *Wildlife News*, Vol. 7, No. 1 (Jan. 1960).

[31]*Ibid.*, p. 2.

[32]*Arizona Revised Statutes* (1955), 45-141.

[33]*Ibid.*, 45-147.

[34]Dale Slocum, "Water Filings — Water Grab," *Arizona Wildlife-Sportsman*, March, 1958.

[35]"Wildlife Water Rights Challenge," *Outdoor Life*, Vol. 2, No. 1 (Jan. 1957), pp. 1ff.

[36]For a breakdown of sources of funds, see Arizona Game & Fish Commission, *Annual Report, 1955-1956*, pp. 6-7.

[37]*Arizona Revised Statutes* (1955), 17-261.

[38]*Ibid.*, 17-201; and *Session Laws*, 1929, Ch. 84, par. 2.

[39]*Arizona Revised Statutes* (1955), 17-231.

[40]*Ibid.*, 17-211.

[41]Arizona Game & Fish Dept. *Biennial Report, 1947-1949*, p. 15.

[42]*A Plan for the Operation of the Arizona Game and Fish Commission*, Arizona Game & Fish Dept. (1960).

[43]*Ibid.*

[44]Arizona Game & Fish Dept., *Annual Report, 1955-1956*, p. 22.

[45]*Ibid.*, p. 16.

[46]*Arizona Survey of the Wildlife and Fisheries Resources of the Lower Colorado River*, Arizona Game & Fish Dept. (1952).

[47]*Arizona Revised Statutes* (1955), 17-238.

[48]AGPA, *The Principles Supported by the Arizona Game Protective Association* (mimeo, undated).

[49]AGPA, *By-Laws* (Revised Sept. 2, 1956), Art. 1, Sec. 10.

[50]*Outdoor News*, Vol. 1, No. 3 (Oct. 1956), pp. 1ff.

[51]This, and much information regarding the activities of the Fish and Wildlife Service in Arizona was contained in a letter from Mr. John C. Gatlin, Regional Director, F&WS, Region 2 (Jan. 1957).

[52]Letter from Mr. Gatlin, *supra;* also F&WS, Region 2, *Kofa Game Range* (typewritten, undated).

[53]*Duck Stamps and Wildlife Refuges,* Circular 37, USDI, F&WS (Washington, 1955), p. 22; USDI, F&WS, *Imperial National Wildlife Refuge* (typewritten, undated.)
[54]*Federal Aid in Fish and Wildlife Restoration,* Wildlife Management Institute (Washington, D.C., 1955), p. 2.
[55]USDI, *Annual Report of the Secretary,* 1956, p. 290.
[56]Lyle K. Sowls, *Wildlife Conservation Through Cooperation* (Tucson: University of Arizona Press, 1956), p. 5.
[57]*Ibid.*
[58]*Ibid.,* p. 23.
[59]*Ibid., passim.*
[60]*National Survey of Fishing and Hunting,* USDI, F&WS (Washington, D.C., 1956), *passim.*
[61]*Wildlife News,* Vol. 6, No. 1 (Winter, 1959), p. 28.
[62]William C. Davis, *Values of Hunting and Fishing in Arizona, 1960,* University of Arizona, Bureau of Business and Public Research, Special Studies No. 21 (April 1962), pp. 4, 9, and 24.
[63]*Ibid.,* p. iv.

Chapter 13

[1]*Reviews of Research on Arid Zone Hydrology,* UNESCO, Advisory Committee on Arid Zone Research, Arid Zone Programme I (Paris, 1953), p. 7.
[2]Higbee, p. 251.
[3]Ray K. Linsley, "Report on the Hydrological Problems of the Arid and Semi-Arid Areas of the United States and Canada," *Reviews of Research on Arid Zone Hydrology,* p. 129.
[4]*Ibid.,* pp. 129-140.
[5]U.S. President's Water Resources Policy Commission, *A Water Policy for the American People,* pp. 107-108.
[6]*Ibid.,* pp. 107-108.
[7]U.S. President's Advisory Committee on Water Resources Policy, *Water Resources Policy,* p. 8. It recommended an increase from approximately $40,000,000 to $80,000,-000, an increase of only 2 percent of the amount being expended on natural resources.
[8]*A Water Policy for the American people, pp.* 337-72 *passim.*
[9]*Folio of Geologic and Mineral Maps of Arizona.* Arizona Bureau of Mines (Tucson: University of Arizona Press, 1962).

Chapter 14

[1]AAES, *Fiftieth Annual Report,* pp. 4-5.
[2]Personal communication from the Arizona Agricultural Experiment Station.
[3]AAES, *Fiftieth Annual Report,* p. 20.
[4]W.H. Fuller, *Effects of Kinds of Phosphate Fertilizer and Method of Placement on Phosphorus Absorption by Crops Grown on Arizona Calcareous Soil.* AAES Tech. Bull. 128 (June 1953); W.S. Fuller, N.C. Gomness, and L.V. Sherwood, *The Influence of Soil Aggregate Stabilizers on Stand, Composition, and Yield of Crops on Calcareous Soils of Southern Arizona.* AAES Tech. Bull. 129 (July 1953); W.T. McGeorge, E.L. Breazeale, and J.L. Abbott, *Polysulfides as Soil Conditioners.* AAES Tech. Bull. 131 (June 1956); W.H. Fuller and Catherine G. Padgett, *The Effects of Discing, Rototilling and Water Action on the Structure of Some Calcareous Soils.* AAES Tech. Bull. 134 (July 1958); W.H. Fuller, *Soil Composition.* AAES Report 168 (April 1958).
[5]See the following articles in *Progressive Agriculture*: R.H. Hilgeman, "Irrigating Oranges at Three Moisture Levels," Vol. 9, No. 1, p. 13; W.D. Pew, "Yellowing of Lettuce," Vol. 3, No. 3, p. 10; C.O. Stanberry, "Alfalfa Irrigation," Vol. 6, No. 2, p. 9; Karl Harris, "Irrigation of Alfalfa," Vol. 3, No. 4, p. 4.
[6]AAES, *Fiftieth Annual Report,* p. 22.

[7]Oscar E. Meinzer and F.C. Kelton, *Geology and Water Resources of Sulphur Spring Valley, Arizona,* AAES Bull. 72 (1913); G.E.P. Smith, *Groundwater Supply and Irrigation in Rillito Valley,* AAES Bull. 64 (1910), *The Physiography of Arizona Valleys and the Occurrence of Ground Water,* AAES Tech. Bull. 77 (1938), and *The Groundwater Supply of the Eloy District of Pinal County, Arizona,* AAES Tech. Bull. 87 (June 1940).

[8]C.K. Cooperrider, *The Relationship of Stream Flow to Precipitation on the Salt River Watershed Above Roosevelt Dam,* AAES Tech. Bull. 76 (1938); Harold C. Schwalen, *Rainfall and Runoff in the Upper Santa Cruz River Drainage Basin,* AAES Tech. Bull. 95 (Sept. 1942).

[9]See Schwalen and Shaw.

[10]AAES, *Research Progress, 1955,* pp. 21-22.

[11]*Ibid.,* p. 25.

[12]See Ch. 15 for further discussion of the possibilities of ground-water recharge through wells.

[13]See the following AAES Bulletins: Frank A. Gulley, *Pumping Water for Irrigation* (#3, 1891); A.J. McClatchie, *Irrigation at Station Farm* (#41, 1902); G.E.P. Smith, *Use and Waste of Irrigation Water* (#88, 1919); W.E. Code, *Design and Construction of Small Concrete Lined Canals* (#97, 1923); H.C. Schwalen, *The Stovepipe or California Method of Well Drilling as Practiced in Arizona* (#112, 1924); Rex Rehnberg, *Irrigation Ditch Management on Arizona Irrigation Farms* (#255, 1951), and *The Cost of Pumping Irrigation Water, Pinal County, 1951* (#246, 1953); and H.C. Schwalen, *Sprinkler Irrigation* (#250, 1953).

[14]W.T. McGeorge, E.L. Breazeale, and A. Mark Bliss, *The Salinity Problem — Safford Experiment Farm Field Experiments,* AAES Tech. Bull. 124 (Feb. 1952).

[15]Personal communication to the author from Sol Resnick, Head, Institute of Water Utilization, September 24, 1960.

[16]Bruce L. Branscomb, "Vegetation Changes Continuously on Arizona Ranges," *Progressive Agriculture,* Vol. 9, No. 2 (July-Sept. 1957), p. 11.

[17]See the following AAES publications: W.G. McGinnies and F.J. Arnold, *Relative Water Requirements of Arizona Range Plants.* Tech. Bull. 80 (1939); J.W. Toumey, *Range Grasses of Arizona.* Bull. 2 (1891); A.A. Nichol, *The Natural Vegetation of Arizona.* Bull. 68 (1937) and later revision Bull. 127 (1952); and Robert R. Humphrey, *Arizona Range Grasses.* Bull. 298 (July 1958).

[18]D. Anderson, L.P. Hamilton, H.G. Reynolds, and Robert R. Humphrey, *Reseeding Desert Grassland Ranges in Southern Arizona.* AAES Bull. 249 (1953).

[19]J.J. Thornber, *The Grazing Ranges of Arizona.* AAES Bull. 65 (1910); Robert R. Humphrey, "The Desert Grassland, Past and Present," *Journal of Range Management,* Vol. 6, No. 3 (May 1953), p. 159.

[20]R.B. Streets and E.B. Stanley, *Control of Mesquite and Noxious Shrubs on Southern Arizona Grassland Ranges.* AAES Tech. Bull. 74 (1938); F.H. Tschirley, "Chaparral —Still a Problem," *Progressive Agriculture,* Vol. 6, No 1(April-June 1954), p. 8; Robert R. Humphrey, "Major Aspects of the Woody Plant Problem in Arizona," *Arizona Cattlelog,* Vol. 12, No. 2 (Oct. 1956), p. 38.

[21]R.A. Darrow, *Arizona Range Resources and Their Utilization. I: Cochise County.* Tech. Bull. 103 (Oct. 1944); Robert R. Humphrey, *Arizona Range Resources: Yavapai County.* Bull. 229 (July 1950); *Forage Production on Arizona Ranges. III: Mohave County.* Bull. 244 (Feb. 1953), and *Forage Production on Arizona Ranges. IV: Coconino, Navajo, and Apache Counties — A Study of Range Conditions.* Bull. 226 (Oct. 1955); David Griffiths, *A Protected Stock Range in Arizona.* USDA, Bureau of Plant Industry, Bull. 177 (April 1910).

[22]See AAES, *Research Progress, 1955,* various articles; also Robert R. Humphrey, *The Desert Grassland.* AAES Bull. 299 (Dec. 1958).

[23]Humphrey, "Paved Drainage Basins as a Source of Water for Livesotck or Game," *Journal of Range Management,* Vol. 10, No. 2 (March 1957), p. 59.

[24]24 Stat. 440, sec. 1.

[25]49 Stat. 436, sec. 1.

[26]*Arizona Republic,* August 15, 1957.

[27]W.S. Stallings, Jr., *Dating Prehistoric Ruins by Tree-Rings,* Laboratory of Anthropology, General Series No. 8 (Santa Fe, N.M., 1939); Edmund Schulman, "The Tree-

Ring Laboratory of the University of Arizona," *Chronica Botanica*, Vol. 6, No. 3 (Nov. 1940), pp. 63-64.

[28]A.E. Douglass, "Accuracy in Dating, I," *Tree-Ring Bulletin*, Vol. 1, No. 2 (Oct. 1934), p. 10, and "Accuracy in Dating, II: The Presentation of Evidence," *Tree-Ring Bulletin*, Vol. 1, No. 3 (Jan. 1935), p. 19.

[29]Edmund Schulman, "Some Propositions in Tree-Ring Analysis," *Ecology*, Vol. 22, No. 2 (April 1941), p. 193.

[30]Schulman, *Dendroclimatic Changes in Semiarid America* (Tucson: University of Arizona Press, 1956), p. 7. Hereafter cited as *Dendroclimatic Changes*.

[31]*Ibid.*

[32]See A.E. Douglass, "The Secret of the Southwest Solved by Talkative Tree Rings," *National Geographic Magazine*, Vol. 56 (Dec. 1929), p. 737. For a survey of the work in this field, see Terah L. Smiley, *A Summary of Tree-Ring Dates from Some Southwestern Archaeological Sites*, University of Arizona Laboratory of Tree-Ring Research Bull. No. 5 (Oct. 1951).

[33]Schulman, "Tree-Rings Work for Science," *Arizona Alumnus*, Vol. 15, No 2 (No. 1, 1937), pp. 2ff.

[37]Schulman, *Dendroclimatic Changes*, p. 69.

[35]*Ibid.*

[36]Schulman, "Definitive Dendrochronologies: A Progress Report," *Tree-Ring Bulletin*, Vol. 18, No. 2/3 (Oct. 1951-Jan. 1952), p. 17.

[37]Schulman, *Dendroclimatic Changes*, p. 69.

[38]*Ibid.*, p. 67.

[39]See Roy Lasseter, "The Value of Tree-Ring Analysis in Engineering," *Tree-Ring Bulletin*, Vol. 5, No. 2 (Oct. 1938), p. 13.

[40]Schulman, *Dendroclimatic Changes*, p. 4.

[41]Schulman, *Tree-Ring Hydrology of the Colorado River Basin*, University of Arizona Laboratory of Tree-Ring Research Bull. 2 (Oct. 1945), p. 49.

[42]Ben Avery, "Rain-Making in Arizona," *Reclamation Era*, Vol. 36, No. 5 (May 1950), p. 89. An estimated generation of 11,990 acre-feet was claimed one year.

[43]Institute of Atmospheric Physics, U of A, *General Information on the Institute of Atmospheric Physics* (April 15, 1955), p. A1.

[44]*Ibid.*, p. A3.

[45]Inst. of Atmos. Physics, U of A, *Proceedings of the Conference on the Scientific Basis of Weather Modification Studies* (Tucson, April 10-12, 1956).

[46]Inst. of Atmos. Physics, U of A, *Progress Report No. 2* (Nov. 30, 1955), p. 1; also *Reference List of Research Projects in the Institute of Atmospheric Physics as of January 1957* (Tucson, Jan. 1957), pp. 9-10. Hereafter cited as *Reference List*.

[47]Inst. of Atmos. Physics, U of A, *First Annual Progress Report, Cooperative Punchcard Climatological Program* (Tucson, Nov. 15, 1955).

[48]*Ibid.*, pp. 28-45.

[49]Louis J. Battan and A. Richard Kassander, Jr., "Possibilities of Increasing Water Yields by Cloud Seeding," *Watershed and Related Water Management Problems*, Proceedings, Fourth Annual Watershed Symposium (Sept. 21, 1960), p. 40.

[50]Battan and Kassander, *Seeding of Summer Cumulus Clouds*, Inst. of Atmos. Physics, U of A, Scientific Report No. 10 (July 1, 1959), pp. 2 and 6, and *Evaluation of Effects of Airborne Silver-Iodide Seeding of Convective Clouds*, Inst. of Atmos. Physics, U of A, Scientific Report No. 18 (March 1, 1962).

[51]Kassander and Lee L. Sims, *Cloud Photogrammetry with Ground-Located K-17 Aerial Cameras*, Inst. of Atmos. Physics, U of A, Scientific Report No. 2 (June 15, 1956).

[52]Inst. of Atmos. Physics, U of A, *Reference List*, p. 2.

[53]*General Information on the Institute of Atmospheric Physics*, p. A3.

[54]See Bernice Ackerman, *Characteristics of Summer Radar Echoes in Arizona, 1956*, Inst. of Atmos. Physics, U of A, Scientific Rept. No. 11 (July 8, 1959); and Battan, Kassander and Sims, *Randomized Seeding of Orogeophic Cumuli, 1959, Part II*, Inst. of Atmos. Physics, U of A, Scientific Rept. No. 9 (Oct. 1, 1958).

[55]See the following Institute of Atmospheric Physics Technical Reports on the Meteorology and Climatology of Arid Regions: Clayton H. Reitan, *The Role of Precipitable Water Vapor in Arizona's Summer Rains* (No. 2, Jan. 31, 1957); Bryson, *The Annual March of Precipitation in Arizona, New Mexico, and Northwestern Mexico* (No. 6, June 7, 1957); Christine R. Green, *Arizona Statewide Rainfall* (No. 7, Nov. 30, 1959); Green, *Probabilities of Drought and Rainy Periods for Selected Points in*

284   THE POLITICS OF WATER IN ARIZONA

the *Southwestern United States* (No. 8, Jan. 31, 1960). See also William D. Sellers, *Distribution of Relative Humidity and Dew Point in the Southwestern United States,* Scientific Rpt. No. 13 (Feb. 1, 1960).

[50]University of Arizona, *The Utilization of Arid Lands, A Research Proposal Submitted to the Rockefeller Foundation* (Feb. 6, 1958), and *First Annual Report on an Interdisciplinary Study of the Utilization of Arid Lands* (June 11, 1959).

Chapter 15

[1]USDI, *Annual Report of the Secretary, 1956,* p. 131.

[2]Wilbur E. Heckler, *History of Run Off Investigations and Development of Storage in Arizona* (typewritten, undated talk).

[3]USDI, *Official Organization Handbook* (1951), p. 70.

[4]Arthur P. Davis, *Irrigation Near Phoenix, Arizona,* USGS Water Supply and Irrigation Paper No. 2 (1897); Joseph B. Lippincott, *Storage of Water on the Gila River, Arizona,* USGS Water Supply and Irrigation Paper No. 33 (1900).

[5]Arthur P. Davis, *Water Storage on Salt River, Arizona,* USGS Water Supply and Irrigation Paper No. 73 (1903).

[6]A.T. Schwennesen, *Geology and Water Resources of the Gila and San Carlos Valleys in the San Carlos Indian Reservation, Arizona,* USGS Water Supply Paper No. 450 (1919).

[7]E.C. Murphy et al., *Destructive Floods in the United States in 1904,* USGS Water Supply and Irrigation Paper No. 147 (1905), and *Destructive Floods in the United States in 1905,* USGS Water Supply and Irrigation Paper No. 162 (1906).

[8]E.C. LaRue, *Colorado River and Its Utilization,* USGS Water Supply Paper No. 395 (1916).

[9]For example, La Rue suggested 5,580,000 acres as the maximum for irrigation in the Colorado River Basin; the Bureau of Reclamation now estimates a maximum of 4,753,000 acres including diversions outside the natural basin.

[10]See *Compilation of Records of Surface Water of the United States through September 1950, Part 9, Colorado River Basin.*

[11]Project Planning Report (Nov. 1952) and *Memorandum Supplement* (Nov. 1953).

[12]Douglas A. Lewis, "Cottonwood Wash Project: Water Use by Channel Vegetation," *Progress in Watershed Management,* p. 100.

[13]In preface to A.P. Davis, *Irrigation Near Phoenix, Arizona,* p. 11.

[14]USDI, *Official Organization Handbook* (1951), p. 70.

[15]Willis T. Lee, *The Underground Waters of the Gila Valley, Arizona,* USGS Water Supply and Irrigation Paper No. 104 (1904), and *The Underground Waters of the Salt River Valley,* USGS Water Supply and Irrigation Paper No. 136 (1905). O.E. Meinzer and A.J. Ellis, *Ground Water in Paradise Valley, Arizona,* USGS Water Supply Paper No. 375-B (1915).

[16]Kirk Bryan, *The Papago Country, Arizona,* USGS Water Supply Paper No. 499 (1925); Herbert E. Gregory, *The Navajo Country,* USGS Water Supply Paper No. 380 (1916); Clyde P. Ross, *Routes to Desert Watering Places in the Lower Gila Region, Arizona,* USGS Water Supply Paper No. 490 (1922).

[17]See ASLD, *Thirty-second Annual Report,* p. 5.

[18]L.C. Halpenny, J.D. Hem, and I.I. Jones, *Definitions of Geologic, Hydrologic, and Chemical Terms Used in Reports on the Ground-Water Resources and Problems of Arizona,* USGS (mimeo, March 18, 1947).

[19]*Ibid.*

[20]*Pumpage and Ground-Water Levels in Arizona in 1955,* pp. 1-2.

[21]These studies were printed in mimeograph form by the state office of the USGS and have been collected by the University of Arizona as the *Arizona Water Supply Papers.*

[22]This estimate was given by J. W. Harshbarger, then State Director of the Ground Water Branch in an interview (Sept. 16, 1957).

[23]The first was published in 1956. See *Pumpage and Ground-Water Levels in Arizona in 1955.*

[24]Harris R. McDonald and Harold D. Padgett, Jr., *Geology and Ground-Water*

Resources of the Verde River Valley near Fort McDowell, Arizona, USGS, Ground Water Branch, Tucson (mimeo, Nov. 1, 1945); G.E. Hazen and S.F. Turner, Geology and Ground-Water Resources of the Upper Pinal Creek Area, Arizona, USGS (mimeo, Dec. 1946).

[25]See J.W. Harshbarger, C.A. Repenning, and R.L. Jackson, Jurassic Stratigraphy of the Navajo Country, USGS (Holbrook, Ariz., mimeo, Aug. 1951); L.C. Halpenny, Preliminary Report on the Ground-Water Resources of the Navajo and Hopi Indian Reservations, Arizona, New Mexico, and Utah, USGS (Holbrook, mimeo, Aug. 1951).

[26]S.F. Turner, and L.C. Halpenny, "Ground-Water Inventory in the Upper Gila Valley, New Mexico and Arizona: Scope of Investigation and Methods Used," Transactions of the American Geophysical Union, Vol. 22, Part 3, (1941); J.D. Hem, Quality of Water of the Gila River Basin Above Coolidge Dam, Arizona, USGS Water Supply Paper No. 1104 (1950).

[27]H.M. Babcock, Memorandum on Ground-Water Supply of the Joseph City Irrigation District, USGS (Tucson, mimeo, Aug. 1955).

[28]D.C. Metzger, Geology and Ground-Water Resources of the Northern Part of the Ranegras Plain Area, Yuma County, Arizona, USGS (Tucson, mimeo, Feb. 1951).

[29]H.W. Wolcott, H.E. Skibitzke, and L.C. Halpenny, "Water Resources of Bill Williams River Valley Near Alamo, Arizona," Contributions to the Hydrology of the United States, USGS Water Supply Paper No. 1360-D (1956).

[30]For a list, see Pumpage and Ground-Water Levels in Arizona in 1955, pp. 4-6.

[31]H.M. Babcock, and E.M. Cushing, "Recharge to Ground-Water from Floods in A Typical Desert Wash, Pinal County, Arizona," Transactions of the American Geophysical Union, Vol. 23, Part 1 (1942).

[32]J.S. Gatewood, et al., Use of Water by Bottom-Land Vegetation in Lower Safford Valley, Arizona, USGS Water Supply Paper No. 1103 (1950).

[33]Pumpage and Ground-Water Levels in Arizona in 1955, p. 3; see also R.L. Cushman and L.C. Halpenny, "Effect of Western Drought on the Water Resources of Safford Valley, Arizona," Transactions of the American Geophysical Union, Vol. 36 (1955), pp. 87-94.

[34]USDI, Official Organization Handbook, p. 70.

[35]see Hem.

[36]Quality of Surface Waters for Irrigation, Western United States, 1952, USGS Water Supply Paper No. 1362 (1955), pp. 122-150.

[37]Ibid., p. 1.

[38]Quality of Surface Waters for Irrigation, Western United States, 1952, pp. 1-2; see also, R.A. Krieger, J.L. Hatchett, and J.L. Poole, Preliminary Survey of the Saline-Water Resources of the United States, USGS Water Supply Paper No. 1374 (1957). Studies are also being conducted on the suitability of water for industrial use; see The Industrial Utility of Public Water Supplies in the United States, 1962; part 2, States West of the Mississippi, USGS Water Supply Paper No. 1300 (1954).

[39]Harshbarger, "Use of Ground Water in Arizona," Climate and Man in the Southwest, (Tucson: University of Arizona Press, 1958), p. 16.

[40]Interview in Tucson, Sept. 16, 1957.

[41]Note the work being done on this practice by Harold Schwalen of the Arizona Agricultural Experiment Station.

[42]Babcock and Cushing, p. 56.

[43]Harshbarger, "Capturing Additional Water in the Tucson Area," Progress in Watershed Management, p. 25. The city of Tucson in particular has been unwilling to provide its share of financial support.

Chapter 16

[1]Organization of the Federal Government for Scientific Activities, U.S. National Science Foundation (Washington, D.C., 1956), p. 38.

[2]Toward Meeting Soil and Water Conservation Needs. I, USDA, ARS (Washington, D.C., Dec. 1955). In spite of the broad field of needed study, the President's Water Resources Policy Commission reported that less than 10 percent of Forest Service funds went for research, with resulting waste in forest management. A Water Policy for the American People, p. 135.

[3]For a summary of Forest Service watershed management research in Arizona, see Hudson G. Reynolds, "Current Watershed Management Research by the U.S. Forest Service in Arizona," *Progress in Watershed Management*, p. 63, and *Watershed Management Research in Arizona*, Progress Report, 1959, USDA, FS, Rocky Mtn. Forest & Range Exper. Sta. (1960).

[4]*The Sierra Ancha Experimental Watersheds*, USDA, FS, Southwestern Forest and Range Experiment Station (Tucson, June 1953), pp. 3-4.

[5]*Ibid.*, pp. 6-14.

[6]*Ibid.*, pp. 14-31.

[7]USDA, FS, Rocky Mtn. Forest & Range Experiment Station, *Annual Report, 1956*, (Fort Collins, Colo., 1957), p. 7; see also Hudson G. Reynolds in *Progress in Watershed Management*, p. 69.

[8]Fred H. Kennedy, "National Forest Watershed Projects in Arizona," *Progress in Watershed Management*, p. 53.

[9]For a brief summary, see the Rocky Mountain Station's *Annual Report, 1956*, p. 2 and the *Arizona Farmer-Ranchman*, "Experiment at Beaver Creek," Vol. 36, No. 9 (April 27, 1957). One hydrologist who approved of the project felt that the results might not be too significant since the measurements will be relatively "coarse."

[10]L.R. Rich, "Preliminary Effects of Forest Tree Removal on Water Yields and Sedimentation," *Watershed and Related Water Management Problems*, Proceedings of Fourth Annual Watershed Symposium [Arizona Watershed Program], (Sept. 21, 1960), p. 13.

[11]H.E. Brown and E.G. Dunford, *Streamflow in Relation to Extent of Snow Cover*, USDA, FS, Rocky Mtn. Forest & Range Experiment Station Paper 24 (June 1956); Fred H. Kennedy in *Progress in Watershed Management*, p. 61.

[12]USDA, FS, Rocky Mtn. Forest & Range Exper. Station, *Annual Report, 1956*, p. 8; Kennedy, in *Progress in Watershed Management*, p. 61.

[13]Reynolds in *Progress in Watershed Management*, p. 76.

[14]D.E. Le Crone, "Corduroy and Cibecue Watershed Projects: Fort Apache Indian Reservation," *Progress in Watershed Management*, p. 94.

[15]H.C. Fletcher and Harold B. Elmendorf, "Phreatophytes — A Serious Problem in the West," *Water: The Yearbook of Agriculture, 1955*, p. 426.

[16]Reports of researchers working on the saltcedar problem are found in J.S. Horton, "Ecology of Saltcedar," *Watershed and Related Water Management Problems*, Proceedings of Fourth Annual Watershed Symposium, p. 19; T.E.A. van Hylckama, "Measuring Water Use in Saltcedar," *ibid.*, p. 22, and H.F. Arle, "Saltcedar Control with Chemicals," *ibid.*, p. 27.

[17]USDA, FS, Rocky Mtn. Forest & Range Exper. Station, *Annual Report, 1955*, p. 64.

[18]See Harold Weaver, "A Preliminary Report on Prescribed Burning in Virgin Ponderosa Pine," *Journal of Forestry*, Vol. 50, No. 9 (Sept. 1952), p. 662. Also, Harry R. Kallender, "Controlled Burning in Ponderosa Pine Stands of the Fort Apache Indian Reservation in Arizona," *Arizona Cattlelog*, Vol. 11, No. 12 (Aug. 1956), p. 26.

[19]Harry R. Kallender, Harold Weaver, and Edward M. Gaines, "Additional Information on Prescribed Burning in Virgin Ponderosa Pine in Arizona," *Journal of Forestry*, Vol. 53, No. 10 (Oct. 1955), p. 730.

[20]Weaver, f.n. 18 *supra*, p. 664. H.H. Biswell and A.M. Schultz, "Surface Runoff and Erosion as Related to Prescribed Burning," *Journal of Forestry*, Vol. 55, No. 5 (May 1957), p. 374.

[21]W.H. Fuller, Stanton Shannon, and P.S. Burgess, "Effect of Burning on Certain Forest Soils of Northern Arizona," *Forest Science*, Vol. 1 (1955), pp. 44-50.

[22]Kallender, Weaver, and Gaines, p. 731.

[23]*Ibid.*

[24]USDA, FS, Rocky Mtn. Forest & Range Exper. Station, *Annual Report, 1956*, p. 56.

[25]*Ibid.*, p. 63.

[26]*Ibid.*, p. 65.

[27]Humphrey, "The Desert Grassland, Past and Present," *Journal of Range Management*, Vol. 6, No. 3 (May 1953), p. 159, and "Fire as a Means of Controlling Velvet Mesquite, Burrweed and Cholla on Southern Arizona Ranges," *Journal of Range Management*, Vol. 2, No. 4 (Oct. 1949), p. 175.

[28]See Humphrey, "Major Aspects of the Woody Plant Problem in Arizona," *Arizona Cattlelog*, Vol. 12, No. 2 (Oct. 1956), pp. 38-40.

[29]*Arizona Daily Star*, July 17, 1957.

[30]USDA, FS, Rocky Mtn. Forest & Range Exper. Station, *Annual Report, 1955*, p. 67. See also p. 41.

[31]M.E. Roach and G.E. Glendening, "Response of Velvet Mesquite in Southern Arizona to Airplane Spraying of 2,4,5-T," *Journal of Range Management*, Vol. 9, No. 3 (March 1956), pp. 70-73.

[32]H.G. Reynolds, F. Lavin, and H.W. Springfield, *Preliminary Guide for Range Reseeding in Arizona and New Mexico*, USDA, FS, Southwest Forest & Range Exper. Station Research Report No. 7 (July 1949); F. Lavin, *Intermediate Wheatgrass for Reseeding Southwestern Ponderosa Pine and Upper Woodland Ranges in the Southwest*, USDA, FS, SW Forest & Range Exper. Station Research Report No. 9 (March 1953).

[33]USDA, FS, Santa Rita Experimental Range, *Annual Report, 1956*, pp. 57-60.

[34]Reynolds, "Meeting Drought on Southern Arizona Rangelands," *Journal of Range Management*, Vol. 7, No. 1 (Jan. 1954), p. 33.

[35]USDA, FS, Rocky Mtn. Forest & Range Exper. Sta., *Annual Report, 1956*, p. 106.

[36]*Toward Meeting Soil and Water Conservation Needs*, p. 191.

[37]*Arizona Republic*, April 19, 1957.

[38]*Ibid.*, April 22, 1957.

[39]USDA, *Report of the Secretary, 1954*, pp. 10-11.

[40]*Organization of the Federal Government for Scientific Activities*, p. 7.

[41]*Directory of Organization and Field Activities of the Department of Agriculture, 1957*, USDA, Agriculture Handbook No. 76 (1957), pp. 54-56, 143-46, 196-98.

[42]*Ibid.*, pp. 54-56.

[43]*Workers in Subjects Pertaining to Agriculture in Land-Grant Colleges and Experiment Stations*, USDA, Agriculture Handbook No. 116 (1957), pp. 4-6.

[44]*Toward Meeting Soil and Water Conservation Needs.*

[45]Robert V. Boyle, "Southwest Gets Thirstier as Water Problem Grows," *Soil Conservation*, Vol. 17, No. 9, p. 196.

[46]*Ibid.*

[47]See R.H. Peebles, G.T. Den Hartog, and E.H. Pressley, *Effect of Spacing on Some Agronomic and Fiber Characteristics of Irrigated Cotton*, USDA Tech. Bulletin 1140 (1956); K. Harris, D.C. Aaepli, and W.D. Pew, *Tillage Practices for Irrigated Soils*, AAES Bulletin 257 (1954); and K. Harris, "Tillage Changes During the Next Ten Years," *Crops and Soils* (Oct. 1954).

[48]C.O. Stanberry, *et al.*, "Effect of Moisture and Phosphate Variables on Alfalfa Hay Production on the Yuma Mesa," *Soil Science Society of America Proceedings*, Vol. 19, No. 3, pp. 303-311; *Toward Meeting Soil and Water Conservation Needs: I*, pp. 100-114, and *II*, pp. 69-74; K. Harris, "Pre-Planting Irrigation," *Progressive Agriculture* (July-Sept. 1954).

[49]See Barry N. Freeman and Robert R. Humphrey, "The Effects of Nitrates and Phosphates upon Forage Production of a Southern Arizona Desert Grassland Range," *Journal of Range Management*, Vol. 9, No. 5 (May 1956), p. 176.

[50]*Toward Meeting Soil and Water Conservation Needs: I*, pp. 179-244, and *II*, pp. 109-148. For example, see Roach and Glendening, *Journal of Range Management*, Vol. 9, No. 2, p. 170. For a summary of ARS work on shrub control and watershed research, see H.A. Rodenhiser, "Progress in Watershed and Brush Control Research, Arizona," *Progress in Watershed Management*, p. 14.

[51]Joseph F. Arnold and W.L. Schroeder, *Juniper Control Increases Forage Production on the Fort Apache Indian Reservation*, USDA, FS, Rocky Mtn. Forest & Range Exper. Station Paper No. 18 (1955); also W.L. Schroeder, "History of Juniper Control on the Fort Apache Reservation," *Arizona Cattlelog*, Vol. 8, No. 10 (1953), p. 18.

[52]Frank S. Boice, "A Southwestern Rancher's Viewpoint of Shrub Control," *Journal of Range Management*, Vol. 8, No. 3 (May 1955), p. 103.

[53]*Toward Meeting Soil and Water Conservation Needs: I*, pp. 15-44, and *II* pp. 13-30.

[54]USDA, SCS, Region VI, *Water Yield and Range Conservation Studies on Walnut Gulch Watershed* (mimeo., Oct. 7, 1953). The project was originally under the jurisdiction of the SCS.

[55]*Ibid.*, p. 3.

[56]C.B. Brown, *Rates of Sediment Production in Southwestern United States*, USDA, SCS (processed, 1945), p. 39.

[57]Lloyd E. Myers, "Program and Facilities of the Southwest [now U.S.] Water Conservation Laboratory," p. 7. Paper presented before the sixteenth annual meeting of the Colorado Water Users Association, Las Vegas, Nevada, Dec. 3, 1959. See also Myers, "Flow Regimes in Surface Irrigation," *Agricultural Engineering*, Vol. 40, No. 11 (Nov. 1959), pp. 676-77, 682-83; and Myers, "Waterproofing Soil To Collect Precipitation," *Journal of Soil and Water Conservation*, Vol. 16, No. 6 (Nov.-Dec. 1961), pp. 281-82.

Chapter 17

[1]R.H. Forbes in AAES, *Fiftieth Annual Report*, p. 10.

[2]See Lester Velie, "Are We Short of Water," *Reclamation Era*, Vol. 34, No. 7 (July 1948), p. 123, reprinted from *Colliers*, May 15, 1948.

[3]Foreword to G.E.P. Smith, *The Groundwater Supply of the Eloy District in Pinal County, Arizona*.

Bibliography

Ackerman, Bernice. *Characteristics of Summer Radar Echoes in Arizona, 1956.* Institute of Atmospheric Physics, U of A, Scientific Report, No. 11. July 8, 1959.

Agricultural Conservation Program: Arizona Handbook for 1956. USDA, Agricultural Conservation Program Service. Washington, 1955.

Agricultural Statistics, 1956. U. S. Department of Agriculture, 1957.

An Analysis of Arizona's Agricultural Productive Capacity. USDA, published in cooperation with the University of Arizona, 1951.

Anderson, D., Hamilton, L. P., Reynolds, H. G., and Humphrey, Robert R. *Reseeding Grassland Ranges in Southern Arizona.* AAES Bulletin 249, 1953.

Appraisal Report, Central Arizona Project. USDI, Bureau of Reclamation, Region 3. Boulder City, Nevada, 1962.

Arizona. *Acts, Resolutions, and Memorials* (also called *Session Laws*).

Arizona (Territory). *Compiled Laws of the Territory of Arizona.* 1877.

_____. *Howell Code.* 1864.

_____. *Revised Statutes of Arizona.* 1887.

Arizona Agricultural Experiment Station. *Annual Reports.*

_____. *Research Progress, 1955.*

Arizona Agricultural Extension Service. *Annual Reports of County Extension Agents.*

Arizona Agriculture. AAES Bulletin 226, 1950, and 252, 1954; Agriculture Extension Service Circular 270, 1959.

Arizona Commission on Indian Affairs. *Annual Reports.*

Arizona-County Base Book. Tucson: University of Arizona Bureau of Business Research, 1954, and 1958.

Arizona Daily Star (Tucson).

Arizona Department of Mineral Resources. *Inventory of Arizona Lands As of June 30, 1961.* Phoenix, August 1962.

Arizona Development Board. *Preliminary Survey and Recommendations Relating to the Establishment of a State Parks and Recreation Board.* Prepared for the Legislative Council. Phoenix, 1955.

_____. *Report on Preliminary Survey of Selected Natural Resources.* Phoenix, 1955.

_____. *Year-End Reports.*

Arizona Farmer (see *Arizona Farmer-Ranchman*).

Arizona Farmer-Ranchman.

Arizona Game and Fish Commission. *Annual Reports.*

Arizona Game Protective Association. *By-Laws* (revised Sept. 2, 1956).

_____. *The Principles Supported by the Arizona Game Protective Association.* Mimeo., undated.

Arizona Interstate Stream Commission. *Annual Reports.*

Arizona, Land of Promise and Fulfillment: Addresses Delivered by Four Former Presidents on November 16, 1951, as a Symposium at the Inauguration of Richard Anderson Harvill as President of the University of Arizona. U of A General Bulletin No. 18, 1955.

Arizona Legislative Council. *Third Annual Report*. Phoenix, 1955.
Arizona Legislature. House of Representatives. House Joint Memorial No. 6, Feb. 17, 1931.
————. House of Representatives. *Journals of the House*.
————. Senate. *Journals of the Senate*.
Arizona Legislature (Territorial). *Laws*. 1885.
Arizona Park, Parkway and Recreation Area Plan: Progress Report. Arizona Highway Commission, Arizona Resources Board, Arizona State College, Tempe, in cooperation with the National Park Service. Phoenix, Feb. 1941.
Arizona Power Authority. *Annual Reports*.
————. *Colorado River Project: Bridge Canyon Development, Marble Canyon Development*. Application for License before the Federal Power Commission. Phoenix, July 1958.
————. *Colorado River Project: Bridge Canyon Development, Marble Canyon Development, Little Colorado River Development*. Amendment to Application for License Project 2248, before the Federal Power Commission. Phoenix, November, 1959.
Arizona Power Survey. Federal Power Commission. March, 1942.
Arizona Republic (Phoenix).
Arizona Resources Board. *Annual Report*. Phoenix, Dec. 31, 1920.
Arizona Revised Statutes. 1939 and 1955.
The Arizona Soil Conservationist. Vol. 8, No. 2 (April 1956).
Arizona State Land Department. *Annual Reports*.
————. "Arizona Land Ownership." Prepared by the Service Division. Sept. 1, 1962, multilithed.
Arizona State Planning Board. *Reports*. Published in cooperation with the National Resources Committee, Works Progress Administration. Phoenix, 1936.
Arizona State Resources and Planning Board, Post-War Planning Committee. *Preliminary Post-War Plans for Arizona Agriculture*. Phoenix, 1944.
Arizona State Water Commissioner. *Biennial Reports*.
Arizona Statistical Review. Valley National Bank, Phoenix. Published annually.
Arizona Stockman.
Arizona Survey of the Wildlife and Fisheries Resources of the Lower Colorado River. Arizona Game and Fish Commission. Phoenix, 1952.
Arizona Tax Research Association. *Newsletters*. Phoenix.
Arizona Water Resources, Hearings on Sen. Res. 304. U.S. Senate, Committee on Irrigation and Reclamation, 78th Cong., 2d Sess., 1944.
Arizona Wildlife-Sportsman.
Arizona's Growth and Its Future. Arizona House of Representatives, Committee on Arizona Development. Phoenix, 1950.
Arizona's National Monuments. Southwestern National Monuments Association, Popular Series No. 2. Santa Fe, N.M., 1945.
Arle, H. F. "Saltcedar Control with Chemicals," *Watershed and Related Water Management Problems* (which see).
Arnold, Joseph F. "Effect of Heavy Selection Logging on the Herbaceous Vegetation in a Ponderosa Pine Forest in Northern Arizona," *Journal of Forestry*, Vol. 51, No. 2 (1953).

Arnold, Joseph F., and Schroeder, W.L. *Juniper Control Increases Forage Production on the Fort Apache Indian Reservation.* USDA, FS, Rocky Mountain Forest & Range Experiment Station Paper No. 18, 1955.

Aston, Rollah E. *Boulder Dam and the Public Utilities.* University of Arizona Master's thesis, 1936.

Average Electric Bills. FPC. Washington, D.C., 1939.

Average Typical Residential Bills. FPC, Electric Rate Survey, Rate Series No. 3. Washington, D.C., Jan., 1935.

Avery, Ben. "Rain-Making in Arizona," *Reclamation Era,* Vol. 36, No. 5 (May 1950).

Babcock, H. M. *Memorandum on Ground-Water Supply of the Joseph City Irrigation District.* USGS, Tucson. Mimeo., August 1955.

_____, and Cushing, E.M. "Recharge to Ground-Water from Floods in a Typical Desert Wash, Pinal County, Arizona," *Transactions of the American Geophysical Union,* Vol. 22 (1952).

Balchin, W. G. V., and Pye, Norman. "The Drought in the Southwestern United States," *Weather,* Vol. 8, No. 8 (1953).

_____, _____. "Recent Economic Trends in Arizona," *Geographical Journal,* Vol. 120, Part 2, (1954).

Banks, Harvey O. "Bases of an Adequate State Water Program," *State Government,* Vol. 33, No. 2 (Spring 1960).

_____. "Utilization of Underground Storage Reservoirs," *Transactions of the American Society of Civil Engineers,* Vol. 118 (1953).

Barsalou, Frank, Gale, John R., and Gillies, James. *Yuma: Its Economic Growth and Land Use Potential.* Stanford Research Institute: [Menlo Park, Calif.], 1956.

Battan, Louis J. and Kassander, A. Richard, Jr. *Evaluation of Effects of Airborne Silver-Iodide Seeding of Convective Clouds.* Institute of Atmospheric Physics, U of A, Scientific Report No. 18, March 1, 1962.

_____, _____. "Possibilities of Increasing Water Yields by Cloud Seeding," *Watershed and Related Water Management Problems* (which see).

_____, _____. *Seeding of Summer Cumulus Clouds.* Institute of Atmospheric Physics, U of A, Scientific Report No. 10. July, 1959.

_____, _____, and Sims, Lee L. *Randomized Seeding of Orogeophic Cumuli, 1959.* Institute of Atmospheric Physics, U of A, Scientific Report No. 9, Part 2. Oct. 1958.

Bernard, Merrill. "The Appraisal of Water Resources in the United States: Analysis and Utilization of Data, Water Supply and Flood Forecasting," in *Proceedings of the United Nations Scientific Conference on the Conservation and Utilization of Resources,* Vol. IV, *Water Resources.* Lake Success, N.Y., 1951.

Biswell, H. H., and Schultz, A. M. "Surface Runoff and Erosion as Related to Prescribed Burning," *Journal of Forestry,* Vol. 55, No. 5 (May 1957).

Blaney, Harry F., and Huberty, Martin R. "Irrigation in the Far West," *Agricultural Engineering,* Vol. 38, No. 6 (1957).

Boice, Frank S. "A Southwestern Rancher's Viewpoint of Shrub Control," *Journal of Range Management,* Vol. 8, No. 3 (May 1955).

Boulder Canyon Project Final Reports. USDI, Bureau of Reclamation. Washington, D.C., 1948.

THE POLITICS OF WATER IN ARIZONA

Boyle, Robert V. "Southwest Gets Thirstier as Water Problem Grows,"
Soil Conservation, Vol. 17, No. 9 (April 1952).

Brannon, V. D. *Employers' Liability and Workmen's Compensation in
Arizona.* University of Arizona Social Science Bulletin No. 7, 1934.

Branscomb, Bruce L. "Vegetation Changes Continuously on Arizona Ranges,"
Progressive Agriculture, Vol. 9, No. 2 (July-Sept. 1957).

Bredo, William. "A Framework for Water Pricing Policy," *Research Needs
and Problems* (which see).

Brown, C.B. *Rates of Sediment Production in Southwestern United States.*
USDA, SCS. Processed, 1945.

Brown, H.E. and Dunford, E.G. *Streamflow in Relation to the Extent of Snow
Cover.* USDA, FS, Rocky Mtn. Forest & Range Experiment Station Paper
24, 1956.

Brown, Virginia L. *Some Aspects of the History of the Arizona Education
Association.* University of Arizona Master's thesis, 1952.

Bryan, Kirk. *The Papago Country, Arizona.* USGS Water Supply Paper 499,
1925.

Bryson, Reid. *The Annual March of Precipitation in Arizona, New Mexico,
and Northwestern Mexico.* Institute of Atmospheric Physics, U of A,
Technical Reports on the Meteorology and Climatology of Arid Regions,
No. 6. June 1957.

_____. *Some Factors in Tucson Summer Rainfall.* Institute of Atmospheric
Physics, U of A, Technical Reports on the Meteorology and Climatology
of Arid Regions, No. 4. May 1957.

"The Buttes Dam Story," *Arizona Farmer-Ranchman.* June 9, 1956.

Carhart, Arthur H. *Water—Or Your Life.* New York: Lippincott, 1951.

Carr, William H. *The Desert Speaks.* Arizona-Sonora Desert Museum Educational Series No. 1. Tucson, 1956.

Carson, Charles A. "Arizona's Interest in the Colorado River," *Rocky Mountain
Law Review,* Vol. 19 (June 1947).

Casaday, L.W. *Arizona—An Economic Report.* University of Arizona Bureau
of Business Research, Special Studies No. 7. Dec. 1953.

Catlin, C. N. *Character of Groundwater Resources of Arizona.* AAES Bulletin 114, 1926.

Census of Manufactures. U.S. Bureau of the Census. Washington, D.C.,
various years.

Census of Population, 1960: Number of Inhabitants. U.S. Bureau of the Census. Washington, D.C., 1961.

Central Arizona Project Association. *News of Arizona's Water Fight.* June
1955.

_____. *Report to Directors and Members,* by John M. Jacobs, President.
Dec. 2, 1955.

The Central Arizona Project, Hearings on H.R. 1500 and H.R. 1501. 2 parts.
U.S. House of Representatives, Committee on Interior and Insular Affairs.
82d Cong., 1st Sess., 1951.

Central Arizona Project. U.S. House of Representatives, 81st Cong., 1st Sess.,
House Doc. 136, 1949.

Clawson, Marion. "The Crisis in Outdoor Recreation," *American Forests*
(March and April, 1959).

_____. *Uncle Sam's Acres.* New York: Dodd, Mead, 1951.

Clawson, Marion and Held, Burnell. *The Federal Lands: Their Use and Management.* Baltimore, Md.: published by The Johns Hopkins Press for Resources for the Future, Inc., 1957.

Code, W. E. *Design and Construction of Small Concrete Lined Canals.* AAES Bulletin 97, 1923.

Colbert, Edwin H. "Rates of Erosion in the Chinle Formation," *Plateau,* Vol. 28, No. 4 (1956).

Collingwood, C. B. *Waters and Water Analysis.* AAES Bulletin 3, 1891.

The Colorado River. USDI, Bureau of Reclamation. Washington, D.C., 1946.

Colorado River Commission of Arizona. *Report.* Feb. 2, 1933 to March 3, 1935.

Colorado River Compact. U.S. Dept. of State Doc. No. 6241. Washington, D.C., 1922.

Colorado River Development Within the State of Arizona: Colorado River Projects. Preliminary Planning Report. Harza Engineering Co., Chicago, 1958. Prepared for the Arizona Power Authority, Phoenix.

Colter, Fred. *Highline Book.* Phoenix, 1934.

Colton, Harold S. "Early Failure to Solve the Water Shortage," *Plateau,* Vol. 29, No. 2 (1956).

_____. "Some Notes on the Original Condition of the Little Colorado River: A Side Light on the Problems of Erosion," *Plateau,* Vol. 10, No. 6 (1937).

Compilation of Records of Surface Water of the United States Through September 1950, Part 9: Colorado River Basin. USGS Water Supply Paper 1313, 1954.

Contributions to the Hydrology of the United States. USGS Water Supply Paper 1360-D, 1956.

Cooke, P. St. George. "Report," in Emory, W.H., *Notes of a Military Reconoissance...* (which see).

Cooper, Charles F. *Yield and Value of Wild-land Resources of the Salt River Watershed, Arizona.* University of Arizona Master's thesis, 1956.

Cooperrider, C. K. *The Relationship of Stream Flow to Precipitation on the Salt River Watershed Above Roosevelt Dam.* AAES Technical Bulletin 76, 1938.

Corker, Charles E. "The Issues in Arizona v. California: California's View," in Pollak (which see).

Cowden, Ray. "Brush-Land Control," *Arizona Cattlelog,* Vol. 11, No. 12 (1956).

Cozzens, Samuel W. *The Marvellous Country: Or, Three Years in Arizona and New Mexico, the Apaches' Home.* Boston: Shepard and Gill, 1873.

Craig, Robert G. "Farmers' Home Administration Loans," *Arizona Cattlelog,* Vol. 12, No. 2 (Oct. 1956).

Croft, A. R., and Hoover, Marvin D. "The Relation of Forests to our Water Supply," *Journal of Forestry,* Vol. 49, No. 4 (April 1951).

Cross, Jack L., Shaw, Elizabeth H., and Scheifele, Kathleen. *Arizona: Its People and Resources.* Tucson: University of Arizona Press, 1960.

Current Population Reports: Population Estimates. U.S. Bureau of the Census, Series P-25, No. 160. Aug. 9, 1957.

Cushman, R. L., and Halpenny, L. C. "Effect of Western Drought on the Water Resources of Safford Valley, Arizona," *Transactions of the American Geophysical Union,* Vol. 36 (1955).

Darrow, R. A. *Arizona Range Resources and Their Utilization,* I: *Cochise County.* AAES Technical Bulletin 103, 1944.

D'Autremont, H.H. *More Data on the Colorado River Question.* Tucson: Privately published, 1943.

Davis, Arthur P. *Irrigation Near Phoenix, Arizona.* USGS Water Supply and Irrigation Paper 2, 1897.

————. *Water Storage on Salt River, Arizona.* USGS Water Supply and Irrigation Paper 73, 1903.

Davis, Hiram S. "New Manufacturing Plants Strengthen the Phoenix Economy," *Arizona Business and Economic Review,* Vol. 6, No. 2 (Feb. 1957).

Davis, William C. *Values of Hunting and Fishing in Arizona, 1960.* University of Arizona, Bureau of Business and Public Research Special Studies No. 21. April 1962.

DeVoto, Bernard. "Our Great West — Boom or Bust," *Colliers,* Vol. 132 (Dec. 25, 1953).

Directory of Organization and Field Activities of the Department of Agriculture, 1957. USDA, Agriculture Handbook No. 76. Washington, D.C., 1957.

Dobkins, D. A. "Soil Conservation in Arizona," *Arizona Highways,* Vol. 32, No. 8 (August 1956).

Dobyns, Henry F. "The Plight of the Papagos," *Frontier,* Vol. 3, No. 5 (March 1952).

————. "Shift in Arizona," *Frontier,* Vol. 7, No. 7 (May 1956).

Donnelly, Thomas C. *Rocky Mountain Politics.* Albuquerque: University of New Mexico Press, 1940.

Douglass, A. E. "Accuracy in Dating, I," *Tree-Ring Bulletin,* Vol. 1, No. 2 (Oct. 1934).

————. "Accuracy in Dating, II: The Presentation of Evidence," *Tree-Ring Bulletin,* Vol. 1, No. 3 (Jan. 1935).

————. *Precision of Ring Dating in Tree-Ring Chronologies.* University of Arizona Laboratory of Tree-Ring Research Bulletin No. 3. July 1946.

————. *Researches in Dendrochronology.* University of Utah Bulletin, Vol. 37, No. 2, Biological Series, Vol. 10, No. 1 (1946).

————. "The Secret of the Southwest Solved by Talkative Tree Rings," *National Geographic Magazine,* Vol. 56 (Dec. 1929).

————. *Tree-Rings and Chronology.* University of Arizona Physical Science Bulletin No. 1. Oct. 1937.

Elder, James B. *Utilization of Man-Made Water-Holes in Southern Arizona.* University of Arizona Master's thesis, 1954.

Electric Power and Government Policy. Twentieth Century Fund. New York, 1948.

Emory, William H. *Notes of a Military Reconnoissance, from Fort Leavenworth in Missouri to San Diego in California. . . .* 30th Cong., 1st Sess., House Exec. Doc. 41. Washington, D.C.: Wendell and Van Benthuysen, 1848.

Englebert, Ernest A. "Political Aspects of Future Resources Development in the West," in *Research Needs and Problems* (which see).

Estimated Use of Water In the United States, 1955. USGS Circular 398, 1957.

Facts About the Agricultural Conservation Program. USDA, Agricultural Conservation Program Service, PA-272. Washington, 1955.

Federal Aid in Fish and Wildlife Restoration. Annual reports of the Dingell-Johnson and Pittman-Robertson programs for the fiscal year ending June 30, 1955. Wildlife Management Institute and Sport Fishing Institute, Washington D.C. (n.d.).

Fergusson, Erna. *Our Southwest.* New York: Knopf, 1940.

Feth, J. H., White, N. D., and Hem, J. D. *Preliminary Report of Investigations of Springs in the Mogollon Rim Region, Arizona.* USGS, Tucson (mimeo), 1954.

Fletcher, H. C. and Elmendorf, Harold B. "Phreatophytes—A Serious Problem. in the West," in *Water: The Yearbook of Agriculture, 1955* (which see).

—————, and Rich, L. R. "Classifying Southwestern Watersheds on the Basis of Water Yields," *Journal of Forestry,* Vol. 53, No. 3 (March 1955).

Folio of Geologic and Mineral Maps of Arizona. Arizona Bureau of Mines. Tucson: University of Arizona Press, 1962.

Forbes, R. H. *Irrigating Sediments and Their Effects.* AAES Bulletin 53, 1906.

—————. *Irrigation and Agricultural Practice in Arizona.* AAES Bulletin 63, 1911.

—————. *The River-Irrigating Waters of Arizona — Their Character and Effects.* AAES Bulletin 44, 1902.

Forecast of Total Visits, 1966-1970. USDI, NPS. Mimeo [1960?].

Forer, Lois G. "Water Supply: Suggested Federal Regulation," *Harvard Law Review,* Vol. 75, No. 2 (Dec. 1961).

Foster, Gene "A Brief Archaeological Survey of Glen Canyon," *Plateau,* Vol. 25, No. 2 (1952).

Fox, Kel M., and Perkins, R. W. (Bob). "Cattlemen Say It's Time for a Change," *Arizona Farmer,* Vol. 26, No. 20 (1947).

Freeman, Barry N. and Humphrey, Robert R. "The Effects of Nitrates and Phosphates upon Forage Production of a Southern Arizona Desert Grassland Range," *Journal of Range Management,* Vol. 9, No. 5 (May 1956).

French, Norman R. "A New Look at Erosion Control in Arizona," *Journal of Range Management,* Vol. 10, No. 5 (Sept. 1957).

—————. "Silt Is a Major Thief." Typewritten. June 13, 1957.

Fuller, W.H. *Effects of Kinds of Phosphate Fertilizer and Method of Placement on Phosphorus Absorption by Crops Grown on Arizona Calcareous Soils.* AAES Technical Bulletin 128. June 1953.

—————. *Soil Composition.* AAES Report 168. Spring 1958.

—————, Gomness, N. C., and Sherwood, L. V. *The Influence of Soil Aggregate Stabilizers on Stand, Composition, and Yield of Crops on Calcareous Soils of Southern Arizona.* AAES Technical Bulletin 129. 1953.

—————, and Padgett, Catherine G. *The Effects of Discing, Rototilling and Water Action on the Structure of Some Calcareous Soils.* AAES Technical Bulletin 134. July 1958.

—————, Shannon, Stanton, and Burgess, P. S. "Effect of Burning on Certain Forest Soils of Northern Arizona," *Forest Science,* Vol. 1 (1955).

Future Needs for Reclamation in the Western United States. USDI, Bureau of Reclamation. Committee Print No. 14 for the Senate Select Committee on National Water Resources, 86th Cong., 2d Sess., 1960.

Garnsey, Morris E. America's New Frontier: the Mountain West. New York: Knopf, 1950.

———. "Water: West," The Annals of the American Academy of Political and Social Science, Vol. 281 (1952).

Gatewood, J. S., Robinson, T. W., Colby, P. R., Hem, J. D., and Halpenny, L. C. Use of Water by Bottom-Land Vegetation in Lower Safford Valley, Arizona. USGS Water Supply Paper 1103, 1930.

Geology and Ground-Water Resources of the Salt River Valley Area, Maricopa and Pinal Counties, Arizona. USGS. Mimeo., Feb. 4, 1947.

Gila Project. USDI, Bureau of Reclamation, Region 3. Boulder City, Nevada, 1953.

"Glen Canyon Dam," Reclamation Era, Vol. 44, No. 1 (Feb. 1958).

Gramm, Warren S. "Economics and Politics: Inseparables in Water Resource Development," Proceedings of the Thirty-First Annual Conference of the Western Economic Association. Los Angeles, Aug. 30-31, 1956.

Green, Christine R. Arizona Statewide Rainfall. Institute of Atmospheric Physics, U of A, Report No. 7. Nov. 1959.

———. Probabilities of Drought and Rainy Periods for Selected Points in the Southwestern United States. Institute of Atmospheric Physics, U of A, Report No. 8. Jan. 1960.

Gregory, Herbert E. The Navajo Country. USGS Water Supply Paper No. 380, 1916.

Griffenhagen and Associates. Report on General State Organization. Prepared for the Arizona Senate, Special Legislative Committee on State Operations, 19th Legislature, 1st Special Sess. 3 vols. Chicago, 1949.

Griffiths, David. A Protected Stock Range in Arizona. USDA, Bureau of Plant Industry, Bulletin 177. April 1910.

Griffiths, V. S. State Regulation of Railroad and Electric Rates in Arizona to 1925. University of Arizona Master's thesis, 1931.

Ground Water in the Gila River Basin and Adjacent Areas, Arizona—A Summary. USGS, in cooperation with the Underground Water Commission of Arizona, 1952.

Groundwater Regions of the United States—Their Storage Facilities. Vol. III of The Physical and Economic Foundation of Natural Resources. U.S. House of Representatives, Committee on Interior and Insular Affairs, 1952.

Gulley, Frank A. Pumping Water for Irrigation. AAES Bulletin 3, 1891.

Hackenberg, Robert A. Economic and Political Change Among the Gila River Pimas. University of Arizona, Bureau of Ethnic Research. Report to the John Hay Whitney Foundation. Mimeo., 1954.

Halpenny, L. C. Preliminary Report on the Ground-Water Resources of the Navajo and Hopi Indian Reservations, Arizona, New Mexico, and Utah. USGS, Holbrook, Ariz. Mimeo., August 1951.

———, Hem, J. D., and Jones, I. I. Definitions of Geologic, Hydrologic, and Chemical Terms Used in Reports on the Ground-Water Resources and Problems of Arizona. USGS. Mimeo., March 18, 1947.

Halseth, Odd S. "Arizona's 1500 Years of Irrigation History," *Reclamation Era*, Vol. 33, No. 12 (Dec. 1947).

Hamilton, Patrick. *Arizona, For Homes, For Health, For Investments*. (A revision with alterations of *The Resources of Arizona*.) Phoenix, 1886.

————. *The Resources of Arizona*. San Francisco: A. L. Bancroft, 1884.

Hamman, Charles L. "Water Policy and Western Industrial Development," *Research Needs and Problems* (which see).

Hardin, Charles. *Politics of Agriculture*. Glencoe, Ill.: Free Press, 1952.

Harding, S. T. "Statutory Control of Ground Water in the Western United States," *Transactions of the American Society of Civil Engineers*, Vol. 120 (1955).

Harris, Karl. "Irrigation of Alfalfa," *Progressive Agriculture*, Vol. 3, No. 4 (Jan.-March 1952).

————. "Pre-Planting Irrigation," *Progressive Agriculture*, Vol. 6, No. 2 (July-Sept. 1954).

————. "Tillage Changes During the Next Ten Years," *Crops and Soils* (Oct. 1954).

————, Aepli, D. C., and Pew, W. D. *Tillage Practices for Irrigated Soils*. AAES Bulletin 257, 1954.

Harris, R. McDonald, and Padgett, Harold D. *Geology and Ground-Water Resources of the Verde River Valley near Fort McDowell, Arizona*. USGS. Mimeo., Nov. 1, 1945.

Harshbarger, J. W., "Capturing Additional Water in the Tucson Area," in *Progress in Watershed Management* (which see).

————. "Use of Ground Water in Arizona," in Smiley, ed., *Climate and Man in the Southwest* (which see).

————, Repenning, C.A., and Callahan, J.T. "The Navajo Country, Arizona-Utah-New Mexico," in *Subsurface Facilities of Water Management and Patterns of Supply—Type Area Studies* (which see).

————, ————, and Jackson, R. L. *Jurassic Stratigraphy of the Navajo Country*. USGS, Holbrook, Arizona. Mimeo., Aug. 1951.

Harvill, Richard A. "The Economy of Arizona," *Arizona Business and Economic Review*, Vol. 6, No. 1 (Jan. 1957).

Hastings, James R. "Vegetation Change and Arroyo Cutting in Southeastern Arizona During the Past Century," *Journal of the Arizona Academy of Science*, Vol. 1, No. 2 (Oct. 1959).

Hazen, G. E., and Turner, S. F. *Geology and Ground-Water Resources of the Upper Pinal Creek Area, Arizona*. USGS. Mimeo., Dec. 1946.

Heckler, Wilbur E. [State Director, USGS Surface Water Branch, Arizona.] *History of Run Off Investigations and Development of Storage in Arizona*. Typewritten, undated speech.

Hem, J. D. *Quality of Water of the Gila River Basin Above Coolidge Dam, Arizona*. USGS Water Supply Paper 1104, 1950.

Herber, B. P. *An Analysis of the Economy of Arizona*. University of Arizona Master's thesis, 1955.

Higbee, Edward. *The American Oasis*. New York: Knopf, 1957.

Hilgeman, R. H. "Irrigating Oranges at Three Moisture Levels," *Progressive Agriculture*, Vol. 9, No. 1 (April-June 1957).

Hill, Ralph R. "Wildlife Management in Relation to Multiple Use," *Journal of Forestry*, Vol. 53, No. 3 (1955).

Hinton, Richard J. The Hand-book of Arizona: Its Resources, History, Towns, Mines, Ruins, and Scenery. Tucson: Arizona Silhouettes, 1954. Reprint.

Hirshleifer, Jack, DeHaven, James C., and Milliman, Jerome. Water Supply: Economics, Technology, and Policy. Chicago: University of Chicago Press, 1960.

Hobart, Charles, and Harris, Karl. Fitting Cropping Systems to Water Supplies in Central Arizona. AAES Circular 127. November 1950.

Hodge, H. C. Arizona As It Is. New York, 1877.

Hoover, J. W. "Ground-Water Problems in Arizona and Neighboring States," Geographical Review, Vol. 29, No. 2 (1939).

Horton, J.S. "Ecology of Saltcedar," in Watershed and Related Water Management Problems (which see).

Houghton, N. D. "Arizona's Experience with the Initiative and Referendum," New Mexico Historical Review, Vol. 29, No. 3 (July 1954).

――――. "The 1950 Elections in Arizona," Western Political Quarterly, Vol. 4, No. 1 (March 1951).

――――. "The 1954 Elections in Arizona," Western Political Quarterly, Vol. 7, No. 4 (Dec. 1954).

――――. "The 1956 Elections in Arizona," Western Political Quarterly, Vol. 10, No. 4 (1957).

――――. "Problems of the Colorado River as Reflected in Arizona Politics," Western Political Quarterly, Vol. 4, No. 4 (Dec. 1951).

――――. "Problems of Public Power Administration in the Southwest— Some Arizona Applications," Western Political Quarterly, Vol. 4, No. 1 (March 1951).

How Reclamation Pays. USDI, Bureau of Reclamation. Washington, D.C., 1947.

Hudspeth, T. J. "A Supplement to 'Tin Roof or Solid Mat?'" Arizona Cattlelog, Vol. 7, No. 2 (1951).

Huffman, Roy E. "Public Water Policy for the West," Journal of Farm Economics, Vol. 35, No. 5 (1953).

Humphrey, Robert R. Arizona Range Grasses. AAES Bulletin 298, July 1958.

――――. Arizona Range Resources: Yavapai County. AAES Bulletin 229, July 1950.

――――. The Desert Grassland. AAES Bulletin 299, December 1958.

――――. "The Desert Grassland, Past and Present," Journal of Range Management, Vol. 6, No. 3 (May 1953).

――――. "Fire as a Means of Controlling Velvet Mesquite, Burroweed, and Cholla on Southern Arizona Ranges," Journal of Range Management, Vol. 2, No. 4 (Oct. 1949).

――――. Forage Production on Arizona Ranges, III: Mohave County. AAES Bulletin 244, Feb. 1953.

――――. Forage Production on Arizona Ranges, IV: Coconino, Navajo, and Apache Counties—A Study of Range Conditions. AAES Bulletin 266, Oct. 1955.

――――. "Major Aspects of the Woody Plant Problem in Arizona," Arizona Cattlelog, Vol. 12, No. 2 (Oct. 1956).

――――. "Paved Drainage Basins as a Source of Water for Livestock or Game," Journal of Range Management, Vol. 10, No. 2 (1957).

Hutchins, Wells A. *Selected Problems in the Law of Water Rights in the West.* USDA Misc. Pub. No. 418, 1942.

————. "Trends in the Statutory Law of Ground Water in the Western States," *Texas Law Journal,* Vol. 34, No. 2 (1955).

Hyatt, Gilbert E. "Western Water Rights: May They Be Taken Without Compensation?" *Montana Law Review,* Vol. 13 (Spring 1952).

Hydrology of Stock-Water Reservoirs in Arizona. USGS Circular 110, 1951.

The Industrial Utility of Public Water Supplies in the United States, 1952; States West of the Mississippi. USGS Water Supply Paper 1300, 1954.

Inskeep, Lester N. "State to Embark on Park Development," *Arizona Daily Star,* April 15, 1957.

Institute of Atmospheric Physics, U of A. *First Annual Progress Report, Cooperative Punchcard Climatological Program.* Nov. 15, 1955.

————. *General Information on the Institute of Atmospheric Physics.* April 15, 1955.

————. *Proceedings of the Conference on the Scientific Basis of Weather Modification Studies.* Tucson, April 10-12, 1956.

————. *Progress Report No. 2.* Nov. 30, 1955.

————. *Reference List of Research Projects in the Institute of Atmospheric Physics as of Jan. 1, 1957.*

Irrigating in Arizona. Arizona Agricultural Extension Service Circular 123. June 1944.

Irrigation and Flood Protection: Papago Indian Reservation, Arizona. U.S. Senate, 62d Cong., 2d Sess., Sen. Doc. 973, 1913.

Ives, Joseph C. *Report Upon the Colorado River of the West.* 36th Cong., 1st Sess., House Exec. Doc. No. 90, 1861.

James, Harold C. "Keep the Desert Living," *Pacific Discovery,* Vol. 9, No. 2 (March-April 1956).

Johnson, Rich. "Conservation and a Farm Editor," *Arizona Wildlife-Sportsman,* Vol. 27, No. 12 (1956).

————. "Transfer of the Public Domain," *Arizona Farmer,* Vol. 26, No. 23 (Nov. 15, 1947).

Kallender, Harry. "Controlled Burning in Ponderosa Pine Stands of the Fort Apache Indian Reservation in Arizona," *Arizona Cattlelog,* Vol. 11, No. 12 (Aug. 1956).

————, Weaver, Harold, and Gaines, Edward M. "Additional Information on Prescribed Burning in Virgin Ponderosa Pine in Arizona," *Journal of Forestry,* Vol. 53, No. 10 (Oct. 1955).

Karie, Jack. "Apaches' New Lake First Step in Creating Vacation Paradise," *Arizona Republic,* June 9, 1957.

Kassander, A. Richard, Jr., and Sims, Lee L. *Cloud Photogrammetry with Ground-Located K-17 Aerial Cameras.* Institute of Atmospheric Physics, U of A, Scientific Report No. 2. June 15, 1956.

————, ————, and McDonald, James E. *Observations of Freezing Nuclei over the Southwestern United States.* Institute of Atmospheric Physics, U of A, Scientific Report No. 3, 1956.

Kelly, Desmond G. "California and the Colorado River," *California Law Review,* Vol. 38, No. 4 (Oct. 1950).

Kelly, George H. (comp.). *Legislative History of Arizona, 1864-1912.* Phoenix: Manufacturing Stationers Inc., 1926.

300   THE POLITICS OF WATER IN ARIZONA

Kelly, William H., Kunstadter, Peter, and Hackenberg, Robert A. *Social and Economic Resources Available for Indian Health Purposes in Five Southwestern States.* 4 books. University of Arizona, Bureau of Ethnic Research. Prepared under contract with the U.S. Public Health Service. Mimeo., June 15, 1956.

Kelso, Paul. "The Arizona Ground Water Act," *Western Political Quarterly,* Vol. 1, No. 2 (1948).

_____. "The 1952 Elections in Arizona," *Western Political Quarterly,* Vol. 6, No. 1 (1953).

Kennedy, Fred H. "National Forest Watershed Projects in Arizona," in *Progress in Watershed Management* (which see).

Keppel, R. V. and Fletcher, Joel E. "Runoff from Rangelands of the Southwest," Agricultural Research Service. Typewritten, 1960.

Kinney, Clesson S. *A Treatise on the Law of Irrigation.* San Francisco, 1912.

Kirkwood, Marion Rice. "Appropriation of Percolating Water," *Stanford Law Review,* Vol. 1 (Nov. 1948).

Kluckhohn, Clyde and Leighton, Dorothea. *The Navajo.* Cambridge, Mass.: Harvard University Press, 1946.

Krieger, R. A., Hatchett, J. L., and Poole, J. L. *Preliminary Survey of the Saline-Water Resources of the United States.* USGS Water Supply Paper 1374, 1957.

Krutch, Joseph Wood. *The Desert Year.* New York: Sloane, 1952.

_____. "Man is the Enemy," *Frontier,* Vol. 6, No. 1 (1954).

_____. *The Voice of the Desert.* New York: Sloane, 1954.

Krutilla, John V., and Eckstein, Otto. *Multiple-Purpose River Development.* Baltimore, Md.: Johns Hopkins Press, 1958.

Langbein, Walter B. *Water Yield and Reservoir Storage in the United States.* USGS Circular 409, 1959.

Langley, Dana. "Hopi Trek to the Land of 'Big Water,'" *Desert,* Vol. 9, No. 8 (1946).

LaRue, E. C. *Colorado River and Its Utilization.* USGS Water Supply Paper 395, 1916.

Lassen, Leon, and Frank, Bernard. "Forest Management for More and Better Water," *Journal of Forestry,* Vol. 48, No. 12 (1950).

Lasseter, Roy. "The Value of Tree-Ring Analysis in Engineering," *Tree-Ring Bulletin,* Vol. 5, No. 2 (Oct. 1938).

Lauver, Mary E. *A History of the Use and Management of the Forested Lands of Arizona, 1862-1936.* University of Arizona Master's thesis, 1938.

Lavin, F. *Intermediate Wheatgrass for Reseeding Southwestern Ponderosa Pine and Upper Woodland Ranges in the Southwest.* USDA, FS, Southwest Forest & Range Experiment Station Research Report No. 9. March 6, 1953.

Le Crone, D. E. "Corduroy and Cibecue Watershed Projects: Fort Apache Indian Reservation," in *Progress in Watershed Management* (which see).

Lee, J. Karl. "Economic Analysis of Water Resources Policy," in *Research Needs and Problems* (which see).

Lee, Willis T. *The Underground Waters of the Gila Valley, Arizona.* USGS Water Supply and Irrigation Paper 104, 1904.

_____. *The Underground Waters of the Salt River Valley, Arizona.* USGS Water Supply and Irrigation Paper 136, 1905.

Lemon, Carroll. "Arizona's Wildlife Research Unit," *Arizona Wildlife-Sportsman,* Vol. 27, No. 11 (1956).

Leopold, Luna. *Conservation and Protection.* USGS Circular 414-A, 1960.

————. *The Conservation Attitude.* USGS Circular 414-C, 1960.

————. "Vegetation of Southwestern Watersheds in the Nineteenth Century," *Geographical Review,* Vol. 41, No. 2 (1951).

Lewis, Douglas A. "Cottonwood Wash Project: Water Use by Channel Vegetation," in *Progress in Watershed Management* (which see).

Linsley, Ray K. "Report on the Hydrological Problems of the Arid and Semi-Arid Areas of the United States and Canada," *Reviews of Research on Arid Zone Hydrology* (which see).

Lippincott, Joseph B. *Storage of Water on the Gila River, Arizona.* USGS Water Supply and Irrigation Paper 33, 1900.

Long Range Program of Soil and Water Conservation Operations: Arizona. USDA, SCS. 1956.

Lord, Russell. *To Hold This Soil.* USDA Misc. Pub. 321, 1938.

Los Angeles Examiner.

Lyons, John D. "Weather or Not," *Arizona Quarterly,* Vol. 8, No. 1 (Spring, 1952).

Maass, Arthur. *Muddy Waters.* Cambridge, Mass.: Harvard University Press, 1952.

MacKichen, K.A. and Kammerer, J.C. *Estimated Use of Water in the United States, 1960.* USGS Circular 456, 1961.

Major Irrigation Developments in Region 3. USDI, Bureau of Reclamation, Region 3. Boulder City, Nevada, June 1953.

Manessier, Hugh. "Pollution on the Colorado River," *Arizona Wildlife-Sportsman,* Vol. 26, No. 8 (1955).

Mann, Dean E. "The Legislative Committee System in Arizona," *Western Political Quarterly,* Vol. 14, No 4 (Dec. 1961).

McClatchie, A. J. *Irrigation at Station Farm.* AAES Bulletin 41, 1902.

McDonald, Harris R. and Padgett, Harold D., Jr. *Geology and Ground-Water Resources of the Verde River Valley near Fort McDowell, Arizona.* USGS Ground Water Branch, Tucson. Mimeo., Nov. 1, 1945.

McGeorge, W. T., Breazeale, E. L., and Abbott, J. L. *Polysulphides as Soil Conditioners.* AAES Technical Bulletin 131, 1956.

————, ————, and Bliss, A. Mark. *The Salinity Problem — Safford Experiment Farm Field Experiments.* AAES Technical Bulletin 124, 1952.

McGinnies, W. G., and Arnold, J. F. *Relative Water Requirements of Arizona Range Plants.* AAES Technical Bulletin 80, 1939.

McGinnis, Tru A. *The Influence of Organized Labor on the Making of the Arizona Constitution.* University of Arizona Master's thesis, 1930.

McMullin, R.J. "Rehabilitation and Betterment Pays a Dividend," *Reclamation Era,* Vol. 41, No. 1 (Feb. 1955).

Meinzer, Oscar E. (ed.). *Hydrology.* New York: Dover, 1949.

————, and Ellis, A. J. *Ground-Water in Paradise Valley, Arizona.* USGS Water Supply Paper 375-B, 1915.

————, and Kelton, F. C. *Geology and Water Resources of Sulphur Spring Valley, Arizona.* AAES Bulletin 72, 1913.

Memorandum of Understanding between the Soil Conservation Service, the Forest Service, and the Agricultural Research Service relating to Inter-Agency Coordination of Programs in Watersheds as Authorized by Section 6 of Public Law 566, 83d Congress. USDA. Mimeo., Feb. 2, 1956.

Memorandum of Understanding between the Soil Conservation Service, U.S.D.A., and the Bureau of Land Management, U.S.D.I., Relative to Inter-Agency Cooperation with Soil Conservation Districts in Arizona. USDA, SCS. Mimeo., Aug. 1, 1956.

Metzger, D. C. Geology and Ground-Water Resources of the Northern Part of the Ranegras Plain Area, Yuma County, Arizona. USGS, Tucson. Mimeo., Feb. 1951.

Meyer, Louis S. State of Arizona: A Brief Analysis of Land Status and Utilization. Arizona State University, Bureau of Government Research, Research Study No. 6. Tempe, 1962.

Middleton, James E. Water Management. Arizona Agricultural Extension Service Circular 205. Oct. 1952.

Miller, Cecil. "Multiple Use of Range Lands in Arizona," Arizona Cattlelog, Vol. 7, No. 10 (June 1952).

Miller, Joseph. The Arizona Story. New York: Hastings House, 1952.

Minette, Bill. "Cows on a Hot Tin Roof," Arizona Wildlife-Sportsman, Vol. 27, No. 12 (Dec. 1956).

————. "Farm Editor Lost in the Wilderness," Arizona Wildlife-Sportsman, Vol. 28, No. 5 (1957)

Mission 66 for the National Park System. USDI, NPS. Washington, D.C., 1957.

Morrissey, Richard J. "The Early Range Cattle Industry in Arizona," Agricultural History, Vol. 24, No. 3 (July 1950).

Mowry, Sylvester. Arizona and Sonora. New York: Harper, 1864.

Muns, Edward N. "Yield and Value of Water From Western National Forests," Journal of Forestry, Vol. 50, No. 5 (June 1952).

Murbarger, Nell. "Dam in Glen Canyon," Desert, Vol. 20, No. 4 (April 1957).

Murphy, E. C. Destructive Floods in the United States in 1905. USGS Water Supply and Irrigation Paper 162, 1906.

————, et al. Destructive Floods in the United States in 1904. USGS Water Supply and Irrigation Paper 147, 1905.

Myers, Lloyd E. "Flow Regimes in Surface Irrigation," Agricultural Engineering, Vol. 40, No. 11 (Nov. 1959).

————. "Program and Facilities of the Southwest Water Conservation Laboratory." Paper presented before the sixteenth annual meeting of the Colorado Water Users Association, Las Vegas, Nevada, Dec. 3, 1959.

————. "Waterproofing Soil to Collect Precipitation," Journal of Soil and Water Conservation, Vol. 16, No. 6 (Nov.-Dec. 1961).

National Forest Facts: Southwestern Region. USDA, FS (mimeo.). Various years.

National Forest Progress: Southwestern Region. USDA, FS. Various years.

National Power Survey: Principal Electric Utility Systems in the United States. FPC, Power Series No. 2. Washington, D.C., 1935.

National Survey of Fishing and Hunting. USDI, F&WS. Washington, D.C., 1956.

The Navajo: Long-Range Program for Navajo Rehabilitation. Report of J. A. Krug, Secretary of the Interior, USDI, BIA, 1948.

Nichol, A. A. *The Natural Vegetation of Arizona.* AAES Bulletin 68, 1937.
————. *The Natural Vegetation of Arizona.* AAES Bulletin 127, 1952.

Niehuis, Gladys and Charles. "Greed and Grass," *Arizona Teacher-Parent,* Dec. 1946.

Olson, Reuel L. *The Colorado River Compact.* Los Angeles: The Author, 1926.

Organization of the Federal Government for Scientific Activities. National Science Foundation. Washington, D.C., 1956.

Ostrom, Vincent. *Water and Politics.* Los Angeles: Haynes Foundation, 1953.
————. "State Administration of Natural Resources in the West," *American Political Science Review,* June 1953.

Outdoor Life.

Papago Tribal Council. *The Papago Development Program: 1949.* Sells, Ariz., 1949.

Parker, Kenneth W. and Martin, S. Clark. *The Mesquite Problem on Southern Arizona Ranges.* USDA, Circular 908, 1952.

Parsons, Malcolm B. *The Colorado River in Arizona Politics.* University of Arizona Master's thesis, 1947.
————. "Party and Pressure Politics in Arizona's Opposition to Colorado River Development," *Pacific Historical Review,* Vol. 19, No. 1 (Feb. 1950).

Pearson, G.A. *The Fort Valley Experiment Station.* USDA, FS, Southwest Forest & Range Experiment Station, 1942.

Peebles, R.H., Den Hartog, G.T., and Pressley, E.H. *Effects of Spacing on Some Agronomic and Fiber Characteristics of Irrigated Cotton.* USDA, Technical Bulletin 1140, 1956.

Peffer, E. Louise. *The Closing of the Public Domain.* Stanford, Calif.: Stanford University Press, 1951.

Penny, J. Russell, and Clawson, Marion. "Economic Possibilities of the Public Domain," *Land Economics,* Vol. 29, No. 3 (Aug. 1953).

Pew, W. D. "Yellowing of Lettuce," *Progressive Agriculture,* Vol. 3, No. 3 (Oct.-Dec. 1951).

The Pima Indians and the San Carlos Irrigation Project. U.S. House of Representatives, Committee on Indian Affairs, 68th Cong., 1st Sess., 1924.

Piper, Arthur M. "The Nationwide Water Situation," in *Subsurface Facilities of Water Management and Patterns of Supply—Type Area Studies* (which see).

A Plan for the Operation of the Arizona Game and Fish Commission. Arizona Game and Fish Department. Phoenix, 1960.

Pollak, Franklin S. (ed.). *Resources Development: Frontiers for Research.* Western Resources Conference, 1959. Boulder: University of Colorado Press, 1960.

Power Market Survey—Colorado River Storage Project. Federal Power Commission. Washington, D.C., 1958.

"Pre-Planting Irrigation," *Progressive Agriculture,* Vol. 6, No. 2 (July-Sept. 1954).

Preservation of Natural and Wilderness Values in the National Parks. USDI, NPS. March 1957.

The Problems of Imperial Valley and Vicinity. 67th Cong., 2d Sess., Sen. Doc. 142, 1922.

Proceedings of the United Nations Scientific Conference on the Conservation and Utilization of Resources. 8 vols. Lake Success, N.Y., 1950-1953. Vol. IV, Water Resources, 1951.

Program for the National Forests. USDA, FS. May, 1959.

Progress in Watershed Management. Proceedings of the Third Annual Meeting [Arizona Watershed Program], Sept. 21, 1959.

"Public Domain States Knuckled Under," Arizona Stockman, Vol. 14, No. 1 (1948).

Public Use: National Parks and Related Areas. USDI, NPS. Washington, D.C., 1960.

Pumpage and Ground-Water Levels in Arizona in 1955, by P.W. Johnson, N.D. White, and J.M. Cahill of the USGS. Arizona State Land Department, Phoenix, 1956.

Quality of Surface Waters for Irrigation, Western United States, 1952. USGS Water Supply Paper 1362, 1955.

Reclamation Project Data. USDI, Bureau of Reclamation. Washington, D.C., 1948.

Recovering Rainfall: More Water for Irrigation. 2 vols. Tucson: Arizona Watershed Program, 1956.

Rehnberg, Rex. The Cost of Pumping Irrigation Water, Pinal County, 1951. AAES Bulletin 246, 1953.

————. Irrigation Ditch Management on Arizona Irrigated Farms. AAES Bulletin 225, 1951.

Reich, Charles A. Bureaucracy and the Forests. Occasional Papers, Center for the Study of Democratic Institutions. Santa Barbara, Calif., 1962.

Reitan, Clayton H. The Role of Precipitable Water Vapor in Arizona's Summer Rains. Institute of Atmospheric Physics, U of A, Tech. Reports on Meteorology and Climatology of Arid Regions, No. 2, 1957.

Report of Land Use and Ownership in Soil Conservation Districts: Arizona Summary—All Districts. USDA, SCS, Arizona State Office. Phoenix, June 30, 1956.

Report on the Gila River and Tributaries Below Gillespie Dam. 81st Cong., 1st Sess., House Doc. 331, 1949.

Research Needs and Problems. Report No. 1 of Water Resources and Economic Development of the West. Western Agricultural Economics Research Council, Committee on the Economics of Water Resources Development. Berkeley, Calif., 1953.

Report on Water Supply of the Lower Colorado River Basin. US Bureau of Reclamation, Project Planning Report. November, 1952.

Resources for Freedom. U.S. President's Materials Policy Commission. 5 vols. Washington, D.C., 1952.

Resume of the Use of Electric Power and Energy in the State of Arizona. APA, annual power survey for the year 1954. Mimeo., Aug. 31, 1956.

Reviews of Research on Arid Zone Hydrology. UNESCO, Advisory Committee on Arid Zone Research, Arid Zone Programme I. Paris, 1953.

Reynolds, Hudson G. "Current Watershed Management Research by the U.S. Forest Service in Arizona," in Progress in Watershed Management (which see).

Reynolds, Hudson G. "Meeting Drought on Southern Arizona Rangelands," *Journal of Range Management,* Vol. 7, No. 1 (Jan. 1954).

————. *Watershed Management Research in Arizona, Progress Report, 1959.* USDA, FS, Rocky Mountain Forest and Range Experiment Station, 1960.

————, Lavin, F., and Springfield, H.W. *Preliminary Guide for Range Reseeding in Arizona and New Mexico.* USDA, FS, Southwest Forest & Range Experiment Station, Research Report No. 7. July 1949.

Rich, L.R. "Preliminary Effects of Forest Tree Removal on Water Yields and Sedimentation," in *Watershed and Related Water Management Problems* (which see).

Riggs, R. E. *Administrative Reorganization in Arizona.* University of Arizona Master's thesis, 1955.

Roach, M. E., and Glendening, G. E. "Response of Velvet Mesquite in Southern Arizona to Airplane Spraying with 2,4,5-T," *Journal of Range Management,* Vol. 9, No. 3 (March 1956).

Robertson, G. Gordon. "Percolating Waters—Ownership Rule Restated in Arizona," *Rocky Mountain Law Review,* Vol. 26 (1954).

Rodenhiser, H.A. "Progress in Watershed and Brush Control Research, Arizona," in *Progress in Watershed Management* (which see).

Ross, Clyde P. *Routes to Desert Watering Places in the Lower Gila Region, Arizona.* USGS Water Supply Paper 490, 1922.

Ross, P.H. *Twenty Years of Agricultural Extension Work in Arizona.* Arizona Agricultural Extension Service, Extension Project Circular 15. June 1935.

Rowland, Edward. "The Public Domain and the Bureau of Land Management: Brief History and Problems." Speech given at Arizona State College, Flagstaff, June, 1956.

Salmond, G.R. and Croft, A.R. "The Management of Public Watersheds," in *Water: The Yearbook of Agriculture, 1955* (which see).

Salt River Project: Major Facts in Brief. Salt River Valley Water Users' Association. Phoenix, 1956.

Salt River Valley Water Users' Association. *Annual Reports.* Phoenix, Arizona.

————. *Articles of Incorporation.* Phoenix, 1903.

Sampson, Arthur W. *Range Management.* New York: Wiley and Sons, 1952.

The San Carlos Apache Indian Reservation. Stanford Research Institute [Stanford, Calif.?], 1954.

Saunderson, Mont H. *Western Land and Water Use.* Norman: University of Oklahoma Press, 1951.

————. *Western Stock Ranching.* Minneapolis: University of Minnesota, 1950.

Schroeder, Albert H. *A Brief Survey of the Lower Colorado River from Davis Dam to the International Border.* NPS, Region 3. Reproduced by Bureau of Reclamation Reproduction Unit, Region 3. Boulder City, Nevada (1952?).

Schroeder, W. L. "History of Juniper Control on the Fort Apache Reservation," *Arizona Cattlelog,* Vol. 8, No. 10 (1953).

Schulman, Edmund. "Definitive Dendrochronologies: A Progress Report," *Tree-Ring Bulletin,* Vol. 18, No. 2, 3 (Oct. 1951, Jan. 1952).

306   THE POLITICS OF WATER IN ARIZONA

Schulman, Edmund. *Dendroclimatic Changes in Semi-Arid America.* Tucson: University of Arizona Press, 1956.
————. "Dendroclimatic Changes in Semi-Arid Regions," *Tree-Ring Bulletin,* Vol. 20, No. 3, 4 (Jan.-April 1954).
————. "Selection of Trees for Climatic Study," *Tree-Ring Bulletin,* Vol. 3, No. 3 (Jan. 1937).
————. *Tree-Ring Hydrology of the Colorado River Basin.* U of A Laboratory of Tree-Ring Research Bulletin No. 2. Oct. 1945.
————. "The Tree-Ring Laboratory of the University of Arizona," *Chronica Botanica,* Vol. 6, No. 3 (Nov. 1940).
————. "Tree-Rings Work for Science," *Arizona Alumnus,* Vol. 15, No. 2 (1937).
Schwalen, Harold C. "Arizona's Water Problem," *Progressive Agriculture,* Vol. 6, No. 2 (July-Sept. 1954).
————. *Rainfall and Runoff in the Upper Santa Cruz River Drainage Basin.* AAES Technical Bulletin 95, 1942.
————. *Sprinkler Irrigation.* AAES Bulletin 250, 1953.
————. *The Stovepipe or California Method of Well Drilling as Practiced in Arizona.* AAES Bulletin 112, 1924.
————, and Shaw, R. J. *Groundwater Supplies of the Santa Cruz Valley of Southern Arizona Between Rillito Station and the International Boundary.* AAES Bulletin 288, 1957.
Schwennesen, A. T. *Geology and Water Resources of the Gila and San Carlos Valleys in the San Carlos Indian Reservation, Arizona.* USGS Water Supply Paper 450, 1919.
Sears, Paul B. "Comparative Costs of Restoration and Reclamation of Land," *The Annals of the American Academy of Political and Social Science,* Vol. 281 (May 1952).
————. "Science and Natural Resources," *American Forests,* Vol. 63, No. 3 (March 1957).
Sellers, William D. (ed.). *Arizona Climate.* Tucson: University of Arizona Press, 1960.
————. *Distribution of Relative Humidity and Dew Point in the Southwestern United States.* Institute of Atmospheric Physics, U of A, Scientific Report No. 13. February 1, 1960.
Seltzer, Raymond. "Cooperative Financing Possibilities in Arizona," in *Water: Preliminary Economic Considerations of the Arizona Watershed Program* (which see).
Shadegg, Stephen. *The Phoenix Story: An Adventure in Reclamation.* Phoenix, 1958.
Shirer, John S. and Weissmiller, Josef C. "Arizona's Energy Base, 1949-1952," *Arizona Business and Economic Review,* Vol. 3, No. 3 (March 1954).
Sieker, John H. "Planning for Recreational Use of Waters: A Plea," in *Water, The Yearbook of Agriculture, 1955* (which see).
The Sierra Ancha Experimental Watersheds. USDA, FS, Southwestern Forest and Range Experiment Station. Tucson, June 1953.
Sitgreaves, Lorenzo. *Report of an Expedition Down the Zuni and Colorado Rivers.* 32d Cong., 2d Sess., Sen. Exec. Doc. 59, 1853.
Smiley, Terah L. *A Summary of Tree-Ring Dates from Some Southwestern Archaeological Sites.* U of A Laboratory of Tree-Ring Research Bulletin No. 5. Oct. 1951.

Smiley, Terah L. (ed.). *Climate and Man in the Southwest.* University of Arizona Program in Geochronology, Contribution No. 6 [a symposium held before the 33d annual meeting of the Southwestern and Rocky Mountain Division of the American Association for the Advancement of Science, Tucson, April 30, 1957]. Tucson: University of Arizona Press, 1958.

Smith, Anthony Wayne. "Campaign for the Grand Canyon," *National Parks Magazine* (April 1962).

Smith, G. E. P. *Cement Pipe for Small Irrigation Systems.* AAES Bulletin 55, 1907.

————. "Future Water Supply and Irrigated Agriculture in Arizona," Part 2 of *More Data on the Colorado River Question* by H.H. D'Autremont (which see).

————. *Groundwater Law in Arizona and Neighboring States.* AAES Technical Bulletin 65, 1936.

————. *Groundwater Supply and Irrigation in Rillito Valley.* AAES Bulletin 64, 1910.

————. *The Groundwater Supply of the Eloy District in Pinal County, Arizona.* AAES Technical Bulletin 87, 1940.

————. *The Physiography of Arizona Valleys and the Occurrence of Ground Water.* AAES Technical Bulletin 77, 1938.

————. *Use and Waste of Irrigation Water.* AAES Bulletin 88, 1919.

Smith, H. V. *The Chemical Composition of Representative Arizona Waters.* AAES Bulletin 225, 1949.

————. *The Climate of Arizona.* AAES Bulletin 279, 1956.

Soil and Water Conservation in Arizona. USDA, SCS. Phoenix, 1961.

Soil Conservation in Arizona. USDA, SCS, Region Six. Albuquerque, N.M., 1947.

Some Economic Implications of the Tourist Industry for Northern Arizona. Stanford Research Institute. Phoenix, Feb. 19, 1954.

Sowls, Lyle K. *Wildlife Conservation Through Cooperation.* Tucson: Published for the Arizona Cooperative Wildlife Research Unit by the University of Arizona Press, 1956.

Spicer, Edward H. *Cycles of Conquest: The Impact of Spain, Mexico, and the United States on the Indians of the Southwest, 1533-1960.* Tucson: University of Arizona Press, 1962.

Stallings, W.S., Jr. *Dating Prehistoric Ruins by Tree Rings.* Laboratory of Anthropology, General Series No. 8. Santa Fe, N.M., 1939.

Stanberry, C. O. "Alfalfa Irrigation," *Progressive Agriculture,* Vol. 6, No. 2 (July-Sept. 1954).

————, Converse, C.D., Hoise, H.R., and Kelley, O.J. "Effect of Moisture and Phosphate Variables on Alfalfa Hay Production on the Yuma Mesa," *Soil Science Society of America Proceedings,* Vol. 19, No. 3 (1955).

State Parks, Areas, Acreages and Accommodations, 1960. USDI, NPS. March 1961.

Statistical Abstract of the United States.

Statistical Summary of Major Range and Farm Practices Completed Under the Agricultural Conservation Program for the Years 1936-1955. USDA, Agricultural Stabilization and Conservation Service, Arizona. Typewritten, 1956.

308   *THE POLITICS OF WATER IN ARIZONA*

Stocker, Joseph. "Arizona's Maverick Conservative," *Frontier,* Vol. 1, No. 16 (July 1, 1950).

————. "Arizona's New Liberal Movement," *Frontier,* Vol. 1, No. 18 (Aug. 1, 1950).

————. "The Big Grab in Arizona," *Frontier,* Vol. 1, No. 7 (Feb. 15, 1950).

————. *A Guide to Easier Living.* New York: Harper, 1955.

————. "Saying No to Three-Cents-An-Acre," *Survey* (Sept. 1949).

Stong, Benton J. "Washington Report," *Frontier,* Vol. 1, No. 9 (1950).

Strauss, Michael. *Why Not Survive?* New York: Simon and Schuster, 1955.

Streets, R. B., and Stanley, E. B. *Control of Mesquite and Noxious Shrubs on Southern Arizona Grassland Ranges.* AAES Technical Bulletin 74, 1938.

Struckmeyer, Fred C., and Butler, Jeremy E. *A Review of Water Rights in Arizona.* Phoenix, 1960.

A Study of the Park and Recreation Problem of the United States. USDI, NPS, Washington, D.C., 1941.

Subsurface Facilities of Water Management and Patterns of Supply—Type Area Studies. Vol. IV of *The Physical and Economic Foundation of Natural Resources.* U.S. House of Representatives, Committee on Interior and Insular Affairs, 1953.

A Summary of Ground Water Law in Arizona. ASLD Bulletin 302. Phoenix, 1957.

A Summary of Ground Water Legislation in Arizona. ASLD Bulletin 301. Phoenix, 1954.

A Survey of the Recreational Resources of the Colorado River Basin. USDI, NPS. Washington, D.C., 1950.

Ten Rivers in America's Future. Vol. II of the *Report* of the President's Water Resources Policy Commission. Washington, D.C., 1950.

Thiele, Heinrich J. "Groundwater in Arizona," *Arizona Cattlelog,* Vol. 10, No. 12 (1955).

"30 Years of Progress," *Wildlife News,* Vol. 7, No. 1 (Jan. 1960).

Thornber, J. J. *The Grazing Ranges of Arizona.* AAES Bulletin 65, 1910.

Thornwaite, C. Warren, Sharpe, C.F. Steward, and Dosch, Earl F. *Climate and Accelerated Erosion in the Arid and Semi-Arid Southwest with Special Reference to the Polacca Wash Drainage Basin, Arizona.* USDA Technical Bulletin 808. May 1942.

Timber Resources Review. USDA, FS. Washington, D.C., 1955.

Timmons, John F. "Theoretical Considerations of Water Allocations Among Competing Uses and Users," *Journal of Farm Economics,* Vol. 38, No. 5 (1956).

Toumey, J. W. *Range Grasses of Arizona.* AAES Bulletin 2, 1891.

Toward Meeting Soil and Water Conservation Needs: I. USDA, ARS. Washington, D.C., 1955.

Toward Meeting Soil and Water Conservation Needs: II. USDA, ARS. Washington, D.C., 1956.

Trelease, Frank J. "Preferences to the Use of Water," *Rocky Mountain Law Review,* Vol. 27 (Feb. 1955).

Tschirley, F. H. "Chaparral—Still a Problem," *Progressive Agriculture,* Vol. 6, No. 1 (April-June 1954).

Tucson Daily Citizen.

Turner, Samuel F. *Further Investigations of the Ground-Water Resources of the Santa Cruz Basin, Arizona.* USGS Duplicate Report, 1947 [Arizona Water Supply Paper No. 20].

_____, and Halpenny, L. C. "Ground-Water Inventory in the Upper Gila Valley, New Mexico and Arizona; Scope of Investigations and Methods," *Transactions of the American Geophysical Union,* 22d annual meeting, part 3 (Aug. 1941).

Turney, O.A. "Prehistoric Irrigation," *Arizona Historical Review,* Vol. 2, Nos. 1,2,3,4 (April, July, Oct. 1929 and Jan. 1930).

The Underground Water Resources of Arizona: A Report. Arizona Underground Water Commission. Phoenix, 1953.

U.S. Army. *Annual Reports of the Chief of Engineers.*

_____. Corps of Engineers, South Pacific Division. *Water Resources Development by the Army Corps of Engineers in Arizona.* Jan. 1, 1961.

U.S. Department of Agriculture. *Reports of the Secretary of Agriculture.*

_____. Forest Service. *Reports of the Chief of the Forest Service.*

_____. Forest Service. Santa Rita Experimental Range. *Annual Reports.*

_____. Forest Service. Southwest Forest and Range Experiment Station. *Annual Reports.*

_____. Forest Service. Rocky Mountain Forest and Range Experiment Station. *Annual Reports.*

_____. Soil Conservation Service, *Administrator's Memorandum.* SCS-71. Mimeo., Dec. 2, 1954.

_____. Soil Conservation Service. *Administrator's Memorandum.* SCS-84. Mimeo., May 12, 1955.

_____. Soil Conservation Service. *Administrator's Memorandum.* SCS-109.

_____. Soil Conservation Service, Region VI. *Water Yield and Range Conservation Studies on Walnut Gulch Watershed.* Mimeo., Oct. 7, 1953.

U.S. Department of the Interior. *Official Organization Handbook, 1951.*

_____. Bureau of Indian Affairs, Phoenix Area Office. *Soil and Moisture Conservation Annual Report, Colorado River Reservation.* 1956.

_____. Bureau of Indian Affairs, Phoenix Area Office. *Soil Conservation Activities, Branch of Land Operations.* Phoenix, 1955.

_____. Bureau of Indian Affairs, Phoenix Area Office, Branch of Land Operations. *Outline of Work Necessary to Complete Soil and Moisture Conservation Program: Colorado River Indian Agency.* Parker, Ariz., Dec. 21, 1954.

_____. Bureau of Indian Affairs, Phoenix Area Office, Sells Agency (Papago). *Soil Conservation Program: 20 Year Period.* 1954.

_____. Bureau of Land Management. *Report of the Director of the Bureau of Land Management, 1956: Statistical Appendix.*

_____. Bureau of Land Management. *Soil and Moisture Conservation Operations.* Undated.

_____. Bureau of Land Management, Area 2, Arizona. *Annual Narrative Report—Range Management.*

_____. Bureau of Land Management, Area 2, Arizona. *Detailed Plan: Railroad Wash Community Watershed,* LM-2-0-6-4; LM-29-0-6-A4 (1956).

_____. Bureau of Land Management, Area 2, Arizona. *State Summary of 20-Year Program for Sub-basins* (1956).

U. S. Department of the Interior. Bureau of Reclamation. *Davis Dam and Power Plant.* 1955.

—————. Bureau of Reclamation. *General Information Concerning the Salt River Project, Arizona.* April 15, 1941.

—————. Bureau of Reclamation. "News Releases."

—————. Bureau of Reclamation. *Proposed Report of the Bureau of Reclamation.* Washington, D.C., Jan. 26, 1948 (approved by Julius Krug Feb. 5, 1948).

—————. National Park Service. *Mission 66 for Chiricahua National Monument.* Mimeo., n.d.; *Mission 66 for Organ Pipe Cactus National Monument.* Mimeo., n.d.; *Mission 66 for Pipe Springs National Monument.* Mimeo., n.d.; *Mission 66 for Saguaro National Monument.* Mimeo., n.d.

—————. Office of the Solicitor. Opinion M-36263, Feb. 23, 1955.

—————. U.S. Reclamation Service. *Salt River Irrigation Project.* Oct. 1, 1909.

U.S. National Resources Board. *State Planning: A Review of Activities and Progress.* Washington, D.C., 1935.

U.S. Senate. Committee on Interior and Insular Affairs. *Water Rights Settlement Act,* Hearings Before the Subcommittee on Irrigation and Reclamation on S. 863. 84th Cong., 2d Sess., 1956.

—————. Select Committee on National Water Resources. *Population Projections and Economic Assumptions.* 86th Cong., 2d Sess., Committee Print No. 5. 1960.

—————. Select Committee on National Water Resources. *Report.* 87th Cong., 1st Sess., Report No. 29, 1961.

U.S. *Statutes at Large.*

University of Arizona. *First Annual Report on an Interdisciplinary Study of the Utilization of Arid Lands.* June 11, 1959.

—————. *The Utilization of Arid Lands, A Research Proposal Submitted to the Rockefeller Foundation.* Feb. 6, 1958.

Velie, Lester. "Are We Short of Water?" *Reclamation Era,* Vol. 34, No. 7 (July 1948), reprinted from *Colliers* (May 15, 1948).

Wagoner, J.J. *History of the Cattle Industry in Southern Arizona, 1540-1940.* University of Arizona Social Science Bulletin No. 20, 1952.

Waltz, Waldo. E. "Arizona: A State of New-Old Frontiers," in Donnelly, *Rocky Mountain Politics* (which see).

Water for Arizona. Central Arizona Project Association. Phoenix, Spring, 1954.

Water Law With Special Reference to Ground Water. USGS Circular 117, 1951.

Water Levels and Artesian Pressures in Observation Wells in the United States. USGS Water Supply Paper 1270, 1956.

A Water Policy for the American People. Vol. I of the *Report* of the President's Water Resources Policy Commission. Washington, D.C., 1950.

Water: Preliminary Economic Considerations of the Arizona Watershed Program. Proceedings of the Second Annual Meeting of the Arizona Watershed Program, Sept. 22, 1958.

Watershed and Related Water Management Problems. Proceedings, Fourth Annual Watershed Symposium [Arizona Watershed Program], Sept. 21, 1960.

Water, The Yearbook of Agriculture, 1955. U.S. Department of Agriculture, 1955.

Waugh, Robert. "Tourism—A Billion Dollar Industry," *Arizona Business and Economic Review,* Vol. 8, No. 4 (April 1959).

Weaver, Harold. "A Preliminary Report on Prescribed Burning in Virgin Ponderosa Pine," *Journal of Forestry,* Vol. 50, No. 9 (Sept. 1952).

Wheeler, George M. *Preliminary Report Concerning Explorations and Surveys Principally in Nevada and Arizona.* U.S. War Department, 1872.

White, Gilbert F. *The Future of Arid Lands.* American Association for the Advancement of Science, Publication 43. Washington, D. C., 1956.

Wiel, Samuel C. *Water Rights in the Western States.* San Francisco, 1912.

Wildlife News.

"Wildlife Water Rights Challenge," *Outdoor Life,* Vol. 2, No. 1 (Jan. 1957).

Williams, D.A. "Urbanization of Productive Farmland," *Soil Conservation,* Vol. 22, No. 3 (Oct. 1956).

Williams, Jack. "Water," *Arizona Cattlelog,* Vol. 13, No. 4 (1956).

Wilm, H. G., *et al.* "The Training of Men in Forest Hydrology and Watershed Management," *Journal of Forestry,* Vol. 55, No. 4 (April 1957).

Wingfield, Kenneth. "The Stockman's Viewpoint on Brushland Control," *Arizona Cattlelog,* Vol. 11, No. 9 (May 1956).

Wolcott, H. N., Skibitzke, H. E., and Halpenny, L. C. "Water Resources of Bill Williams River Valley Near Alamo, Arizona," in *Contributions to the Hydrology of the United States* (which see).

Wolman, Abel. "Utilization of Surface, Underground and Sea Water," in *Proceedings of the United Nations Scientific Conference on the Conservation and Utilization of Resources,* Vol. IV, *Water Resources.*

Womer, Stanley. "Arizona Today—What We Have and What We Need," *Arizona Business and Economic Review,* Vol. 5, No. 4 (April 1956).

Wooten, H.H. *Supplement to Major Uses of Land in the United States.* USDA Technical Bulletin 1082 (Oct. 1953).

Woosley, Edward. "Our Public Domain," *Arizona Cattlelog,* Vol. 11, No. 7 (March 1956).

Workers in Subjects Pertaining to Agriculture in Land-Grant Colleges and Experiment Stations. USDA, Agriculture Handbook No. 116. Washington, D.C., 1957.

Wright, Frank Lloyd. "Plan for Arizona State Capitol," *Oasis,* Scottsdale [?], Ariz., Feb. 17, 1957.

Young, Robert W., comp., *The Navajo Yearbook of Planning in Action.* USDI, BIA, Navajo Agency, Window Rock, Arizona. Issued yearly, title varies; also called *Navajo Yearbook,* and *Planning in Action.*

Yuma: Federal Reclamation Project. USDI, Bureau of Reclamation, 1936.

Zon, Raphael. "Forestry Mistakes and What They Taught Us," *Journal of Forestry,* Vol. 49, No. 3 (1951).

Cases: Arizona

Arizona v. *Anway,* 87 Arizona 206, 349 P. 2d 774 (1960).

Boquillas Land and Cattle Company v. *Curtis et al.,* 11 Arizona 128, 89 Pac. 504 (1907).

312   THE POLITICS OF WATER IN ARIZONA

Brewster v. *Salt River Valley Water Users' Association,* 27 Arizona 23, 229 P.2d 929 (1924).
Bristor v. *Cheatham,* 73 Arizona 228, 240 P.2d 185 (1952).
Bristor v. *Cheatham,* 75 Arizona 227, 225 P.2d 173 (1953).
Campbell v. *Willard,* 45 Arizona 221, 42 P.2d 403 (1935).
Clough v. *Wing,* 2 Arizona 371 (1888).
Ernst v. *Collins,* 81 Arizona 178, 302 P.2d 941 (1956).
Ernst v. *Superior Court of Apache County,* 82 Arizona 17, 307 P.2d 911 (March 5, 1957).
Ethington et al. v. *Wright et al.,* 66 Arizona 382, 189 P.2d 209 (1948).
Fourzan v. *Curtis,* 43 Arizona 140, 29 P.2d 722 (1934).
Gould v. *Maricopa Canal Company,* 8 Arizona 429, 76 Pac. 598 (1904).
Howard v. *Perrin,* 8 Arizona 347, 76 Pac. 465 (1904).
Hurley v. *Abbott et al.,* No. 4564 Decree (Mar. 1, 1910).
McKenzie v. *Moore,* 20 Arizona 1, 176 Pac. 568 (1918).
Maricopa County Municipal Water Conservation District v. *Southwest Cotton Company et al.,* 39 Arizona 65, 4 Pac. 369 (1926).
Parker et al. v. *McIntyre et al.,* 47 Arizona 484, 56 P.2d 1337 (1936).
Proctor v. *Pima Farms Company,* 30 Arizona 96, 245 Pac. 369 (1926).
Salt River Valley Water Users' Association v. *Norviel,* 29 Arizona 360, 241 Pac. 583 (1925).
Slosser v. *Salt River Valley Canal Company,* 7 Arizona 376, 65 Pac. 332 (1901).
Southwestern Engineering Company v. *Ernst,* 79 Arizona 376, 291 P.2d 764 (1955).
Stewart v. *Verde River Irrigation and Power District,* 49 Arizona 531, 68 P.2d 329 (1937).
Stuart v. *Norviel,* 26 Arizona 493, 226 P.2d 908 (1924).
Vance v. *Lassen,* 82 Arizona 188, 310 P.2d 510 (1957).

Cases: Federal

Arizona v. *California,* 283 U. S. 423 (1931).
Arizona v. *California,* 292 U. S. 341 (1934).
Arizona v. *California,* 298 U. S. 338 (1936).
Arizona v. *California, et al.,* October Term (1952).
Boquillas Land and Cattle Company v. *Curtis et al.,* 29 Sup. Ct. 493, 213 U. S. 339 (1909).
California Oregon Power Company v. *Beaver Portland Cement Company,* 295 U. S. 142 (1936).
Federal Power Commission v. *Oregon,* 340 U. S. 345 (1955).
Howard v. *Perrin,* 200 U. S. 71 (1906).
United States v. *Arizona,* 295 U. S. 174 (1935).
Wattson v. *United States,* 260 F. 506 (1919).
Wormser et al. v. *Salt River Valley Canal Company* (1892), Federal District Court for Arizona, Phoenix.

Index

Administrative agencies, state: 117-18
Administrative reorganization, proposed: 24-25
   and Water Division: 25
Agricultural acreage, decline and expansion: 65
Agricultural acreage, future needs: 19
Agricultural Conservation Program: 190
Agricultural Experiment Station: 213 ff.
   research in soil and water problems: 215 ff.
Agricultural Extension Service: 180-81
Agricultural Research Service: 214, 248-52
   Soil and Water Conservation Division: 248-49
   U. S. Water Conservation Laboratory: 252
Agricultural use of water: 9-10, 43-44
   future requirements: 19
Agriculture, expansion of and water supply: 20-21
Apache Reservations: 174-75
Appropriative rights *see* Prior Appropriation doctrine
Appropriative rights, legal determination of: 33-34
Arid Lands Program: 227
Arizona Association of Soil Conservation Districts: 192
Arizona Cooperative Wildlife Research Unit: 206-207
Arizona Corporation Commission *see* Corporation Commission
Arizona Development Board: 26
Arizona Game and Fish Department: 72, 201, 203
   and Arizona Game Protective Association: 203-204
   water management activities: 20, 201-203
Arizona Game Protective Association: 72, 203-204
Arizona Interstate Stream Commission *see* Interstate Stream Commission
Arizona Legislative Council: 26
Arizona Power Authority: 100 ff.
   and private utilities: 113-14
   creation of: 101-102
   criticism of Hoover Dam operation: 113
   internal conflicts: 113-114
   opposition to: 102
   political troubles: 103-105
   position enhanced: 105
   powers of: 102-103

   private utility attacks on: 104-105
   work of: 103-104
Arizona Public Service Company: 106
Arizona Resources Board: 23
Arizona State Land Department *see* State Land Department
Arizona State Parks Board: 199
Arizona State Planning Board: 23
Arizona State Resources and Planning Board: 24
Arizona Tax Research Association: 73
Arizona v. *California see* Colorado River controversy with California
Arizona Water Resources Committee: 26-27, 126
Arizona Watershed Program: 12-13, 126, 157-60, 204
   opinions on: 158-60
Army Engineer Corps: 144-45
Artesian water: 48
Atmospheric physics *see* Climate research

Barr, George: 219
Barr Report *see* Arizona Watershed Program
Beneficial use: xii, 30, 34, 38
Boulder Canyon Project Act: 84, 85, 99
Bridge Canyon Project: 111-12, 140-41, 142, 194
Bristor v. *Cheatham:* 17, 54-55, 56-57, 58, 60-61
Bureau of Indian Affairs: 145-46, 169 ff. 242
   soil and water conservation programs: 170 ff.
Bureau of Land Management: 160 ff.
   and grazing lands: 162 ff.
   conservation work: 165-66
Bureau of Reclamation: 133 ff.
Burning, prescribed: 243-46

Campbell, Thomas E.: 82, 83
Canal companies and water rights: 35-36
Central Arizona Project: 88-89, 137-38
   efforts for federal authorization: 138-40
   estimated costs: 142-43
   feasibility studies: 140-41
   impossibility of state financing: 130-31, 141
   Interstate Stream Commission work on: 129, 130
   revised proposals of 1962: 142

Tyler Printing Company set the text of *The Politics of Water in Arizona* in Garamond, a modern type face derived from the designs of Claude Garamond, a Parisian type-cutter of the early sixteenth century who has been called the "father of type-founders." Chapter titles and sub-heads are in Spartan Bold. The book was printed by Tyler on S.D. Warren's sixty-pound University Eggshell textstock, and bound by Arizona Trade Bindery in Joanna bright blue vellum parchment. Douglas Peck was the designer. Cartography is by Don Bufkin.

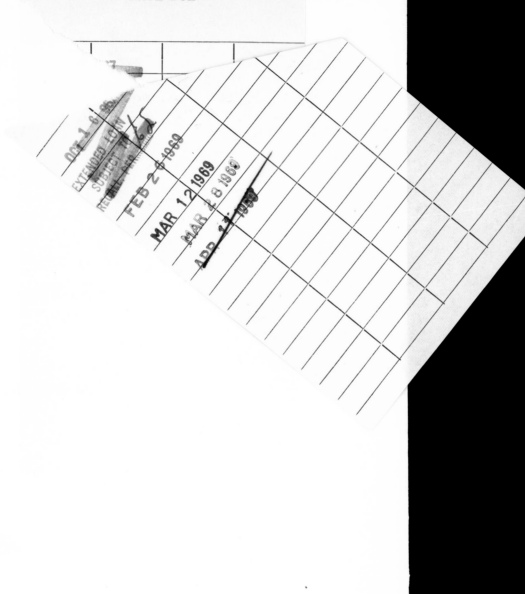